AF782679

Remote Sensing in Ecology and Conservation

Remote Sensing in Ecology and Conservation

Editor:

Maria Tattaris, Ph.D.

ArclerPress

www.arclerpress.com

Remote Sensing in Ecology and Conservation

Editor: Maria Tattaris, Ph.D.

© 2017 Arcler Press LLC

708 3rd Avenue, 6th Floor
New York
NY 10017
United States of America
www.arclerpress.com

ISBN: 978-1-68094-482-2

Library of Congress Control Number: 2016951341

This book contains information obtained from highly regarded resources. Reprinted material sources are indicated. Copyright for individual articles remains with the authors as indicated. A Wide variety of references are listed. Reasonable efforts have been made to publish reliable data and views articulated in the chapters are those of the individual contributors, and not necessarily those of the editors or publishers. Editors or publishers are not responsible for the accuracy of the information in the published chapters or consequences of their use. The publisher believes no responsibility for any damage or grievance to the persons or property arising out of the use of any materials, instructions, methods or thoughts in the book. The editors and the publisher have attempted to trace the copyright holders of all material reproduced in this publication and apologize to copyright holders if permission has not been obtained. If any copyright holder has not been acknowledged, please write to us so we may rectify.

Notice: Registered trademark of products or corporate names are used only for explanation and identification without intent of infringement.

Arcler Press LLC publishes wide variety of books and eBooks. For more information about Arcler Press and its products, visit our website at www.arclerpress.com

About the Editor

Maria Tattaris, Ph.D.

Maria Tattaris has a background in Mathematics and a PhD in remote sensing within the area of atmospheric sciences, awarded in 2012. Her interests include field spectroscopy, image processing, spectral analysis, field data collection, and investigating new data analysis methods for remote sensing techniques, all within the areas of environmental, plant, atmospheric and related sciences.

Contents

List of Contributors

Mirjana Bevanda
Biogeographical Modelling, Bayreuth Center for Ecology and Environmental Research BayCEER, University of Bayreuth, Universitaetsstr. 30, 95447 Bayreuth, Germany

Ned Horning
American Museum for Natural History, Central Park West at 79th Street, NY 10024-5192 New York, USA

Bjoern Reineking
Biogeographical Modelling, Bayreuth Center for Ecology and Environmental Research BayCEER, University of Bayreuth, Universitaetsstr. 30, 95447 Bayreuth, Germany
Unité de recherche écosystèmes montagnards, Irstea, 2 rue de la Papeterie-BP 76, 38402 St-Martin-d'Hères, France

Marco Heurich
Bavarian Forest National Park, Department of Research and Documentation, Freyunger Str. 2, 94481 Grafenau, Germany

Martin Wegmann
Department of Remote Sensing, Remote Sensing for Biodiversity Unit, University Wuerzburg, Oswald Kuelpe Weg 86, 97074 Wuerzburg, Germany

Joerg Mueller
Bavarian Forest National Park, Department of Research and Documentation, Freyunger Str. 2, 94481 Grafenau, Germany

Dimitris Stratoulias
Balaton Limnological Institute, Centre for Ecological Research, Hungarian Academy of Sciences, Klebelsberg Kuno u. 3., Tihany, H-8237, Hungary
University of Leicester, Centre for Landscape and Climate Research, Bennett Building, University Road, Leicester, LE1 7RH, UK

Heiko Balzter
University of Leicester, Centre for Landscape and Climate Research, Bennett Building, University Road, Leicester, LE1 7RH, UK

Olga Sykioti
National Observatory of Athens, Institute for Astronomy, Astrophysics, Space Applications and Remote Sensing, Metaxa and Vas. Pavlou Str., Penteli, Athens GR-15236, Greece

András Zlinszky
Balaton Limnological Institute, Centre for Ecological Research, Hungarian Academy of Sciences, Klebelsberg Kuno u. 3., Tihany, H-8237, Hungary

Viktor R. Tóth
Balaton Limnological Institute, Centre for Ecological Research, Hungarian Academy of Sciences, Klebelsberg Kuno u. 3., Tihany, H-8237, Hungary

Anna K Schweiger
Remote Sensing Laboratories, Department of Geography, University of Zurich, Winterthurerstrasse 190, 8057 Zürich, Switzerland
Research Unit Community Ecology, Swiss Federal Institute for Forest, Snow and Landscape Research WSL, Zürcherstrasse 111, 8903 Birmensdorf, Switzerland
Department of Research and Geoinformation, Swiss National Park, Chastè Planta-Wildenberg, 7530 Zernez, Switzerland

Martin Schütz
Research Unit Community Ecology, Swiss Federal Institute for Forest, Snow and Landscape Research WSL, Zürcherstrasse 111, 8903 Birmensdorf, Switzerland

Pia Anderwald
Department of Research and Geoinformation, Swiss National Park, Chastè Planta-Wildenberg, 7530 Zernez, Switzerland

Michael E Schaepman
Remote Sensing Laboratories, Department of Geography, University of Zurich, Winterthurerstrasse 190, 8057 Zürich, Switzerland

Mathias Kneubühler
Remote Sensing Laboratories, Department of Geography, University of Zurich, Winterthurerstrasse 190, 8057 Zürich, Switzerland

Rudolf Haller
Department of Research and Geoinformation, Swiss National Park, Chastè Planta-Wildenberg, 7530 Zernez, Switzerland

Anita C Risch
Research Unit Community Ecology, Swiss Federal Institute for Forest, Snow and Landscape Research WSL, Zürcherstrasse 111, 8903 Birmensdorf, Switzerland

Krishna Bahadur K.C.
Department of Farming and Rural System in the Tropics and Subtropics, University of Hohenheim, Institute 490c, 70593 Stuttgart, Germany

Talib Oliver-Cabrera
Division of Marine Geosciences, University of Miami, Miami, FL 33149, USA

Shimon Wdowinski
Division of Marine Geosciences, University of Miami, Miami, FL 33149, USA

Samuel M. Jantz
Department of Geographical Sciences, University of Maryland, College Park , MD 20742, USA

Lilian Pintea
Conservation Science Department, the Jane Goodall Institute, Vienna, VA 22182, USA

Janet Nackoney
Department of Geographical Sciences, University of Maryland, College Park , MD 20742, USA

Matthew C. Hansen
Department of Geographical Sciences, University of Maryland, College Park , MD 20742, USA

Trevor G. Jones
Blue Ventures Conservation, Villa Bella Fiharena, Rue Gambetta, Lot 259, Toliara, Madagascar
Dynamic Ecosystems and Landscapes Lab, Department of Environmental Science and Management, Portland State University, Portland, OR 97201, USA

Leah Glass
Blue Ventures Conservation, Villa Bella Fiharena, Rue Gambetta, Lot 259, Toliara, Madagascar

Samir Gandhi
Blue Ventures Conservation, Villa Bella Fiharena, Rue Gambetta, Lot 259, Toliara, Madagascar

Lalao Ravaoarinorotsihoarana
Blue Ventures Conservation, Villa Bella Fiharena, Rue Gambetta, Lot 259, Toliara, Madagascar

Aude Carro
Blue Ventures Conservation, Villa Bella Fiharena, Rue Gambetta, Lot 259, Toliara, Madagascar

Lisa Benson
Blue Ventures Conservation, Villa Bella Fiharena, Rue Gambetta, Lot 259, Toliara, Madagascar

Harifidy Rakoto Ratsimba
Department of Forestry, University of Antananarivo, Antananarivo 101, Madagascar

Chandra Giri
United States Geological Survey, Earth Resources Observation and Science Center, Duke University, Durham, NC 27708, USA

Dannick Randriamanatena
WWF Madagascar West Indian Ocean Programme Office, Antsakaviro, B.P. 738, Antananarivo 101, Madagascar

Garth Cripps
Blue Ventures Conservation, Villa Bella Fiharena, Rue Gambetta, Lot 259, Toliara, Madagascar

Paul H. Evangelista
Natural Resource Ecology Laboratory, Colorado State University, Fort Collins, Colorado 80523-499, USA

Thomas J. Stohlgren
U.S. Geological Survey, Fort Collins Science Center, Fort Collins, Colorado 80526, USA

Jeffrey T. Morisette
U.S. Geological Survey, Fort Collins Science Center, Fort Collins, Colorado 80526, USA

Sunil Kumar
Natural Resource Ecology Laboratory, Colorado State University, Fort Collins, Colorado 80523-499, USA

Usha Chirala
Department of Geo-Engineering, College of Engineering (A), Andhra University, Visakhapatnam, India

Murali Krishna Gurram
GIS Technology & Applications, GIS Division, Xinthe Technologies Pvt. Ltd., Visakhapatnam, India

Nooka Ratnam Kinthada
Department of Geology, College of Science and Technology, Adikavi Nannaya University, Rajahmundry, India

Shaun R. Levick
Max Planck Institute for Biogeochemistry, Hans-Knoell-Str. 10, Jena 07745, Germany

Samantha A. Setterfield
Research Institute for Environment and Livelihoods, Charles Darwin University, Darwin, NT 0909, Australia

Natalie A. Rossiter-Rachor
Research Institute for Environment and Livelihoods, Charles Darwin University, Darwin, NT 0909, Australia

Lindsay B. Hutley
Research Institute for Environment and Livelihoods, Charles Darwin University, Darwin, NT 0909, Australia

Damien McMaster
Research Institute for Environment and Livelihoods, Charles Darwin University, Darwin, NT 0909, Australia

Jorg M. Hacker
Airborne Research Australia, Flinders University, Salisbury South, SA 5106, Australia

Christian Hüttich
Department of Remote Sensing, University of Wuerzburg, Wuerzburg, Germany

Ursula Gessner
Department of Remote Sensing, University of Wuerzburg, Wuerzburg, Germany

Martin Herold
Friedrich-Schiller-University, ESA GOFC-GOLD Project Office, Jena, Germany

Ben J. Strohbach
National Botanical Research Institute of Namibia, Windhoek, Namibia

Michael Schmidt
Department of Remote Sensing, University of Wuerzburg, Wuerzburg, Germany
German Remote Sensing Data Center, German Aerospace Center, Oberpfaffenhofen, Germany

Manfred Keil

Stefan Dech
Department of Remote Sensing, University of Wuerzburg, Wuerzburg, Germany
German Remote Sensing Data Center, German Aerospace Center, Oberpfaffenhofen, Germany

Wiebke Neumann
Department of Forest and Wildlife Ecology, University of Wisconsin-Madison, 1630 Linden Drive, Madison WI 53706, USA
Department of Wildlife, Fish, and Environmental Studies, Swedish University of Agricultural Sciences, Umeå SE-90183, Sweden

Sebastian Martinuzzi
Department of Forest and Wildlife Ecology, University of Wisconsin-Madison, 1630 Linden Drive, Madison WI 53706, USA

Anna B Estes
Department of Forest and Wildlife Ecology, University of Wisconsin-Madison, 1630 Linden Drive, Madison WI 53706, USA
The Huck Institutes of the Life Sciences, Pennsylvania State University, Pennsylvania, USA

Anna M Pidgeon
Department of Forest and Wildlife Ecology, University of Wisconsin-Madison, 1630 Linden Drive, Madison WI 53706, USA

Holger Dettki
Department of Wildlife, Fish, and Environmental Studies, Swedish University of Agricultural Sciences, Umeå SE-90183, Sweden

Göran Ericsson
Department of Wildlife, Fish, and Environmental Studies, Swedish University of Agricultural Sciences, Umeå SE-90183, Sweden

Volker C Radeloff
Department of Forest and Wildlife Ecology, University of Wisconsin-Madison, 1630 Linden Drive, Madison WI 53706, USA

Quazi K. Hassan
Department of Geomatics Engineering, Schulich School of Engineering, University of Calgary, 2500 University Dr. NW Calgary, Alberta, T2N 1N4, Canada

Charles P.-A. Bourque
Faculty of Forestry and Environmental Management, University of New Brunswick, Fredericton, New Brunswick, E3B 6C2, Canada

Somayeh Dodge
Department of Civil, Environmental & Geodetic Engineering, The Ohio State University, 475 Hitchcock Hall, 2070 Neil Avenue, Columbus, USA

Gil Bohrer
Department of Civil, Environmental & Geodetic Engineering, The Ohio State University, 475 Hitchcock Hall, 2070 Neil Avenue, Columbus, USA

Rolf Weinzierl
Max Planck Institute for Ornithology, Radolfzell, Germany

Sarah C Davidson
Department of Civil, Environmental & Geodetic Engineering, The Ohio State University, 475 Hitchcock Hall, 2070 Neil Avenue, Columbus, USA
Max Planck Institute for Ornithology, Radolfzell, Germany

Roland Kays
NC Museum of Natural Sciences & NC State University, Raleigh, USA

David Douglas
U.S. Geological Survey, Alaska Science Center, Juneau, USA

Sebastian Cruz
Max Planck Institute for Ornithology, Radolfzell, Germany

Jiawei Han
University of Illinois at Urbana-Champaign, Urbana, USA

David Brandes
Lafayette College, Easton, USA

Martin Wikelski
Max Planck Institute for Ornithology, Radolfzell, Germany
Department of Biology, University of Konstanz, Konstanz, Germany

Kai Wang
Department of Geography and Planning, University of Saskatchewan, Saskatoon, SK S7N 5C8, Canada

Steven E. Franklin
Trent University, Peterborough, ON K9J 7B8, Canada

Xulin Guo
Department of Geography and Planning, University of Saskatchewan, Saskatoon, SK S7N 5C8, Canada

Marc Cattet
Department of Veterinary Pathology, University of Saskatchewan, Saskatoon, SK S7N 5B4, Canada

Preface

Ecology and conservation are both equally important fields of research. With concerns such as the changing climate, and the decline in biodiversity, the need to improvement our understanding of these two subject areas is always ongoing. Particularly so that natural resources and ecosystems can be managed in an efficient way and the sustainability of biodiversity can be maintained. In, general the study of these two areas requires manually collecting data in the field. Although this type of data collection is considered as highly accurate, it can be costly in terms of labour and time, and there can also exist restrictions in accessibility to target areas, therefore this approach is not considered practical for large scale studies. Remote sensing can overcome these restrictions, for example, remote sensing measurements are non-intrusive, the problem of accessibility is eliminated and measurements can be quick and made at large scales. With their constant development and improvement in technologies and data analysis, remote sensing techniques can be key in the areas of ecology and conservation. Further in this book, the ways in which remote sensing is used as a tool are presented, for example, land classification, the derivation of vegetation and other spectral indices, the mapping and monitoring of numerous processes (such as habitat use and erosion) and time series of various processes and temporal dynamics of landscapes (such as vegetation phenology). Numerous remote sensing data types are described as the inputs for the above techniques, such as multispectral spaceborne imagery, radar datasets, airborne LiDAR and hyperspectral imagery.

Editor
Maria Tattaris, Ph.D.

INTRODUCTION

Remote sensing (RS), also called earth observation, refers to obtaining information about objects or areas at the Earth's surface without being in direct contact with the object or area. Humans accomplish this task with aid of eyes or by the sense of smell or hearing; so, remote sensing is day-today business for people. Reading the newspaper, watching cars driving in front of you are all remote sensing activities. Most sensing devices record information about an object by measuring an object's transmission of electromagnetic energy from reflecting and radiating surfaces.

Remote sensing techniques allow taking images of the earth surface in various wavelength region of the electromagnetic spectrum (EMS). One of the major characteristics of a remotely sensed image is the wavelength region it represents in the EMS. Some of the images represent reflected solar radiation in the visible and the near infrared regions of the electromagnetic spectrum, others are the measurements of the energy emitted by the earth surface itself i.e. in the thermal infrared wavelength region. The energy measured in the microwave region is the measure of relative return from the earth's surface, where the energy is transmitted from the vehicle itself. This is known as active remote sensing, since the energy source is provided by the remote sensing platform. Whereas the systems where the remote sensing measurements depend upon the external energy source, such as sun are referred to as passive remote sensing systems.

Principles of Remote Sensing

Detection and discrimination of objects or surface features means detecting and recording of radiant energy reflected or emitted by objects or surface material (Fig. 1). Different objects return different amount of energy in different bands of the electromagnetic spectrum, incident upon it. This depends on the property of material (structural, chemical, and physical), surface roughness, angle of incidence, intensity, and wavelength of radiant energy. The Remote Sensing is basically a multi-disciplinary science which includes a combination of various disciplines such as optics, spectroscopy, photography, computer, electronics and telecommunication, satellite launching etc. All these technologies are integrated to act as one complete system in itself, known as Remote Sensing System. There are a number of stages in a Remote Sensing process, and each of them is important for successful operation.

Stages in Remote Sensing

- Emission of electromagnetic radiation, or EMR (sun/self- emission)
- Transmission of energy from the source to the surface of the earth, as well as absorption and scattering
- Interaction of EMR with the earth's surface: reflection and emission
- Transmission of energy from the surface to the remote sensor
- Sensor data output
- Data transmission, processing and analysis

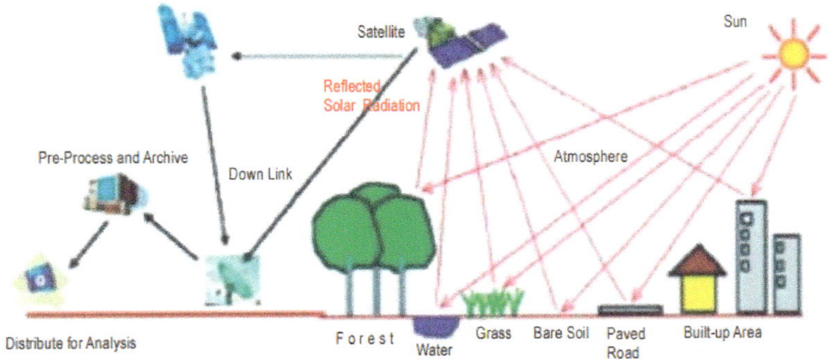

Figure 1: Remote Sensing process.

At temperature above absolute zero, all objects radiate electromagnetic energy by virtue of their atomic and molecular oscillations. The total amount of emitted radiation increases with the body's absolute temperature and peaks at progressively shorter wavelengths. The sun, being a major source of energy, radiation and illumination, allows capturing reflected light with conventional (and some not-so-conventional) cameras and films. The basic strategy for sensing electromagnetic radiation is clear. Everything in nature has its own unique distribution of reflected, emitted and absorbed radiation. These spectral characteristics, if ingeniously exploited, can be used to distinguish one thing from another or to obtain information about shape, size and other physical and chemical properties.

Types of Remote Sensing

Remote sensing can be either passive or active. ACTIVE systems have their own source of energy (such as RADAR) whereas the PASSIVE systems depend upon external source of illumination (such as SUN) or self-emission for remote sensing.

Interaction of EMR with the Earth's Surface

Radiation from the sun, when incident upon the earth's surface, is either reflected by the surface, transmitted into the surface or absorbed and emitted by the surface (Figure 2). The EMR, on interaction, experiences a number of changes in magnitude, direction, wavelength, polarization and phase. These changes are detected by the remote sensor and enable the interpreter to obtain useful information about the object of interest. The remotely sensed data contain both spatial information (size, shape and orientation) and spectral information (tone, color and spectral signature).

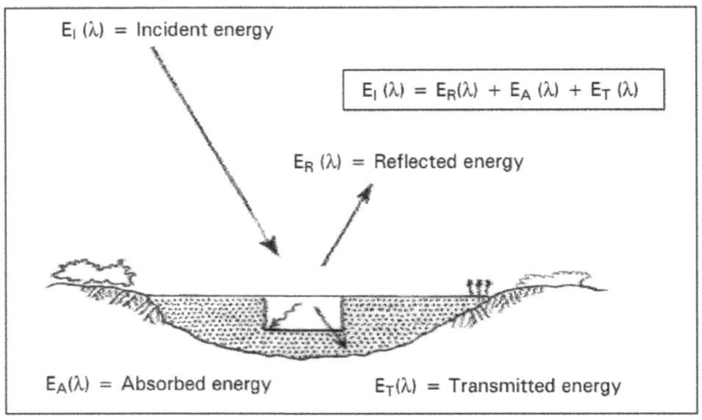

Figure 2: Interaction of Energy with the earth's surface.

From the viewpoint of interaction mechanisms, with the object-visible and infrared wavelengths from 0.3 μm to 16 μm can be divided into three regions. The spectral band from 0.3 μm to 3 μm is known as the reflective region. In this band, the radiation sensed by the sensor is that due to the sun, reflected by the earth's surface. The band corresponding to the atmospheric window between 8 μm and 14 μm is known as the thermal infrared band. The energy available in this band for remote sensing is due to thermal emission from the earth's surface. Both reflection and self-emission are important in the intermediate band from 3 μm to 5.5 μm. In the microwave region of the spectrum, the sensor is radar, which is an active sensor, as it provides its own source of EMR. The EMR produced by the radar is transmitted to the earth's surface and the EMR reflected (back scattered) from the surface is recorded and analyzed. The microwave region can also be monitored with passive sensors, called microwave radiometers, which record the radiation emitted by the terrain in the microwave region.

Reflection

Of all the interactions in the reflective region, surface reflections are the most useful and revealing in remote sensing applications. Reflection occurs when a ray of light is redirected as it strikes a non-transparent surface. The reflection intensity depends on the surface refractive index, absorption coefficient and the angles of incidence and reflection (Figure 3).

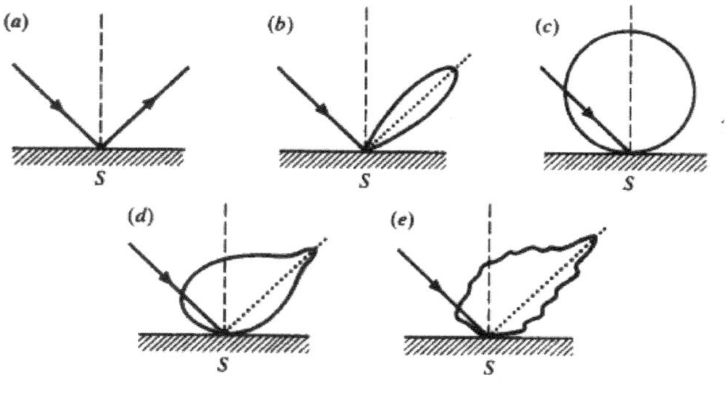

Figure 3: Different types of scattering surfaces (a) Perfect specular reflector (b) Near perfect specular reflector (c) Lambertain (d) Quasi-Lambertian (e) Complex.

Transmission

Transmission of radiation occurs when radiation passes through a substance without significant attenuation. For a given thickness, or depth of a substance, the ability of a medium to transmit energy is measured as transmittance (τ).

$$\tau = \frac{\text{Transmitted radiation}}{\text{Incident radiation}}$$

Spectral Signature Spectral reflectance, [ρ(λ)], is the ratio of reflected energy to incident energy as a function of wavelength. Various materials of the earth's surface have different spectral reflectance characteristics. Spectral reflectance is responsible for the color or tone in a photographic image of an object. Trees appear green because they reflect more of the green wavelength. The values of the spectral reflectance of objects averaged over different, well-defined wavelength intervals comprise the spectral signature of the objects or features by which they can be distinguished. To obtain the necessary ground truth for the interpretation of multispectral imagery, the spectral characteristics of various natural objects have been extensively measured and recorded. The spectral reflectance is dependent on wavelength, it has different values at different wavelengths for a given terrain feature. The reflectance characteristics of the earth's surface features are expressed by spectral reflectance, which is given by:

$$\rho(\lambda) = [E_R(\lambda) \; / \; E_I(\lambda)] \times 100$$

Where,

$\rho(\lambda)$ = Spectral reflectance (reflectivity) at a particular wavelength.

$E_R(\lambda)$ = Energy of wavelength reflected from object

$E_I(\lambda)$ = Energy of wavelength incident upon the object

The plot between $\rho(\lambda)$ and λ is called a spectral reflectance curve. This varies with the variation in the chemical composition and physical conditions of the feature, which results in a range of values. The spectral response patterns are averaged to get a generalized form, which is called generalized spectral response pattern for the object concerned. Spectral signature is a term used for unique spectral response pattern, which is characteristic of a terrain feature. Figure 4 shows a typical reflectance curves for three basic types of earth surface features, healthy vegetation, dry bare soil (grey-brown and loamy) and clear lake water.

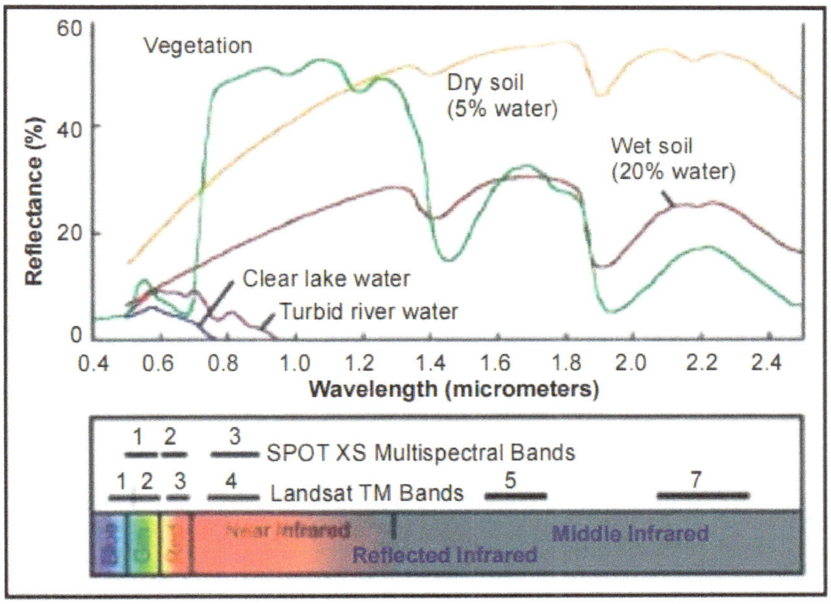

Figure 4: Typical Spectral Reflectance curves for vegetation, soil and water.

Reflectance Characteristics of Earth's Cover types

The spectral characteristics of the three main earth surface features are:

Vegetation: The spectral characteristics of vegetation vary with wavelength. Plant pigment in leaves called chlorophyll strongly absorbs radiation in the red and blue wavelengths but reflects green wavelength. The internal structure of healthy leaves acts as diffuse reflector of near infrared wavelengths. Measuring and monitoring the near infrared reflectance is one way that scientists determine how healthy particular vegetation may be.

Water: Majority of the radiation incident upon water is not reflected but is either absorbed or transmitted. Longer visible wavelengths and near infrared radiation is absorbed more by water than by the visible wavelengths. Thus water looks blue or blue green due to stronger reflectance at these shorter wavelengths and darker if viewed at red or near infrared wavelengths. The factors that affect the variability in reflectance of a water body are depth of water, materials within water and surface roughness of water.

Soil: The majority of radiation incident on a soil surface is either reflected or absorbed and little is transmitted. The characteristics of soil that determine its reflectance properties are its moisture content, organic matter content, texture, structure and iron oxide content. The soil curve shows less peak and valley variations. The presence of moisture in soil decreases its reflectance.

By measuring the energy that is reflected by targets on earth's surface over a variety of different wavelengths, we can build up a spectral signature for that object. And by comparing the response pattern of different features we may be able to distinguish between them, which we may not be able to do if we only compare them at one wavelength. For example, Water and Vegetation reflect somewhat similarly in the visible wavelength but not in the infrared.

Interactions with the Atmosphere

The sun is the source of radiation, and electromagnetic radiation (EMR) from the sun that is reflected by the earth and detected by the satellite or aircraft-borne sensor must pass through the atmosphere twice, once on its journey from the sun to the earth and second after being reflected by the surface of the earth back to the sensor. Interactions of the direct solar radiation and reflected radiation from the target with the atmospheric constituents interfere with the process of remote sensing and are called as "Atmospheric Effects".

The interaction of EMR with the atmosphere is important to remote sensing for two main reasons. First, information carried by EMR reflected/ emitted by the earth's surface is modified while traversing through the atmosphere. Second, the interaction of EMR with the atmosphere can be used to obtain useful information about the atmosphere itself. The atmospheric constituents scatter and absorb the radiation modulating the radiation reflected from the target by attenuating it, changing its spatial distribution and introducing into field of view radiation from sunlight scattered in the atmosphere and some of the energy reflected from nearby ground area. Both scattering and absorption vary in their effect from one part of the spectrum to the other.

The solar energy is subjected to modification by several physical processes as it passes the atmosphere, viz.

- Scattering;
- Absorption, and
- Refraction

Atmospheric Scattering

Scattering is the redirection of EMR by particles suspended in the atmosphere or by large molecules of atmospheric gases. Scattering not only reduces the image contrast but also changes the spectral signature of ground objects as seen by the sensor. The amount of scattering depends upon the size of the particles, their abundance, the wavelength of radiation, depth of the atmosphere through which the energy is traveling and the concentration of the particles. The concentration of particulate matter varies both in time and over season. Thus the effects of scattering will be uneven spatially and will vary from time to time. Theoretically scattering can be divided into three categories depending upon the wavelength of radiation being scattered and the size of the particles causing the scattering. The three different types of scattering from particles of different sizes are summarized below:

Scattering process	Wavelength	Approximate dependence particle size	Kinds of particles
Selective			
• Rayleigh	λ^{-4}	< 1 μm	Air molecules
• Mie	λ^{0} to λ^{-4}	0.1 to 10 μm	Smoke, haze
• Non-selective	λ^{0}	> 10 μm	Dust, fog, clouds

Rayleigh Scattering

Rayleigh scattering predominates where electromagnetic radiation interacts with particles that are smaller than the wavelength of the incoming light. The effect of the Rayleigh scattering is inversely proportional to the fourth power of the wavelength. Shorter wavelengths are scattered more than longer wavelengths. In the absence of these particles and scattering the sky would appear black. In the context of remote sensing, the Rayleigh scattering is the most important type of scattering. It causes a distortion of spectral characteristics of the reflected light when compared to measurements taken on the ground.

Mie Scattering

Mie scattering occurs when the wavelength of the incoming radiation is similar in size to the atmospheric particles. These are caused by aerosols: a mixture of gases, water vapor and dust. It is generally restricted to the lower atmosphere where the larger particles are abundant and dominates under overcast cloud conditions. It influences the entire spectral region from ultra violet to near infrared regions.

Non-selective Scattering

This type of scattering occurs when the particle size is much larger than the wavelength of the incoming radiation. Particles responsible for this effect are water droplets and larger dust particles. The scattering is independent of the wavelength, all the wavelength are scattered equally. The most common example of non-selective scattering is the appearance of clouds as white. As cloud consist of water droplet particles and the wavelengths are scattered in equal amount, the cloud appears as white. Occurrence of this scattering mechanism gives a clue to the existence of large particulate matter in the atmosphere above the scene of interest which itself is a useful data. Using minus blue filters can eliminate the effects of the Rayleigh component of scattering. However, the effect of heavy haze i.e. when all the wavelengths are scattered uniformly, cannot be eliminated using haze filters. The effects of haze are less pronounced in the thermal infrared region. Microwave radiation is completely immune to haze and can even penetrate clouds.

Atmospheric Absorption

The gas molecules present in the atmosphere strongly absorb the EMR passing through the atmosphere in certain spectral bands. Mainly three gases are responsible for most of absorption of solar radiation, viz. ozone, carbon dioxide and water vapour. Ozone absorbs the high energy, short wavelength portions of the ultraviolet spectrum ($\lambda < 0.24$ μm) thereby preventing the transmission of this radiation to the lower atmosphere. Carbon dioxide is important in remote sensing as it effectively absorbs the radiation in mid and far infrared regions of the spectrum. It strongly absorbs in the region from about 13-17.5 μm, whereas two most important regions of water vapour absorption are in bands 5.5 - 7.0 μm and above 27 μm. Absorption relatively reduces the amount of light that reaches our eye making the scene look relatively duller.

Atmospheric Windows

The general atmospheric transmittance across the whole spectrum of wavelengths is shown in Figure 5. The atmosphere selectively transmits energy of certain wavelengths. The spectral bands for which the atmosphere is relatively transparent are known as atmospheric windows. Atmospheric windows are present in the visible part (.4 μm - .76 μm) and the infrared regions of the EM spectrum. In the visible part transmission is mainly effected by ozone absorption and by molecular scattering. The atmosphere is transparent again beyond about λ= 1mm, the region used for microwave remote sensing.

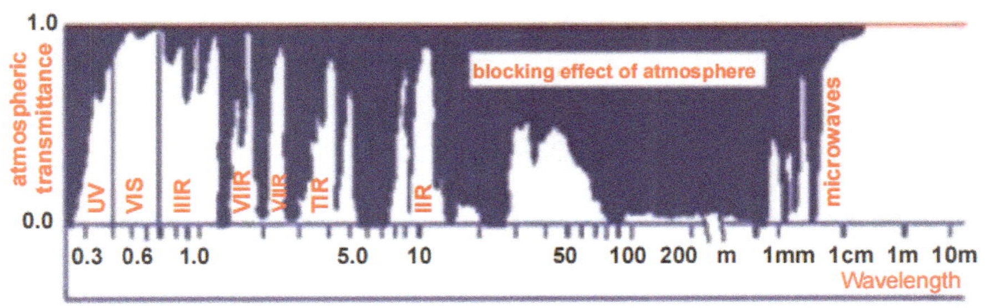

Figure 5: Atmospheric windows.

Refraction

The phenomenon of refraction that is bending of light at the contact between two media also occurs in the atmosphere as the light passes through the atmospheric layers of varied clarity, humidity and temperature. These variations influence the density of atmospheric layers, which in turn, causes the bending of light rays as they pass from one layer to another. The most common phenomena are the mirage like apparitions sometimes visible in the distance on hot summer days.

Applications

Each application itself has specific demands, for spectral resolution, spatial resolution, and temporal resolution. There can be many applications for Remote Sensing, in different fields, as described below. In the body of this introduction of Remote Sensing, some applications relevant for hydrography and oceanography will be given.

Agriculture

Agriculture plays a dominant role in economies of both developed and undeveloped countries. Satellite and airborne images are used as mapping tools to classify crops, examine their health and viability, and monitor farming practices.

Agricultural applications of remote sensing include the following:

- crop type classification
- crop condition assessment
- crop yield estimation
- mapping of soil characteristics
- mapping of soil management practices
- compliance monitoring (farming practices)

Forestry

Forests are a valuable resource providing food, shelter, wildlife habitat, fuel, and daily supplies such as medicinal ingredients and paper. Forests play an important role in balancing the Earth's CO2 supply and exchange, acting as a key link between the atmosphere, geosphere, and hydrosphere.

Forestry applications of remote sensing include the following:

- Reconnaissance mapping: Objectives to be met by national forest/environment agencies include forest cover updating, depletion monitoring, and measuring biophysical properties of forest stands.
- Commercial forestry: Of importance to commercial forestry companies and to resource management agencies are inventory and mapping applications: collecting harvest information, updating of inventory information for timber supply, broad forest type, vegetation density, and biomass measurements.
- Environmental monitoring: Conservation authorities are concerned with monitoring the quantity, health, and diversity of the Earth's forests.

Geology

Geology involves the study of landforms, structures, and the subsurface, to understand physical processes creating and modifying the earth's crust. It is most commonly understood as the exploration and exploitation of mineral and hydrocarbon resources, generally to improve the conditions and standard of living in society.

Geological applications of remote sensing include the following:

- surficial deposit / bedrock mapping
- lithological mapping
- structural mapping
- sand and gravel (aggregate) exploration/ exploitation
- mineral exploration
- hydrocarbon exploration
- environmental geology
- geobotany
- baseline infrastructure
- sedimentation mapping and monitoring
- event mapping and monitoring
- geo-hazard mapping
- planetary mapping

Hydrology

Hydrology is the study of water on the Earth's surface, whether flowing above ground, frozen in ice or snow, or retained by soil

Examples of hydrological applications include:

- wetlands mapping and monitoring,
- soil moisture estimation,
- snow pack monitoring / delineation of extent,
- measuring snow thickness,
- determining snow-water equivalent,
- river and lake ice monitoring,
- flood mapping and monitoring,

- glacier dynamics monitoring (surges, ablation)
- river /delta change detection
- drainage basin mapping and watershed modelling
- irrigation canal leakage detection
- irrigation scheduling

Sea Ice

Ice covers a substantial part of the Earth's surface and is a major factor in commercial shipping and fishing industries, Coast Guard and construction operations, and global climate change studies.

Examples of sea ice information and applications include:

- ice concentration
- ice type / age /motion
- iceberg detection and tracking
- surface topography
- tactical identification of leads: navigation: safe shipping routes/rescue
- ice condition (state of decay)
- historical ice and iceberg conditions and dynamics for planning purposes
- wildlife habitat
- pollution monitoring
- meteorological / global change research

Land Cover & Land Use

Although the terms land cover and land use are often used interchangeably, their actual meanings are quite distinct. Land cover refers to the surface cover on the ground, while Land use refers to the purpose the land serves. The properties measured with remote sensing techniques relate to land cover, from which land use can be inferred, particularly with ancillary data or a priori knowledge.

Land use applications of remote sensing include the following:

- natural resource management
- wildlife habitat protection
- baseline mapping for GIS input
- urban expansion / encroachment
- routing and logistics planning for seismic / exploration / resource extraction activities
- damage delineation (tornadoes, flooding, volcanic, seismic, fire)
- legal boundaries for tax and property evaluation
- target detection - identification of landing strips, roads, clearings, bridges, land/water interface

Mapping

Mapping constitutes an integral component of the process of managing land resources, and mapped information is the common product of analysis of remotely sensed data. Mapping applications of remote sensing include the following:

- Planimetry: Land surveying techniques accompanied by the use of a GPS can be used to meet high accuracy requirements, but limitations include cost effectiveness, and difficulties in attempting to map large, or remote areas. Remote sensing provides a means of identifying and presenting planimetric data in convenient media and efficient manner. Imagery is available in varying scales to meet the requirements of many different users. Defence applications typify the scope of planimetry applications - extracting transportation route information, building and facilities locations, urban infrastructure, and general land cover.
- Digital elevation models (DEM's): Generating DEMs from remotely sensed data can be cost effective and efficient. A variety of sensors and methodologies to generate such models are available and proven for mapping applications. Two primary methods if generating elevation data are 1. Stereogrammetry techniques using airphotos (photogrammetry), VIR imagery, or radar data (radargrammetry), and 2. Radar interferometry.
- Baseline thematic mapping / topographic mapping: As a base map, imagery provides ancillary information to the extracted planimetric or thematic detail. Sensitivity to surface expression makes radar a useful tool for creating base maps and providing reconnaissance abilities for hydrocarbon and mineralogical companies involved in exploration activities.

This is particularly true in remote northern regions, where vegetation cover does not mask the microtopography and generally, information may be sparse. Multispectral imagery is excellent for providing ancillary land cover information, such as forest cover. Supplementing the optical data with the topographic relief and textural nuance inherent in radar imagery can create an extremely useful image composite product for interpretation.

Oceans & Coastal Monitoring

The oceans not only provide valuable food and biophysical resources, they also serve as transportation routes, are crucially important in weather system formation and CO_2 storage, and are an important link in the earth's hydrological balance. Coastlines are environmentally sensitive interfaces between the ocean and land and respond to changes brought about by economic development and changing land-use patterns. Often coastlines are also biologically diverse inter-tidal zones, and can also be highly urbanized.

Ocean applications of remote sensing include the following:

- Ocean pattern identification:
  - currents, regional circulation patterns, shears
  - frontal zones, internal waves, gravity waves, eddies, upwelling zones, shallow water bathymetry ,
- Storm forecasting
  - wind and wave retrieval
- Fish stock and marine mammal assessment
  - water temperature monitoring
  - water quality
  - ocean productivity, phytoplankton concentration and drift
  - aquaculture inventory and monitoring
- Oil spill
  - mapping and predicting oilspill extent and drift
  - strategic support for oil spill emergency response decisions
  - identification of natural oil seepage areas for exploration
- Shipping
  - navigation routing
  - traffic density studies
  - operational fisheries surveillance
  - near-shore bathymetry mapping
- Intertidal zone
  - tidal and storm effects
  - delineation of the land /water interface
  - mapping shoreline features / beach dynamics
  - coastal vegetation mapping · human activity / impact

CHAPTER 1

ADDING STRUCTURE TO LAND COVER – USING FRACTIONAL COVER TO STUDY ANIMAL HABITAT USE

Mirjana Bevanda, Ned Horning, Bjoern Reineking, Marco Heurich, Martin Wegmann and Joerg Mueller

Keywords: Fractional cover, Remote sensing, Land cover classification, Animal movement, Habitat selection, Mixed model

BACKGROUND

Habitat use of animals is assumed to be mainly driven by forage availability and is a complex hierarchical process of behavioural responses and choices [1]. Individuals choose habitat that maximizes resources (e.g. food or shelter) and conditions necessary for survival and reproduction [2], whereas these resources are influenced by temporal and spatial variations of the landscape [3]. Habitat selection is led by behavioural decisions and is shaped by the environment, leading to the observed habitat use [4].

A large majority of animals use certain areas without showing a territorial behaviour, referred to as home range. In contrast to territories, a home range has no defended borders [5]. Home ranges are generally defined as the spatial expression of all behaviours an animal performs in order to survive and reproduce [5]. Since home ranges link individual movement paths to dispersal and population dynamics, understanding why and how home range sizes vary between and among species is a fundamental issue in ecology. The current and prospective availability of large movement data sets and remotely sensed environmental information will allow further detailed analysis [6]. Progress in GPS–sensor receiver technology and satellite telemetry makes it possible to track animals over long time spans with high temporal and spatial resolution and to analyse their habitat requirements and movement paths [7].

By studying variation in home range size and identifying the factors involved in such variation, we can identify how habitat influence individual's habitat use [2] and therefore the variation in home ranges. A number of factors have been adressed for shaping variation in home range sizes, these include the environmental productivity and the heterogeneity of the landscape [8-10]. Especially the availability of forage is a main driver shaping home range sizes [11]. A common trade–off often faced by many large mammals takes places when open habitats provide the best forage, while closed habitats provide shelter against predators and this may vary with different spatio–temporal scales [12].

Typically in habitat use studies the landscape is represented with a categorical habitat map usually derived from a classification [13,14], while in other studies the landscape is represented only by the dominant habitat type [15,16]. A variety of land cover classifications are routinely produced using remotely sensed data such as MODIS and AVHRR [17].

However, the way the landscape is defined is crucial for the analysis of habitat use. In many studies the landscape is defined in land cover categories, containing classes such as "meadows", "forest" and "agriculture" [13,15] and it is common sense that different needs of an animal corresponds to different land cover types, for example "forest" as areas for shelter and therefore resting or hiding sites, and "meadows" as areas for forage sites [12].

However, landscapes rarely contain sharp borders between cover types although that is how they are portrayed using a classical land cover classification approach. Moreover information about spatial variation within an *a–priori* defined land cover class is not provided when using a classification. A forest might vary spatially due to different age classes of the trees or small tree fall gaps which increase spatial

Citation: Mirjana Bevanda, Ned Horning, Bjoern Reineking, Marco Heurich, Martin Wegmann and Joerg Mueller, Adding structure to land cover – using fractional cover to study animal habitat use, DOI 10.1186/s40462-014-0026-1.

heterogeneity. This within land cover variation is not captured by categorical maps.

Therefore we use a continuous land cover approach such as fractional cover for the inclusion of spatial variation within classes for our analyses. Fractional cover is a multiscale analysis combined with spatial prediction. This method is related to spectral unmixing methods [18]. The fractional cover image are typically created using a higher resolution land cover classification image to calculate fractional cover training data for lower resolution imagery. For each pixel of the coarse resolution image the percentage coverage for each land cover class within the high resolution is calculated and used for a spatial prediction of the land cover percentages. The percentage cover for the chosen land cover types per pixel of the coarse resolution image is provided as result.

With this approach a continuous land cover classification can be derived which captures the spatial structure in a fine scale manner and this provides a more realistic and more ecologically meaningful representation of the landscape. Global maps with similar approaches of percentage coverage already exist such as MODIS or AVHRR [19,20] however only at a coarse spatial resolution and not validated in the study area.

Furthermore in many habitat use studies forests have structural attributes like "dense forest" or "light forest" with corresponding functional effects, such as light forest with plentiful food resources due to an established understory as enough sunlight can reach the forest floor. However, these structural attributes are often not validated and instead they are implicitly assumed [21]. With the fractional cover approach these structural attributes can be addressed clearly.

In this study, we investigate the potential of continuous land cover information for habitat use of red deer in the Bohemian Forest. As habitat use leads to differing home range sizes, we investigate the potential of continuous land cover information and its spatial representation for the explanation of their variation in size. We hypothesize larger home ranges with increasing forest cover due to lower density of food resources. We test our hypothesis on different spatial (90%, 70% and 50% isopleths) and temporal scales (monthly, biweekly and weekly) to account for temporal and spatial differences.

METHODS

Study area

The study area is located in Central Europe in the Bohemian Forest, an area belonging to two national parks: the Bavarian Forest National Park on the German side of the border (240 km ^{2}) and the Šumava National Park on the Czech Republic side of the border (640 km ^{2}). These protected areas are embedded within the Bavarian Forest Nature Park (3070 km ^{2}) and the Šumava Landscape Protection Area (1000 km ^{2}). In its entirety, the area is known as the Bohemian Forest Ecosystem. The area is mountainous, with a variation in elevation between 600 and 1450 m.a.s.l.. The mean annual temperature varies between 3°C and 6.5°C along higher elevation and ridges. The mean annual precipitation is between 830 and 2230 mm. Within the park, three major forest types exists: above 1100 m: sub–alpine spruce forests with Norway Spruce (*Picea abies* L.) and some Mountain Ash (*Sorbus aucuparia* L.), on the slopes, between 600 and 1100 m elevation, are mixed montane forests with Norway Spruce, White Fir (*Abies alba* MILL.), European Beech (*Fagus sylvatica* L.), and Sycamore Maple (*Acer pseudoplatanus* L.). In the valley bottoms, spruce forests with Norway Spruce, Mountain Ash, and birches (*Betula pendula* ROTH. and *Betula pubescens* EHRH.) [22]. Since the mid–1990s, the forests of the national park have been affected by massive proliferation of the spruce bark beetle (*Ips typographus*). By 2007, this had resulted in the death of mature spruce stands over an area amounting to 5,600 ha [23,24].

Red deer data

From 2002–2011 red deer were caught during winter, using a procedure approved by the Government of Upper Bavaria, Germany. Red deer were captured and fitted with GPS collars (Vectronic Aerospace, Berlin, Germany) in box traps with side windows after they were lured in with food. Here no immobilization was necessary. A second approach was to tranquillize deer by dart gun where they were attracted by food [25]. We collared 80 deer (39 male, 41 female). Ten individuals were collared two or more times. As animals spend the winter in enclosures, we restricted the analysis temporally from May to the end of September. The most common protocol was to mark red deer in late winter and retrieve the collars after a year by collar drop–off or recapturing, allowing the collars to be used on new individuals. We removed spatial and temporal false fixes (i.e. locations taken only a few seconds apart) beforehand. We defined the samples from the multiple collared animals over the single year as independent. As the schedule of the collars are adjusted to take a location every 15 min for one day of the week we took a random sample of animals with sequences of short time intervals to ensure that all locations have a minimum interval of one hour. The median accuracy of the GPS locations was 16.5 m [26].

Home range estimation

Home ranges were estimated with a commonly used approach, the fixed kernel method [27,28] using the reference method for the smoothing factor h [29]. We used three different home range definitions to include a spatial scale and to investigate the effect on the core area (50% kernel) and a wider range (70% kernel, 90% kernel). In addition, all home range definitions were estimated on three temporal scales: monthly, biweekly and weekly. We only estimated home ranges for individuals with at least ten locations for a given temporal scale, after removing spatial and temporal outliers [30].

Representation of the landscape

For the calculation of fractional cover a high resolution classified image was derived from aerial images and was used for training. The classified image contained 26 categories (different forest types such as coniferous, deciduous and mixed forest, and age classes such as mature, medium, young). Due to used spatial and spectral resolution we grouped those classes to three major categories in order to be able to discriminate them appropriately: forest (containing all forest types and age classes), open areas (e.g. meadows, regeneration areas, clear cut areas) and others (e.g. water, rocks, roads). To create our training data the fractional cover of each class within 30 m Landsat pixels was calculated. The resulting percent cover values for a particular class were used as response variables to train a random forest (RF) regression model [31]. Random forest uses an ensemble of decision trees (in our case regression trees) to model non-linear relations among response variables [32-34]. The resulting RF model was then used to predict percent cover for the cover type being modelled on a Landsat image using pixel spectral values as predictor variables. The number of regression trees used in the random forest model was 1000, the number of predictors tried on each split was set to the algorithm's default value (number of Landsat image bands/3). An unbiased accuracy assessment is provided by RF using "Out Of Bag" statistics calculated from a random selection of 1/3 of the training data [31]. Three cloud free Landsat 5 scenes (path 192, row 26) with bands 1–5 from 2006 (July 15th, October 19th) and 2009 (September 9th) were used for the fractional cover analysis. The three predicted vegetation layers complement each other and sum up to 100%. The class "others" contains only small values in our study area, therefore the major part of the values are split between "forest" and "grassland". Since both layers complement each other we included only the class "forest" in our analysis. Figure 1 shows the categorical map and the fractional cover layers "forest" and "grassland" for the whole study area (upper panels). An enlarged display of a section shows how the formerly categorical representation of the landscape is now split up in continuous values (middle panels). The lower panels show the representation of the categorical values within the fractional cover values in a histogram. The discrete classes are represented by very high cover values within the study area.

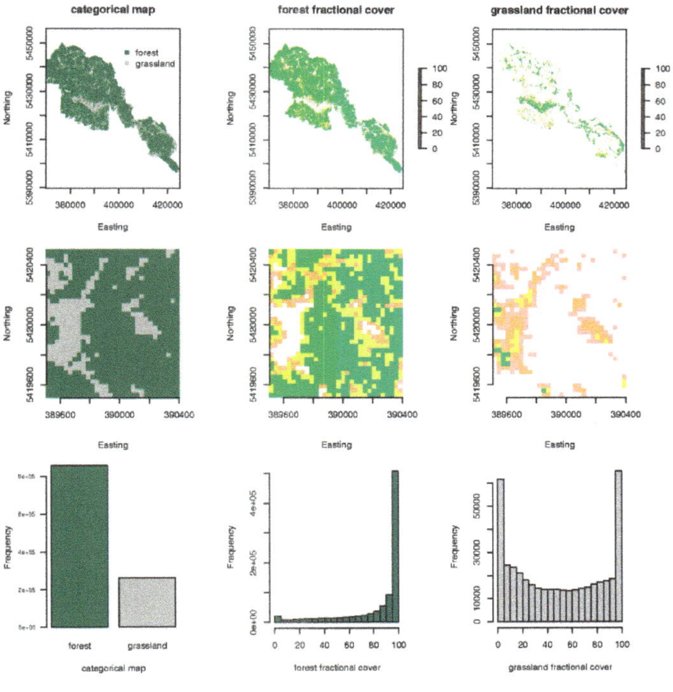

Figure 1: Overview of the landcover and fractional cover values within the study area. The upper panels show the distribution of the categorical (left hand side) and continuous fractional cover values (middel and right hand panel). The second row shows a zoom–in for better representation and the last row shows the distribution of the values for the whole study area.

We extracted all fractional cover values of the forest class within the home ranges and calculated mean, standard deviation and variance. In addition to fractional cover we chose to also calculate texture measures for each home range. Texture metrics were developed by Haralick et al. (1973) [35] and capture habitat structure which can be quantified using the variability of pixel values in a given area. Second–order texture measures are calculated from the gray–level co–occurrence matrix (GLCM) and account for spatial arrangement of pixel values. Haralick et al. (1973) [35] presented a variety of different texture metrics, however he states that these metrics are highly correlated and can be difficult to interpret. To ensure that the chosen texture metric is not size dependent we calculated buffers from 500 to 7000 m in 500 m steps around the home range centres of the 90% kernel isopleths and analysed all texture metrics with regard to their size dependency. We calculated texture measures using all pixel values within the home range. A moving window was used to calculate the texture metric for every pixel relative to its direct neighbours (eight pixels around a centre pixel). We then averaged the resulting texture values to obtain one value for the home range to fit into the mixed model design. We chose to use the texture metric "contrast", as it shows the least size dependency. Throughout the remaining text we will refer to the contrast metric as a texture metric or simply as texture.

We choose to use standard deviation of the forest fractional cover calculated within a home range as a measure for variability and the mean forest fractional cover as an estimate of overall forest fractional cover within each home range. Since variables standard deviation and variance show high collinearity [36], variance is not considered in the analysis. For simplicity we will refer to the standard deviation as variation of fractional cover values.

Furthermore we estimated the mean elevation of the home ranges using the 30 m ASTER Global Digital Elevation Map (GDEM) (http://asterweb.jpl.nasa.gov/gdem.asp).

The chosen variables showed no correlation with each other (Pearson's correlation with the threshold set to 0.7, -0.7 respectively).

Statistical analysis

To investigate the influence of forest fractional cover and texture on home range sizes, we used linear mixed models [37] on the log transformed home range areas (km ^{2}). Afterwards we ran a backfit on the t–values to derive the essential variables [38]. Preliminary analysis showed that the variables texture and elevation have a hump–shaped relationship with home range size in the red deer data and we therefore used a quadratic fit in the models.

Following the framework of Zuur et al. (2009) [39] for mixed effect models, we first identified the best structure for the random effect term. We fitted random intercepts for each individual (ID), different sexes and the year the locations were sampled, using the full model with respect to fixed effects terms and using the REML criterion for fitting. We started with the full random term and then simplified the model. Afterwards we compared the models with an ANOVA and the best model was evaluated with the Akaike Information Criterion (AIC). For variable selection, models were fitted with a maximum likelihood criterion. We considered as fixed effects the mean value of the fractional cover layer forest within a home range, the standard deviation of fractional cover values within a home range, the texture metric contrast and elevation. The final models where fitted using the REML criterion. We derived minimal adequate models by backward stepwise selection using a t–value of 2 as a threshold for inclusion [38]. We repeated the analysis for the three definitions of home range size and for the three definitions of temporal scale.

We used the software tool R version 3.0.1 [40] for all analysis. The package "adehabitatHR" [28] was used for the kernel calculations, "raster" [41], "EBImage" [42] and "randomForest" [43] for creation of the environmental variables and "lmer" [37] and "LMERConvenienceFunctions" [38] were used for the statistical analyses.

RESULTS

The fractional cover approach allows a differentiation of variations within land cover types, compared to categorical classes. The spatial heterogeneity of within class variation is captured by this approach. The fit of the random forest regression model for the forest layer was 70.15%. The diversity of fractional cover values within the home range level can be seen in Figure 2. As outlined in Figure 1, the corresponding categorical values are represented by the very high percentage values within the fractional Cover approach.

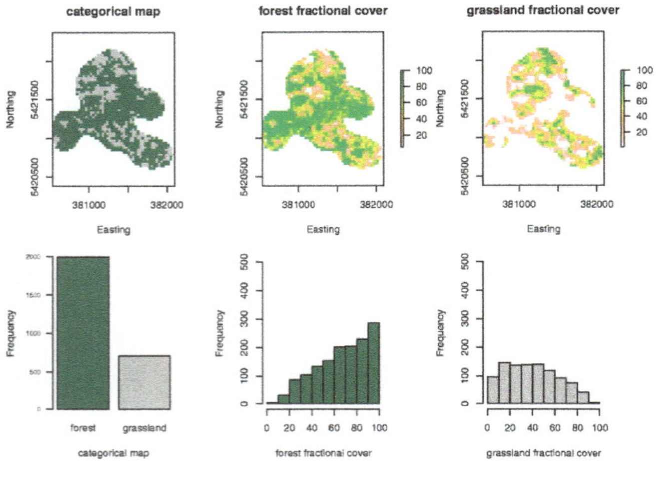

Figure 2: Representation of the landscape for one home range with both approaches, the categorical and the continuous fractional cover. The lower panels show the distribution of the values within the home range for each approach.

Home ranges of red deer show a high variation in size in our study area. We analysed the variation of home range sizes with a mixed model, using mean and standard deviation of the forest fractional cover, as well as the variable elevation and a texture metric. The main random effect in all models was the individual effect (variable ID) with an explained deviance of 0.26–0.38%. The fixed effects of the most parsimonious models explained between 26.88% and 30.88% of the observed variation in home range size for red deer across the different spatio–temporal scales.

In all models the texture metric showed the highest explained deviance (7.98%–14.72%) across scales and was the dominant variable explaining variation in home range size with a hump–shaped relationship. However, this hump–shaped relationship was only pronounced at the monthly time scale, whereas in the biweekly and weekly time scale this relationship changed to a negative linear relationship. The texture metric can be interpreted as an index for spatial heterogeneity in a given area. Hence, at larger temporal scales very homogeneous and very heterogeneous landscapes are leading to small home ranges, while at smaller temporal scales only very heterogeneous landscapes lead to small home ranges.

Furthermore the variation of forest fractional cover (the standard variation) within a home range contributes significantly with an explained deviance of 7.22–11.59% and a positive relationship, leading to larger home ranges where the variation of forest fractional cover values is higher (Figure 3).

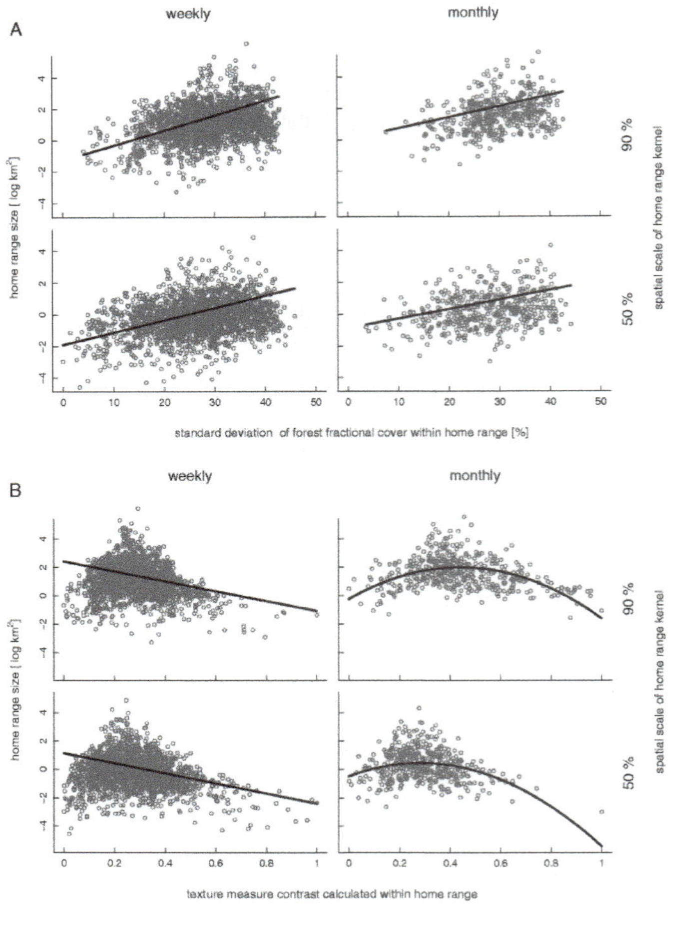

Figure 3: Plot of log–transformed home range sizes (km ^{2}) for red deer in relation to (A) the standard deviation of the forest fractional cover values within each home range and (B) the texture measure calculated within each home range. Home ranges were calculated with the kernel method and the smoothing factor h. Estimates are given for the 90% and 50% kernels and the weekly and monthly time scale. Lines show predicted values and points raw residuals.

Additionally mean showed a positive effect (5.48–7.12% explained deviance), with no effect on the monthly time scale kernel 50% isopleth.

Elevation had a hump–shaped effect on home range size and showed a low explanatory value of 0.35%–6.02%.

DISCUSSION

Many studies of habitat use and home range variation consider the landscape as a categorical map with defined and clearly separated patches [13,14]. This study investigates the use of continuous land cover information, fractional cover, to analyse the within land cover class variation of home ranges over different spatial and temporal scales for red deer in the Bohemian Forest. We demonstrate that small scale variations represented by continuous landscape data provide important information for modelling habitat use.

Red deer as a mixed feeder [44] has the ability to digest a broad spectrum of food items and benefits from forest edges and from the food supply of younger forest stands which show a low forest canopy cover and therefore have a pronounced understory, as sunlight can reach the ground. Mean forest fractional cover shows a positive relationship with home range size meaning that a higher proportion of dense forest will lead to larger home ranges. Whereas in forest patches with less crown cover and therefore more heterogeneous structure, food resources are more abundant which leads to smaller home ranges. This result is in support with other studies [14,45,46]. Mean forest fractional cover is a rather unsuitable derivative, as it averages all pixels within the home range. Nevertheless it shows a significant explanatory value and gives an overview of the overall forest structure within the home range.

The standard deviation of forest fractional cover values captures the variability of values within a home range. High values indicate a wide spectrum of forest fractional cover and therefore a more heterogeneous landscape while small values indicate a more homogeneous landscape within the home range. Tufto et al. (1996) [11] have shown, that female roe deer adjust the size of their home range in response to food supply. In accordance to this study red deer home range sizes increase in our study area with increasing standard deviation and therefore with more heterogeneous forest fractional cover, leading to a higher amount of unfavourable forest habitat within the home range.

The explanatory deviance is largest for the texture metric and also consistent over all spatio–temporal scales with a hump–shaped relationship at larger time scales. Low values of the texture metric correspond to high heterogeneity within the home range, while high values of the texture metric correspond to landscapes which have large aggregated patches. This relationship was detected in a previous study [47] and can be explained by the characteristics of the National Parks. Bark beetle outbreaks in the 90ies affected an area of approximately 5,600 ha especially in the subalpine regions, leading to sunny openings and large regeneration areas characterized by high grass cover, lying dead wood and regrowing vegetation [23]. These areas appear very homogeneous when calculated with a texture metric but offer good habitat for deer, as different resources are provided in a small area, leading to small home ranges, as both requirements, food and cover, are fulfilled at the same spot. Furthermore a heterogeneous landscape, providing many different resources, leads to small home ranges as all the resources needed can be reached within a small distance. The hump–shaped effect flattens in the biweekly and weekly time scale and can only be described with a negative linear trend. However, a pattern towards a hump–shaped distribution can be seen (Figure 3B). This result shows that the temporal scale needs to be accounted for when analysing home ranges as they are likely to change not based on ecological patterns only but on the time scale of the study. The time period of the study is restricted to the summer months, therefore the resource cover can be regarded as static, i.e. not highly changing over the time, while the resource food is dynamic and depleting. Therefore food supply is the main force shaping home range size during summer. When large patches of dense forest occur within the home range, the texture value will increase. These areas provide shelter against predators, but provide only little food resources. Therefore, as food resources are regarded to be a main force shaping home range size, home ranges will increase in size with the inclusion of large patches of dense forest (intermediate values of texture). Furthermore, these regeneration areas are located at higher altitude and are therefore explaining the effect of elevation, reflecting the importance of bark beetle areas in this study. Like the regeneration areas, elevation shows a hump–shaped fit leading to smaller home ranges where important resources are abundant [48].

It is known that other factors, like body mass, age, reproductive status or climatic parameters like temperature or rainfall have an effect on home range size (please see [46] for a more complete list) and it is likely, that by including these parameters, the explanatory value of the models could be increased. However, the best method to estimate home ranges is under debate. While we used at least 10 relocation points [30] to estimate our home ranges other studies suggest at least 20 relocation points [29].

The choice of environmental parameters is important for habitat use modelling. Using classified land cover requires clear definitions of the land cover types but definitions often vary between different maps making them difficult to compare [49]. Moreover do these classes need to reflect the ecological requirements. An increased discrimination of different land cover types is often helpful to better describe a landscape but an increase in the number of land cover classes often results in lower per–class accuracy. Using alternative information such as continuous cover can help to improve how a landscape is represented in a model. Applying remote sensing time–series data can be valuable to further discriminate land cover types and hence allow more fractional cover classes if distinct temporal signature exist for the different targeted land cover types. Applying continuous land cover information for environmental analysis provides detailed information about ecotones and within land cover variation. This research illustrates that fractional cover mapping has potential benefits for ecological research by avoiding categorical values or sharp, most often artificial, boundaries in the landscape. However, the fractional cover approach requires more analytical steps including spatial prediction models and might therefore be potentially biased by the model used.

CONCLUSION

The study demonstrates that continuous land cover information can provide valuable information about spatial within class variation as well as gradual vegetation changes, a feature that is not available when using discrete classes. This is especially relevant in movement ecology where a continuous representation of the landscape might be more ecological appropriate. However, to evaluate the added value of the fractional cover approach with regard to land cover classification or biophysical parameter further analysis are needed. Fractional cover mapping of different land cover types adds information, critical to ecological studies, beyond what traditional land cover categorical mapping can offer. As the synergy between remote sensing and ecology increases improved processing and analysis methods will continue to be developed which will have a positive impact on ecological research. These benefits will be especially important with the growing interest in spatio–temporal movement pattern.

DECLARATIONS

Acknowledgements

This work was financed by the German Federal Environmental Foundation (Deutsche Bundesstiftung Umwelt), the EU-programme INETRREG IV (EFRE Ziel 3), and the Bavarian Forest National Park Administration. We kindly thank Ingo Brauer, Horst Burghart, Rüdiger Fischer, Martin Gahbauer, Helmut Penn, Michael Penn, and Lothar Ertl for technical support. We thank the anonymous reviewers for valuable comments which improved the manuscript considerably.

Authors' contributions

MB performed the statistical analysis and was primarily responsible for writing the manuscript. MB, NH and MW drafted the manuscript and designed the species-environment interaction analysis. BR and JM contributed in the development of the animal movement and resource use approach and its statistical analysis. The movement data within the National Park Bavarian Forest was provided by JM and MH. All authors contributed to the writing of subsequent revisions and approved the final version.

REFERENCES

1. Gaillard J-M, Hebblewhite M, Loison A, Fuller M, Powell R, Basille M, Van Moorter B: Habitat-performance relationships: finding the right metric at a given spatial scale. *Philos Trans R Soc London, Ser B, Biol Sci*2010, 365(1550):2255–2265.
2. Richard E, Said S, Hamann J-L, Gaillard J-M: Toward an identification of resources influencing habitat use in a multi-specific context. *PloS one*2011, 6(12):29048.
3. Gustafson E: Quantifying landscape spatial pattern: what is the state of the art? *Ecosystems*1998, 1:143–156.
4. Johnson DH: The comparison of usage and availability measurements for evaluating resource preference. *Ecology*1980, 61:65–71.
5. Burt W: Territoriality and home range concepts as applied to mammals. *J Mammalogy*1943, 24(3):346–352.
6. Wikelski M, Kays RW, Kasdin NJ, Thorup K, Smith Ja, Swenson GW: Going wild: what a global small-animal tracking system could do for experimental biologists. *J Exp Biol*2007, 210(Pt 2):181–186.
7. Tomkiewicz SM, Fuller MR, Kie JG, Bates KK: Global positioning system and associated technologies in animal behaviour and ecological research. *Philos Trans R Soc London, Ser B*2010, 365(1550):2163–2176.
8. Boyce M, Mao J, Merrill E, Fortin D: Scale and heterogeneity in habitat selection by elk in Yellowstone National Park. *Ecoscience*2003, 10(4):421–431.
9. Nilsen E, Herfindal I, Linnell J: Can intra-specific variation in carnivore home-range size be explained using remote-sensing estimates of environmental productivity? *Ecoscience*2005, 12(1):68–75.
10. Saïd S, Gaillard J-M, Widmer O, Débias F, Bourgoin G, Delorme D, Roux C: What shapes intra-specific variation in home range size? A case study of female roe deer. *Oikos*2009, 118(9):1299–1306.
11. Tufto J, Andersen R, Linnell J: Habitat use and ecological correlates of home range size in a small cervid: the roe deer. *J Animal Ecol*1996, 65(6):715–724.
12. Godvik IMR, Loe LE, Vik JO, Veiberg VO, Langvatn R, Mysterud A: Temporal scales, trade-offs, and functional responses in red deer habitat selection. *Ecology*2009, 90(3):699–710.
13. Torres RT, Virgós E, Santos Ja, Linnell JDC, Fonseca C: Habitat use by sympatric red and roe deer in a Mediterranean ecosystem. *Animal Biol*2012, 62(3):351–366.
14. Massé A, Côté SD: Linking habitat heterogeneity to space use by large herbivores at multiple scales: From habitat mosaics to forest canopy openings. *Forest Ecol Manage*2012, 285:67–76.
15. Börger L, Franconi N, Ferretti F, Meschi F, De Michele G, Gantz A, Coulson T: An integrated approach to identify spatiotemporal and individual-level determinants of animal home range size. *Am Naturalist*2006, 168(4):471–485.
16. Rivrud IM, Loe LE, Mysterud A: How does local weather predict red deer home range size at different temporal scales? *J Animal Ecol*2010, 79(6):1280–1295.
17. Friedl MA, McIver DK, Hodges JCF, Zhang XY, Muchoney D, Strahler AH, Woodcock CE, Gopal S, Schneider A, Cooper A, Baccini A, Gao F, Schaaf C: Global land cover mapping from MODIS: algorithms and early results. *Remote Sensing Environ*2002, 83(1-2):287–302.
18. Asner GP, Heidebrecht KB: Spectral unmixing of vegetation, soil and dry carbon cover in arid regions: Comparing mulitspectral and hyperspectral observations. *Int J Remote Sensing*2002, 23(19):3939–3958.
19. DeFries R, Hansen M, Townshend JRG, Janetos AC, Loveland TR: A new global 1 km data set of percent tree cover derived from remote sensing. *Global Change Biol*2000, 6:247–254.
20. DiMiceli CM, Carroll ML, Sohlberg RA, Huang C, Hansen MC, Townshend JRG: *Annual Global Automated MODIS Vegetation Continuous Fields (MOD44B) at 250 M Spatial Resolution for Data Years Beginning Day 65, 2000 - 2010, Collection 5 Percent Tree Cover*, College Park, MD, USA: University of Maryland; 2011.
21. Debeljak M, Dzeroski S, Jerine K, Kobler A, Adamic M: Habitat suitability modelling for red deer (Cer6us elaphus L.) in South-central Slovenia with classification trees. *Ecol Modell*2001, 138:321–330.
22. Fischer HS, Winter S, Lohberger E, Jehl H, Fischer A: Improving transboundary maps of potential natural vegetation using statistical modeling based on environmental predictors. *Folia Geobotanica*2013, 48(2):115–135.
23. Müller J, Bußler H, Goßner M, Rettelbach T, Duelli P: The European spruce bark beetle Ips typographus in a national park: from pest to keystone species. *Biodivers Conserv*2008, 17(12):2979–3001.
24. Lausch A, Heurich M, Fahse L: Spatio-temporal infestation patterns of Ips typographus (L.) in the

Bavarian Forest National Park, Germany. *Ecol Indicators*2013, 31:73–81.

25. Heurich M: Berücksichtigung von Tierschutzaspekten beim Fang und der Markierung von Wildtieren. In *Internationale Fachtagung zu Fragen Von Verhaltenskunde, Tierhaltung und Tierschutz*. 2011, 142–158.
26. Stache A, Löttker P, Heurich M: Red deer telemetry: dependency of the position acquisition rate and accuracy of GPS collars on the structure of a temperate forest dominated by European beech. *Silva Gabreta*2012, 18(1):35–48.
27. Worton B: Kernel methods for estimating the utilization distribution in home-range studies. *Ecology*1989, 70:164–168.
28. Calenge C: The package "adehabitat" for the R software: a tool for the analysis of space and habitat use by animals. *Ecol Modell*2006, 197(3-4):516–519.
29. Kernohan BJ, Gitzen RA, Millspaugh JJ: Analysis of animal space use and movements. In *Radio Tracking and Animal Populations*. Edited by Millspaugh JJ, Marzluff J. San Diego, California, USA: Academic Press; 2001:126–164.
30. Börger L, Franconi N, De Michele G, Gantz A, Meschi F, Manica A, Lovari S, Coulson T: Effects of sampling regime on the mean and variance of home range size estimates. *J Animal Ecol*2006, 75(6):1393–1405.
31. Breiman L: Random forests. *Machine Learning*2001, 45(1):5–32.
32. Hansen MC, DeFries RS, Townshend JRG, Sohlberg R, Dimiceli C, Carroll M: Towards an operational MODIS continuous field of percent tree cover algorithm: examples using AVHRR and MODIS data. *Remote Sensing Environ*2002, 83(1-2):303–319.
33. Hansen MC, DeFries RS, Townshend JRG, Carroll M, Dimiceli C, Sohlberg Ra: Global percent tree cover at a spatial resolution of 500 meters: first results of the MODIS vegetation continuous fields algorithm. *Earth Interact*2003, 7(10):1–15.
34. Hayes DJ, Cohen WB, Sader Sa, Irwin DE: Estimating proportional change in forest cover as a continuous variable from multi-year MODIS data. *Remote Sensing Env*2008, 112(3):735–749.
35. Haralick R, Shanmugam K, Dinstein I: Textural features for image classification. *IEEE Trans Syst Man Cybernetics*1973, 3(6):610–621.
36. Dormann CF, Elith J, Bacher S, Buchmann C, Carl G, Carré G, Marquéz JRG, Gruber B, Lafourcade B, Leitao PJ, Münkemüller T, McClean C, Osborne PE, Reineking B, Schröder B, Skidmore AK, Zurell D, Lautenbach S: Collinearity: a review of methods to deal with it and a simulation study evaluating their performance. *Ecography*2013, 36:27–46.
37. Bates D, Maechler M, Bolker B: lme4: Linear mixed-effects models using S4 classes. *R package version 0.9*2011.
38. Tremblay A, Ransijn J: LMERConvenienceFunctions: A suite of functions to back-fit fixed effects and forward-fit random effects, as well as other miscellaneous functions. *R package version 1.6.8.3*2011.
39. Zuur AF, Ieno EN, Walker NJ, Saveliev AA, Smith GM: *Mixed Effects Models and Extensions in Ecology with R*: Springer; 2009.
40. Development Core Team R: *R: A Language and Environment for Statistical Computing*: R Foundation for Statistical Computing; 2013.
41. Hijmans RJ: *raster: Geographic data analysis and modeling*; 2013.
42. Pau G, Oles A, Smith M, Sklyar O, Huber W: EBImage - Image processing toolbox for R. *R package version 4.2.1*2013.
43. Liaw A, Wiener M: Classification and regression by random forest. *R News*2002, 2(3):18–22.
44. Albon SD, Langvatn R: Plant phenology and the benefits of migration in a temperate ungulate. *Oikos*1992, 65:502–513.
45. Owen-Smith N, Fryxell JM, Merrill EH: Foraging theory upscaled: the behavioural ecology of herbivore movement. *Philos Trans R Soc London, Ser B, Biol Sci*2010, 365(1550):2267–2278.
46. van Beest FM, Rivrud IM, Loe LE, Milner JM, Mysterud A: What determines variation in home range size across spatiotemporal scales in a large browsing herbivore? *J Animal Ecol*2011, 80(4):771–785.
47. Bevanda M, Fronhofer EA, Heurich M, Müller J, Reineking B: *Landscape configuration is a major determinant of home range size variation*; 2014. in prep.
48. Anderson LO, Shimabukuro YE, Arai E: Cover: Multitemporal fraction images derived from Terra MODIS data for analysing land cover change over the Amazon region. *Int J Remote Sensing*2005, 26(11):2251–2257.
49. Herold M, Mayaux P, Woodcock CE, Baccini A, Schmullius C: Some challenges in global land cover mapping: An assessment of agreement and accuracy in existing 1 km datasets. *Remote Sensing Env*2008, 112(5):2538–2556.

CHAPTER 2 EVALUATING SENTINEL-2 FOR LAKESHORE HABITAT MAPPING BASED ON AIRBORNE HYPERSPECTRAL DATA

Dimitris Stratoulias, Heiko Balzter, Olga Sykioti, András Zlinszky and Viktor R. Tóth

Keywords: Sentinel-2; simulation; hyperspectral; spectral response function; Phragmites; macrophytes; lakeshore vegetation; habitat mapping

INTRODUCTION

Macrophytes have complex spatial mosaic structures and, therefore, require concurrent high spatial and spectral resolution Earth Observation data. Nevertheless, the trade-off between the fixed spectral, spatial, and temporal resolution imposed by most satellite sensors currently in orbit do not allow monitoring of such complex vegetation communities. For instance Landsat-8, one of the most popular satellites for remote sensing, provides multispectral data at 30 m pixel resolution, which is too coarse for fine scale habitat mapping. Worldview-3 on the other hand, a commercial satellite with the finest spatial resolution to date, would lack the temporal resolution of Landsat as the surface coverage potential of the satellite is narrow to pass over the same area in regular intervals and acquire data with a consistent angle of view suitable for monitoring services.

Sentinel-2 is a twin-satellite mission for supporting the Copernicus program developed by the European Space Agency (ESA). The open access and free-of-charge policy adopted for the data products is anticipated to foster the use of remotely sensed data for routine environmental applications. These optical satellites offer high spatial resolution in the spectral domain 443–2190 nm for land monitoring services (including wetlands) and the first satellite, Sentinel-2A, was launched in June 2015. Enhanced specifications include the high spatial resolution, up to 10 m for selected bands, and the integration of four narrowband channels in the red-edge region [1] with the aim to provide enriched capabilities in vegetation studies [2]. One of the main advantages of Sentinel-2, which makes the satellite highly suitable for monitoring and inventorying purposes, is the combination of the wide swath and the frequent revisiting time. In this context, the sensors were designed with the scope to provide continuity to the Landsat mission.

Sentinel-2 will not only enhance the Landsat data archive but also deliver new capabilities in natural habitat mapping. For example, Hedley *et al.* [3] in a study simulating Sentinel-2 data, report an increased performance when comparing with Landsat ETM+ for mapping tropical coral reefs. Richter *et al.* [4] test the potential of Sentinel-2 for estimating Leaf Area Index (LAI) of three contrasting agricultural crops and suggest that Sentinel-2 can estimate biophysical parameters through the SAILH+PROSPECT model. Likewise, it is important to investigate the potential of the data products in anticipation of the full operations of the Sentinel-2 mission in order to assess their capabilities, examine the compatibility with satellites already in orbit, ensure continuity of image provision with similar specifications, and accelerate the operational use of the forthcoming imagery. New sensors, such as the Sentinel-2, are equipped with super-spectral resolution. It is yet to be seen whether the addition of extra bands in combination with the high spatial resolution suffices for fine-scale aquatic vegetation mapping.

This study evaluates the potential of the Sentinel-2 satellites for mapping lakeshore vegetation at the fine scale. A simulation of the high spatial resolution (*i.e.*, 10 m and 20 m) bands of Sentinel-2 on the VNIR (Visible and Near Infrared) region of the spectrum was compiled from airborne hyperspectral imagery acquired with AISA Eagle and Hawk instruments over a reed bed on the shore of Lake Balaton, Hungary. The spectral

Citation: Dimitris Stratoulias, Heiko Balzter, Olga Sykioti, András Zlinszky and Viktor R. Tóth, Evaluating Sentinel-2 for Lakeshore Habitat Mapping Based on Airborne Hyperspectral Data, doi:10.3390/s150922956.

and spatial aggregations are derived separately, as well as the final simulated Sentinel-2 image. The image is classified based on the dominant land cover types encountered in the scene and reed bed subclasses. A comparison of the products with the classification of the original AISA image reveals the effect of spectral and spatial upscaling and the information lost through this process when mapping lakeshore vegetation.

DATASET AND STUDY AREA

This study focuses on the east part of Bozsai Bay on the northwest part of the Tihany Peninsula of Lake Balaton, Hungary (Figure 1). The bay is a nature reserve of the Balaton Uplands National Park and quasi-undisturbed by human activity. The rare and variable vegetation encountered ranging from wetland to arid species is the result of the mild, sub-Mediterranean climate of the peninsula. It encompasses a variety of wetland plants, trees and Pannonic grasslands, however, the main ecological focus is on the dominant species *Phragmites australis* (Cav.) Trin. ex Steud, and a relatively smaller degree on *Typha angustifolia L.*, *Typha latifolia L.* and *Carex sp*. The reed bed has been suffering during the last decades from reed die-back, which is a deterioration of stability and retreat from the relatively deep water [5,6] and, therefore, apart from different species, the habitat contains vegetated patches of different stability, whereas die-back conditions occur locally at the waterward fringe of the reed bed.

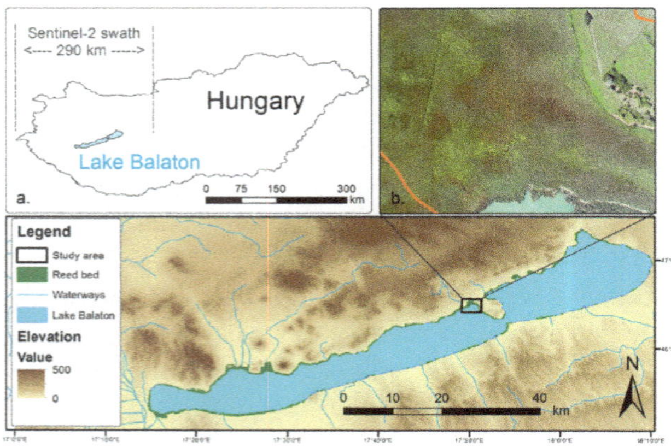

Figure 1: Study area in Lake Balaton, Hungary. (**a**) Position of Lake Balaton and visualization of the 290 km swath of Sentinel-2; and (**b**) true-colour RGB aerial photo (0.25 m) of the study area acquired concurrently with the hyperspectral data.

An airborne hyperspectral image from a Specim AISA dual system (Spectral Imaging Ltd., Oulu, Finland) incorporating the nadir-looking sensors Eagle and Hawk was employed as in similar studies [6,7,8,9,10,11,12,13,14]. The campaign was undertaken at Lake Balaton in August 2010 [15] by the Natural Environment Research Council (NERC) Airborne Research and Survey Facility (ARSF) in the framework of the European Facility for Airborne Research (EUFAR). The sensors recorded incoming radiation cumulatively in 509 bands from 400 nm to 2450 nm with Full-Width Half-Maximum (FWHM) 2.20–2.44 and 6.32 for Eagle and Hawk, respectively, and delivered a spatial resolution of 1.5 m at 1550 m true altitude. The angle between the line of sight of the sensors and the zenith was 180° since they are both nadir looking instruments. The flying orientation of the aircraft was planned to a north-south bearing analogous to the direction of the polar orbit of the Sentinel-2 satellite (inclination 98.5°). It is worth noting the difference in spatial coverage between the AISA and Sentinel-2 systems in Figure 1 and consider the capabilities of the latter for large-scale operational remote sensing applications; the swath width of the airborne imager is small (*i.e.*, 620 m and 990 m for Hawk and Eagle sensors, respectively) due to the low altitude of the aircraft; conversely the 290 km swath width of Sentinel-2 assures that the whole Lake Balaton is included within a maximum of two overpasses.

METHODOLOGY

The methodology is based on the spectral and spatial aggregation of the airborne hyperspectral dataset in order to synthesize an image with characteristics similar to Sentinel-2. AISA specifications are superior to Sentinel-2 specifications [16] as it inherits a higher spectral, spatial and radiometric resolution (Table 1); therefore, simulation of Sentinel-2 is realistic by aggregating the AISA data.

Table 1: Specifications of the Sentinel-2 sensor. The bands simulated in this study are shown in bold

Band No.	Central Wavelength	Bandwidth	Spatial Resolution	Radiometric Resolution
1	443 nm	20 nm	60 m	
2	**490 nm**	**65 nm**	**10 m**	
3	**560 nm**	**35 nm**	**10 m**	
4	**665 nm**	**30 nm**	**10 m**	
5	**705 nm**	**15 nm**	**20 m**	
6	**740 nm**	**15 nm**	**20 m**	
7	**783 nm**	**20 nm**	**20 m**	12-bit
8	**842 nm**	**115 nm**	**10 m**	
8b	**865 nm**	**20 nm**	**20 m**	
9	945 nm	20 nm	60 m	
10	1380 nm	30 nm	60 m	
11	**1610 nm**	**90 nm**	**20 m**	
12	**2190 nm**	**180 nm**	**20 m**	

The AISA dual image (Eagle and Hawk) was first pre-processed by applying radiometric normalization, atmospheric correction, and geo-registration consecutively. Cross-track illumination correction is performed using the ENvironment for Visualising Images (ENVI) 5.2 image processing software to remove sun glint resulting from the perpendicular orientation of the aircraft relative to the sun position at the time of image acquisition. Atmospheric correction was carried out with Fast Line-of-sight Atmospheric Analysis of Spectral Hypercubes (FLAASH) [17] using the mid-latitude atmospheric model without aerosol correction. Geo-registration at a 2 m pixel grid was implemented with the Airborne Processing Library software [18] using bore-sight information recorded concurrently with the imagery. Eagle and Hawk data were spectrally merged to produce a single image covering the spectrum 400–2500 nm. Thereafter the bands outside the spectrum covered cumulatively by the Sentinel-2 channels were removed in order to assure comparability between the hyperspectral and the simulated images. Subsequently the following four input products for classification were derived (Figure 2):

1. The first 10 components of the inversed Minimum Noise Fraction (MNF) transformation. This product contains the entire information from the data and is considered representative of the hyperspectral image in this set-up.
2. Spectrally resampled image. The Spectral Response Function (SRF) is convolved with the hyperspectral data for the spectral resampling. The SRF was provided

by ESA-European Space Research Institute (ESRIN).

3. Spatially aggregated image. The 10 bands as of Table 1 are used. The spatial aggregation is performed individually for each narrow band according to the spatial resolution of the corresponding Sentinel-2 band (*i.e.*, 10 m or 20 m) using bilinear convolution. Bands with coarse resolution (*i.e.*, 20 m) are downscaled to 10 m and then merged with the rest of the bands. Subsequently the first 10 components of the inversed MNF transformation were derived to reduce spectral dimensionality.
4. Consecutive spatial and spectral resampling (steps 3 and 2 consecutively). This product is, in essence, the Sentinel-2 simulated image.

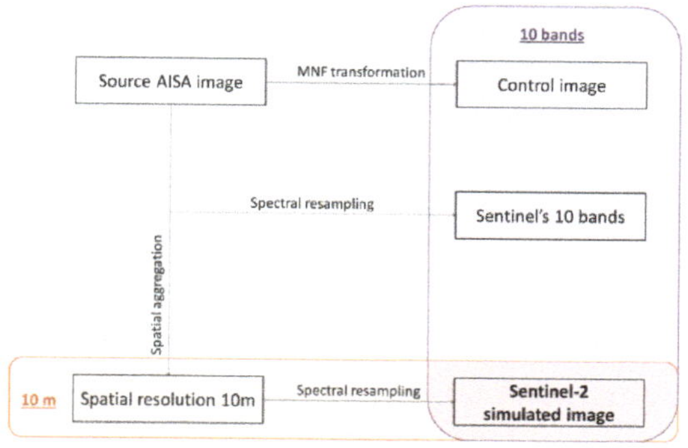

Figure 2: Workflow of the methodology; derivation of the four input datasets from the AISA airborne hyperspectral imagery.

The four products are classified with Support Vector Machines (SVM) using the same parameters and training set. A *Phragmites*-specific classification scheme was set up as the reed bed comprises mainly of *Phragmites* co-occurring with other species. Four categories of *Phragmites* are defined according to the degree of dominance: reed dominant, co-dominant, sub-dominant, and reed die-back. Other macrophyte classes such as *Typha* and *Carex* are considered dominant at the areas where *Phragmites* is sub-dominant. Classes of trees and bare soil were included on the assumption that Sentinel-2 products will be frequently used over large areas and in more generic aquatic environment monitoring.

An assessment of relative accuracy is undertaken, indicating the loss of information when downgrading to the three different thematic maps from the source hyperspectral image, the latter considered as the "truth" data. The comparison is based on the change detection between the AISA hyperspectral classification, considered always as the control image, and each of the three simulated products which are evaluated. Emphasis is placed on the final Sentinel-2 simulated image when discussing the results. The change detection was carried out in ENVI 5.2 (Exelis Visual Information Solutions, Boulder, CO, USA) software and reports pixel values to stress the low frequency of occurrence of some classes, such as in the case of reed die-back.

RESULTS AND DISCUSSION

As demonstrated in Figure 3, the Sentinel-2 bands with 10 m spatial resolution occupy the visible and near-infrared spectrum, while bands with 20 m spatial resolution are concentrated around the red-edge region and longer wavelengths encompassing the hyperspectral vegetation spectrum. From a spectral point of view, the 10 m bands are placed at wavebands containing key information for vegetation studies, as is the chlorophyll absorption characteristics. 20 m bands on the other hand have more diverse characteristics, with wide bands in the near-infrared and four narrow bands distributed in the narrow waveband of the red-edge region. It is worth noting that the red-edge bands are as narrow to include only 3–5 measurements from the corresponding hyperspectral data and, therefore, inherit a spectroscopic nature. Comparing the detail of spectra from the two sources, it becomes apparent that the near-infrared bands are averaging the hyperspectral information available over a spectral region, while in the optical and red-edge domain the waveband covered by the Sentinel-2 channels is rather constant for the case of a pure monospecific reed pixel demonstrated here.

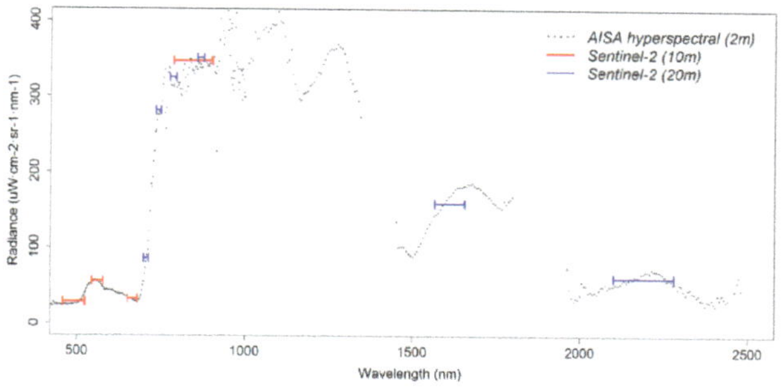

Figure 3: A spectral profile from a pixel of pure monospecific reed from the combined Eagle and Hawk data (grey) and the Sentinel-2 simulated image (colour).

Table 2 presents the Sentinel-2 between-band correlations for the spectrally-resampled image. The three red-edge bands are highly correlated at 0.99 which poses questions about the discriminatory capabilities of the extra bands in regards to vegetation condition and stress. Conversely, Delegido *et al.* [2] have already demonstrated the importance of Sentinel's-2 red-edge bands for the estimation of LAI and chlorophyll content. However, it has to be taken into account that our results derive from simulated data, and the Sentinel-2 sensor might exhibit different robustness in the way radiation is recorded; hence, a real image would be necessary to judge on the inter-band correlation with certain confidence.

Table 2: Correlation between the spectral channels of the simulated Sentinel-2 image. The bands with correlation <0.8 are shown in bold

Central Band	490	560	665	842	705	740	783	865	1610	2190
490 nm	1	0.99	0.91	0.80	**0.52**	**0.50**	**0.49**	**0.49**	**0.60**	**0.62**
560 nm		1	0.88	0.81	**0.58**	**0.55**	**0.55**	**0.55**	**0.61**	**0.60**
665 nm			1	0.92	**0.62**	**0.59**	**0.59**	**0.59**	**0.79**	0.85
842 nm				1	0.87	0.84	0.84	0.84	0.94	0.90
705 nm					1	0.99	0.99	0.99	0.90	**0.74**
740 nm						1	0.99	0.99	0.88	**0.71**
783 nm							1	0.99	0.89	**0.71**
865 nm								1	0.89	**0.72**
1610 nm									1	0.94
2190 nm										1

The results from the classification of the four input images are presented in Figure 4. The transitional steps of spectral or spatial aggregation indicate the level of information loss due to reduction of the spectral dimensionality and spatial upscaling from the airborne hyperspectral dataset respectively. The consecutive spatial and spectral aggregation represents the simulated Sentinel-2 image.

Classification of the spectrally-aggregated image results in a vegetation map very similar to the hyperspectral output. This indicates that the near-complete spectral information inherited by the Sentinel-2 bands contains vital information regarding vegetation targets; indeed, in such a classification set-up the superspectral characteristics of Sentinel-2 are sufficiently suitable for lakeshore vegetation mapping. However, it is worth noting that in such simulation experiments the effects of the atmospheric path, and how well it can be accounted for if a satellite product is used, cannot be simulated. While FLAASH was suitable for hyperspectral data and provides surface reflectance products based on input parameters estimated from scene information [19], the spectral resolution and bandwidth of the satellite channels will not provide adequate information in order to perform such an accurate atmospheric correction; nevertheless, the 60 m bands of Sentinel-2 will supply information on water vapor and aerosols and allow for a broad atmospheric correction based on within-scene information.

Under spatial aggregation, macrophyte-specific classes are not as well delineated as in the hyperspectral image and, therefore, capability of fine-scale mapping appears restricted. This may be attributed to the fact that spatial degradation results in larger pixels and, therefore, more classes are encountered within a single pixel. Consequently, the image is constructed from mixed pixels and the discriminatory capability between similar classes is weakened. For instance, the road network between the bare soil land parcels on the top of the image is not identifiable since the road, which is narrow compared to the pixel size, is integrated in the dominant surrounding class.

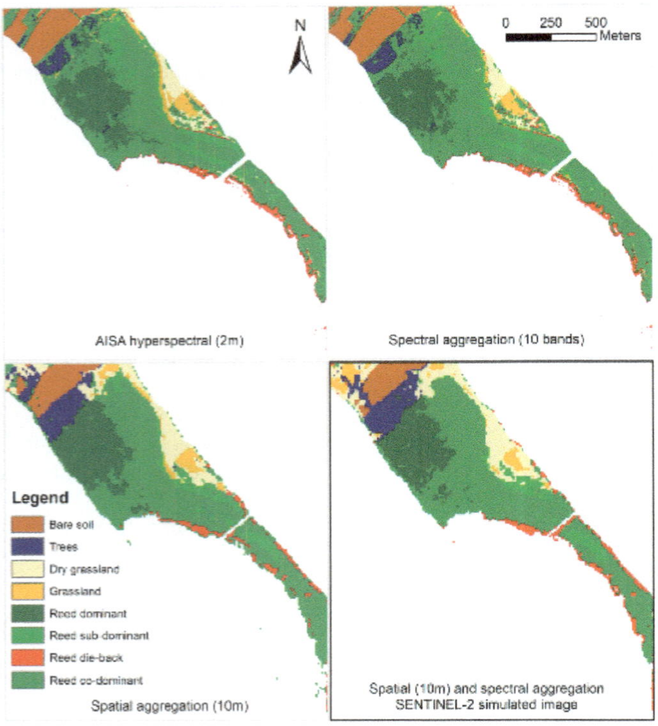

Figure 4: Classification of the four input products; highlighted on the lower right the classification result derived from the Sentinel-2 simulated image.

With regards to the simulated Sentinel-2 image (*i.e.*, lower right inset of Figure 4), separability between generic classes, as well as vegetation species, is maintained and moreover all classes are present. Nevertheless, misclassifications occur between macrophyte classes. Thematic consistency in generic classes such as trees, bare soil, and reed is maintained; however, the discriminatory capability is constrained when mapping classes which inherit higher inter-class spectral variability, such as macrophyte associations, fact indicated by sporadic misclassified patches. The vegetation classes of reed and trees are separated well; nevertheless, sub-dominant reed and grassland are largely being confused in the Sentinel-2 simulated image, as a result of the nature of sub-dominant reed, which most probably includes grassland at the terrestrial part of the reed bed. This suggests that similar classes, such as the aforementioned, are not distinguished as satisfactorily as in the case of the control image.

Reed die-back is classified correctly as slivers at the waterward edge of the reed bed from the Sentinel-2 data at the 10 m resolution and a complete lakeshore vegetation status seems to be feasible based on the simulated data, which would be much needed from a conservation and management perspective. Vegetation-specific applications based on Sentinel-2 imagery are further supported by the fact that mapping of vegetation vigor requires sound atmospheric correction methods, for which Sentinel-2 has dedicated bands.

Table 3: Assessment of relative accuracy between the classification product of the AISA hyperspectral image (control image) and the three simulated products. All values are in number of pixels

	AISA Hyperspectral Classification									
	Bare Soil	**Trees**	**Dry Grassland**	**Grassland**	**Reed Dominant**	**Reed Sub-Dominant**	**Reed Die-Back**	**Reed co-Dominant**	**Row Total**	**Class Total**
Sentinel-2 simulated classification										
Unclassified	3	-	10	1	-	7	497	23	541	1,266,729
Bare soil	**16,918**	17	183	9	42	21	340	721	18,251	18,290
Trees	3208	**10,344**	273	385	29,063	1744	395	9433	54,845	54,985
Dry grassland	6609	1239	**9055**	5244	4343	5105	973	12,717	45,285	46,450
Grassland	9375	108	1265	**3874**	1979	8392	65	4095	29,153	29,165
Reed dominant	-	53	-	1	**22,013**	42	-	7208	29,317	29,330
Reed sub-dominant	197	33	1382	53	333	**3309**	290	6181	11,778	11,800
Reed die-back	422	74	267	7	84	463	**4224**	5778	11,319	15,695
Reed co-dominant	908	1021	1306	121	11,192	10,877	3830	**79,177**	108,432	112,755
Class total	37,640	12,889	13,741	9695	69,049	29,960	10,614	125,333	-	-
Class Changes	20,722	2545	4686	5821	47,036	26,651	6390	46,156	-	-
Image Difference	**-19,350**	**42,096**	**32,709**	**19,470**	**-39,719**	**-18,160**	**5081**	**-12,578**	-	-
Spectral resampled (10 Sentinel bands)										
Unclassified	-	-	-	-	-	4	-	129	133	1,278,528
Bare soil	**36,004**	-	613	-	-	334	-	521	37,472	37,474
Trees	2	**5867**	1	-	805	1069	13	-	7757	7757
Dry grassland	531	-	**9240**	621	-	1049	40	236	11,717	11,717
Grassland	131	-	1439	**5662**	2	308	265	4	7811	7811
Reed dominant	25	6394	7	17	**51,821**	5281	1815	2	65,362	65,362
Reed sub-dominant	174	574	1316	1505	13,784	**109,904**	6924	1644	135,825	135,826
Reed die-back	-	54	75	1888	2636	4997	**20,898**	15	30,563	30,563
Reed co-dominant	773	-	1050	2	1	2387	5	**8067**	12,285	12,583
Class total	37,640	12,889	13,741	9695	69,049	125,333	29,960	10,618	-	-
Class Changes	1636	7022	4501	4033	17,228	15,429	9062	2551	-	-
Image Difference	**-166**	**-5132**	**-2024**	**-1884**	**-3687**	**10,493**	**603**	**1965**	-	-

Spatial aggregation (10 m)										
Unclassified	261	13	26	4	5	307	5	198	819	1,263,059
Bare soil	**21,856**	21	327	22	98	1196	41	549	24,110	24,195
Trees	2424	**11,149**	107	375	30,578	10,948	2493	138	58,212	58,415
Dry grassland	6832	251	**9350**	3948	1831	7453	1480	850	31,995	32,785
Grassland	25	16	603	**3929**	596	592	2055	13	7829	7835
Reed dominant	-	199	11	4	**25,170**	9443	69	-	34,896	34,950
Reed sub-dominant	5500	1232	2093	1190	10,533	**86,738**	18,295	4964	130,545	140,765
Reed die-back	51	-	987	198	129	5313	**4573**	176	11,427	11,475
Reed co-dominant	691	8	237	25	109	3343	949	**3726**	9088	11,720
Class total	37,640	12,889	13,741	9695	69,049	125,333	29,960	10,614	-	-
Class Changes	15,784	1740	4391	5766	43,879	38,595	25,387	6888	-	-
Image Difference	**-13,445**	**45,526**	**19,044**	**-1860**	**-34,099**	**15,432**	**-18,485**	**1106**	-	-

Table 3 presents the relative accuracy of the three simulated products based on the control image, as estimated by change detection between the two products for each instance. For the case of the simulated Sentinel-2 product, upon which the main focus is placed, it is apparent that reed classes, with the exception of reed die-back, are converted to other generic categories. Reed die-back is overestimated due to the small size of the patches, which forces the spatial aggregation of the pixels to assign the reed die-back class to larger pixels. The main misclassification is for the case of trees, due to the homogenization around the patches where individual trees or group or trees are actually found. A comparison of the three tables indicates that the misclassifications for the spectral aggregation are considerably less than the spatial aggregation and the Sentinel-2 product, a fact which indicates that the process of spatial upscaling is the main factor fostering the alteration of the classification outcome. Another observation that can be made apparent is the fact that for crisp classes (*i.e.*, bare soil, trees, dry grassland, and reed dominant) the misclassifications of the Sentinel-2 and the spatial aggregated products are similar; however, for classes with softer spectral boundaries (*i.e.*, grassland, reed sub-dominant, reed-codominant, and reed die-back) the image difference for these two products is reversed, indicating that the misclassifications of Sentinel-2 are not solely attributed to the spatial resampling, but probably to the consecutive execution of spatial and spectral resampling.

Overall, Sentinel-2 performs satisfactorily in classifying wetland ecosystems with high discrimination complexity. The 10 m spatial resolution allows for detecting fragmented patches at the edge of the reed bed, which is the main manifestation of die-back conditions. Confusion between macrophyte classes and associations is low at the individual spectral aggregation and rather high at the spatial resolution, which indicates that mapping of such classes is feasible with the spectral characteristics of Sentinel-2, but at a finer pixel resolution than that of 10 m.

CONCLUSIONS

Sentinel-2 is a twin satellite mission with enhanced spectral and spatial capabilities anticipated to provide further potential in habitat monitoring and classification of environmentally complex areas. In the framework of the recent launch of the first satellite, simulated data are needed to assess the quality of the expected product and fruitfully exploit the imagery.

In this study, a classification from simulated Sentinel-2 imagery was presented, synthesized from hyperspectral airborne data over a nature reserve site on Lake Balaton in order to assess the suitability of the imagery for fine-scale lakeshore vegetation mapping. An SVM supervised classification of a Sentinel-2 simulated image, in comparison to the MNF transformation of the hyperspectral source image, shows that the thematic consistency of generic classes such as trees, bare soil, and reed is maintained; however, when mapping classes inheriting higher inter-class spectral variability (e.g., macrophyte associations) the discrimination capability is constrained. The information is well-preserved in spectral resampling, while spatial upscaling introduces clumping of classes and mixed pixels. This study adds to the already documented high potential of Sentinel-2 for demanding Earth Observational needs [20]. Further research on the usefulness of the introduced narrow-width red-edge bands is required to investigate to what degree they can assist in stress mapping and biophysical parameters estimation.

The very large swath of Sentinel-2 (290 km) provides the opportunity of large scale main categorical mapping of large wetlands and complete lake systems, a foreseen product with high value for the conservation and management of such areas [21]. It is foreseeable that the upcoming Sentinel-2 data will enable users to derive frequent products of aquatic vegetation with wide area coverage, fine spatial resolution and thematic consistency. Moreover, synergistic use of optical data with other Earth Observation sources which have demonstrated capabilities in aquatic mapping, such as Synthetic Aperture Radar (SAR) [22] or Airborne Laser Scanning (ALS) [23] will enhance the vegetation-related information; for instance, SAR provides data associated with the inundation level, biomass, and soil moisture, while ALS relates to plant height, both cases complimentary to optical sensors' information. In an operational context, for example, Sentinel-1, the respective SAR satellite of Sentinel-2, is already streaming free-access data and a more accurate representation of the aquatic vegetation condition could be formed when integrating the two Sentinel sources.

ACKNOWLEDGMENTS

This paper has been published with financial support from the Natural Environment Research Council (NERC). Dimitris Stratoulias has been a Marie Curie Early Stage Researcher

(ESR) with the GIONET program, funded by the European Commission, Marie Curie Programme, Initial Training Networks Grant Agreement number PITN-GA-2010-26450. Heiko Balzter was supported by the Royal Society Wolfson Research Merit Award, 2011/R3 and the NERC National Centre for Earth Observation. The airborne data were collected and provided by the Airborne Research and Survey Facility vested in the Natural Environment Research Council under the EUFAR Contract Number 2271. The Spectral Response Function of Sentinel-2 was kindly provided by Ferran Gascon from ESRIN/ESA.

AUTHOR CONTRIBUTIONS

Dimitris Stratoulias wrote the paper and did the data analysis. Heiko Balzter conceived the general idea. Viktor R. Tóth contributed in the field data collection and interpretation of the results from an ecological point of view. Olga Sykioti assisted in the analysis of the remote sensing data. András Zlinszky organized the airborne data campaign. All authors read, reviewed and approved the manuscript.

REFERENCES

1. Drusch, M.; Del Bello, U.; Carlier, S.; Colin, O.; Fernandez, V.; Gascon, F.; Hoersch, B.; Isola, C.; Laberinti, P.; Martimort, P.; *et al.* Sentinel-2: ESA's optical high-resolution mission for GMES operational services. *Remote Sens. Environ.* 2012, *120*, 25–36.
2. Delegido, J.; Verrelst, J.; Alonso, L.; Moreno, J. Estimation of Sentinel-2 red-edge bands for empirical estimation of green LAI and Chlorophyll content. *Sensors* 2011, *11*, 7063–7081.
3. Hedley, J.; Roelfsema, C.; Koetz, B.; Phinn, S. Capability of the Sentinel 2 mission for tropical coral reef mapping and coral bleaching detection. *Remote Sens. Environ.* 2012, *120*, 145–155.
4. Richter, K.; Atzberger, C.; Vuolo, F.; D'Urso, G. Evaluation of Sentinel-2 spectral sampling for Radiative transfer model based LAI estimation of wheat, sugar beet and maize. *IEEE J. Sel. Top. Appl. Earth Obs. Remote Sens.* 2011, *4*, 458–464.
5. Van der Putten, W.H. Die-back of *Phragmites australis* in European wetlands: An overview of the European Research Programme on Reed Die-Back and Progression (1993–1994). *Aquat. Bot.* 1997, *59*, 263–275.
6. Stratoulias, D.; Balzter, H.; Zlinszky, A.; Tóth, V.R. Assessment of ecophysiology of lake shore reed vegetation based on chlorophyll fluorescence, field spectroscopy and hyperspectral airborne imagery. *Remote Sens. Environ.* 2015, *157*, 72–84.
7. Artigas, F.J.; Yang, J.S. Hyperspectral remote sensing of marsh species and plant vigour gradient in the New Jersey Meadowlands. *Int. J. Remote Sens.* 2005, *26*, 5209–5220.
8. Jensen, R.; Mausel, P.; Dias, N.; Gonser, R.; Yang, C.; Everitt, J.; Fletcher, R. Spectral analysis of coastal vegetation and land cover using AISA+ hyperspectral data. *Geocarto Int.* 2007, *22*, 17–28.
9. Dalponte, M.; Bruzzone, L.; Gianelle, D. Fusion of hyperspectral and LIDAR remote sensing data for classification of complex forest areas. *IEEE Trans. Geosci. Remote Sens.* 2008, *46*, 1416–1427.
10. Shafri, H.Z.M.; Hamdan, N. Hyperspectral imagery for mapping disease infection in oil palm plantation using vegetation indices and red-edge techniques. *Am. J. Appl. Sci.* 2009, *6*, 1031–1035.
11. Yang, J.; Artigas, F.J. Mapping salt marsh vegetation by integrating hyperspectral and LiDAR remote sensing. In *Remote Sensing of Coastal Environments*; Wang, Y., Ed.; CRC Press: Boca Raton, FL, USA, 2009.
12. Burai, P.; Lövei, G.Z.; Csaba, L.; Nagy, I.; Enyedi, P. Mapping aquatic vegetation of the Rakamaz-Tiszanagyfalui Nagy-morotva using hyperspectral imagery. *AGD Landsc. Environ.* 2010, *4*, 1–10.
13. Onojeghuo, A.O.; Blackburn, G.A. Optimising the use of hyperspectral and LiDAR data for mapping reedbed habitats. *Remote Sens. Environ.* 2011, *115*, 2025–2034.
14. Peerbhay, K.Y.; Mutanga, O.; Ismail, R. Commercial tree species discrimination using airborne AISA Eagle hyperspectral imagery and partial least squares discriminant analysis (PLS-DA) in KwaZulu–Natal, South Africa. *ISPRS J. Photogramm. Remote Sens.* 2013, *79*, 19–28.
15. Zlinszky, A.; Tóth, V.R.; Pomogyi, P.; Timár, G. Initial report of the AIMWETLAND project: Simultaneous airborne hyperspectral, LiDAR and photogrammetric survey of the full shoreline of Lake Balaton, Hungary. *Geogr. Tech.* 2011, *1*, 101–117.
16. ESA Sentinel-2 Team. GMES Sentinel-2 Mission Requirements Document. EOP-SM/1163/MR.dr, issue 2.1. Available online: http://esamultimedia.esa.int/docs/GMES/Sentinel-2_MRD.pdf (accessed on 1 September 2015).
17. Module, F. *Atmospheric Correction Module: QUAC and FLAASH User's Guide*; ITT Visual Information Solutions: Boulder, CO, USA, 2009.
18. Warren, M.A.; Taylor, B.H.; Grant, M.G.; Shutler, J.D. Data processing of remotely sensed airborne hyperspectral data using the Airborne Processing Library (APL): Geocorrection algorithm descriptions and spatial accuracy assessment. *Comput. Geosci.* 2014, *64*, 24–34.
19. Griffin, M.K.; Burke, H.K. Compensation of hyperspectral data for atmospheric effects. *Linc. Lab. J.* 2003, *14*, 29–54.
20. Malenovský, Z.; Rott, H.; Cihlar, H.; Schaepman, M.E.; García-Santos, G.; Fernandes, R.; Berger, M. Sentinels for science: potential of Sentinel-1, -2 and -3 missions for scientific observations of ocean, cryosphere, and land. *Remote Sens. Environ.* 2012, *120*, 91–101.
21. Schmieder, K. European lake shores in danger—Concepts for a sustainable development. *Limnologica* 2004, *34*, 3–14.
22. Bourgeau-Chavez, L.L.; Kowalski, K.P.; Mazur, M.L.C.; Scarbrough, K.A.; Powell, R.B.; Brooks, C.N.; Huberty, B.; Jenkins, L.K.; Banda, E.C.; Galbraith, D.M.; *et al.*

Mapping invasive *Phragmites australis* in the coastal Great Lakes with ALOS PALSAR satellite imagery for decision support. *J. Gr. Lakes Res.* 2013, *39*, 65–77.

23. Zlinszky, A.; Mücke, W.; Lehner, H.; Briese, C.; Pfeifer, N. Categorizing wetland vegetation by airborne laser scanning on Lake Balaton and Kis-Balaton, Hungary. *Remote Sens.* 2012, *4*, 1617–1650.

CHAPTER 3

FORAGING ECOLOGY OF THREE SYMPATRIC UNGULATE SPECIES–BEHAVIOURAL AND RESOURCE MAPS INDICATE DIFFERENCES BETWEEN CHAMOIS, IBEX AND RED DEER

Anna K Schweiger, Martin Schütz, Pia Anderwald, Michael E Schaepman, Mathias Kneubühler, Rudolf Haller3and Anita C Risch

Keywords: Biomass, Forage, Imaging spectroscopy, Movement, Nitrogen, Remote sensing, Resource selection, T-LoCoH, Utilisation distribution, Vegetation

BACKGROUND

How ecologically similar species coexist in a shared habitat is a fundamental question in ecology [1]. Resource ecology provides the basis for understanding multi-species assemblages as it deals with plant-nutrient relationships, interactions between consumers and resources and interactions among consumers [2]. Foraging is the central process in resource ecology as it leads to growth, survival and reproduction of the animal and thus, ultimately influences its fitness [2]. Ungulates forage selectively [3] and are not only influenced by vegetation and landscape structure, but are themselves major drivers of landscape heterogeneity [4-8]. Additionally, ungulates affect the abundance and population dynamics of other species, ranging from herbivores [9] to soil decomposers [10] which in turn feed back to vegetation composition and structure. These traits make the spatial distribution and foraging ecology of ungulates an important issue in wildlife management, nature protection and landscape conservation [11].

The diversity of ungulate communities is often explained by differences in their dietary niches [12]. Most studies have investigated forage selection based on plant family or functional type (e.g. graminoids, forbs, shrubs) level and have used either direct observations [13,14], fence experiments [13,15] or micro-histological analysis of faecal pellets [16-20]. Thus, ungulates are traditionally categorised according to their feeding types as grazers, mixed feeders or browsers (concentrate selectors) [3,12]. Previous studies defined chamois (*Rupicapra rupicapra* L.), ibex (*Capra ibex* L.) and red deer (*Cervus elaphus* L.) as mixed feeders, with chamois being closer to browsers, ibex closer to grazers and red deer in between [12,13,21]. Such similarity of dietary niches would imply high potential for competition among the three species, especially when population numbers are high. In the Trupchun valley of the Swiss National Park (SNP) population sizes of sympatric chamois, ibex and red deer are amongst the highest in central Europe. Previous studies in this area investigated the forage composition of the three species using micro-histological analysis of faecal pellets and found no significant differences in the proportions of grasses, sedges, forbs and woody species consumed during spring and summer [22-24].

However, large variations in forage composition at the plant species and plant family level were not only reported between, but also within ungulate species (e.g. for deer species see [3]) suggesting flexibility in their dietary choices. Additionally, plant species within a vegetation type can strongly differ in growth form and nutritious value resulting from small scale heterogeneity of microclimate and soil, which is especially pronounced in alpine landscapes [25]. Thus, vegetation type classifications might conceal the heterogeneity of forage resources [26]. High-resolution remote sensing has demonstrated the potential to detect environmental

Citation: Anna K Schweiger, Martin Schütz, Pia Anderwald, Michael E Schaepman, Mathias Kneubühler, Rudolf Haller and Anita C Risch, Foraging ecology of three sympatric ungulate species – Behavioural and resource maps indicate differences between chamois, ibex and red deer, DOI: 10.1186/s40462-015-0033-x.

heterogeneity [27,28] at a spatial scale fine enough to be relevant for foraging animals [3,26,29]. Advanced observational approaches such as imaging spectroscopy (IS [30,31]) make it possible to detect changes in plant biochemical and biophysical composition [28,32,33], and plant species distribution [34]. Plant biomass and plant nitrogen (N) content are vegetation characteristics important for forage resource selection in ungulates [35-39] and have already been mapped successfully using IS in heterogeneous grassland ecosystems [35,40-42].

The home range (HR) of an animal is the area traversed by the individual during its normal activities of food gathering, mating and caring for offspring [43]. Advances in global positioning system (GPS) technology have made it possible to collect large amounts of location data [44,45] and several HR estimators (polygon methods) - from minimum convex polygons (MCP) to alpha hulls [46], kernel density estimators (KDE) and local convex hulls (LoCoH) [47] - have been proposed.

Traditional HR estimators have been criticised for treating locations as spatially and temporally independent, an assumption that can only be fulfilled when data are collected either at random [48] or at time intervals long enough to allow an animal to move to any place within its HR [49]. However, it has been argued that efforts to handle spatial autocorrelation, which can be an intrinsic data attribute [50], have drawn attention away from more important questions in HR analysis [51]. Instead of removing spatial autocorrelation, which has been shown to be of limited relevance for HR estimators (e.g. KDE) it can be used as a source of biological information and therefore be incorporated in models of animal movement and space use [52].

Similarly, polygon methods have been criticised for giving only limited information about the species' biology when focussing on the perimeter (size and shape) of an HR. Thus, additional insights into the species' biology might be gained (i.e. what the animal did and where) by using spatial and temporal autocorrelation to delineate areas coupled to the animal's behaviour [48]. During the last decade, models of space use incorporating temporal autocorrelation of GPS data became more widely available, including (dynamic) Brownian bridge movement models (BBMM) [53,54], Levy flight movement models [55], movement based kernel density estimators (MKDE) [56] and time geography methods [57,58]. Similarly, behavioural models (models of time-use) that take advantage of the temporal autocorrelation of GPS data, such as cognitive models [59] or state-space models [60] were developed.

One of the few methods that take both spatial and temporal autocorrelation of GPS data into account is the Time Local Convex Hull approach (T-LoCoH) [61]. T-LoCoH introduces time-scaled distance (TSD), which measures the distance between two points in both space (using the x/y coordinates) and time (using timestamps) allowing to calculate time-use metrics, such as directionality of movement, duration of stay or revisitation rate of a specific area [61]. These metrics can be used to generate behavioural maps serving as proxies to delineate migration corridors, resting or foraging areas. Core foraging areas have been defined as regions within an HR that are most heavily used for foraging [62] and have been approximated by taking the 10% to 50% isopleths of an animal's utility distribution (UD; see [63,64]). T-LoCoH's ability to detect behavioural phases using the temporal and spatial autocorrelation structure of GPS data [61] fills an important gap in HR [50] analyses.

Ungulates are mobile, have accurate spatial memory [11], spend most of their time feeding [65] and allocate their time according to the resources available [8,66]. Thus, areas frequently revisited by the animals can be expected to contain important forage resources. Thanks to the SNP's long term monitoring and behavioural studies [67,68] we know that ungulates in the SNP follow daily movement patterns between foraging sites and are active during most time of the day, likely caused by limited disturbances (strong protection status, absence of predators). Red deer are known to follow a bimodal diurnal rhythm (peaks around sunrise and sunset) in areas strongly influenced by human activities [69], while behaving polyphasal (several activity peaks during day and night) when human disturbance is low [70].

The goal of our study was to investigate the potential of combining high-resolution remote sensing data with a HR estimator incorporating the behavioural information contained in GPS data for studying a classical issue in resource ecology, resource partitioning between sympatric species. We used GPS data and T-LoCoH›s revisitation index to delineate the core foraging areas of the three ungulate species, chamois, ibex and red deer, co-occurring at high population densities in the Trupchun valley of the SNP and airborne IS data to map plant biomass (forage quantity) and plant N content (forage quality) at 2 m × 2 m spatial resolution. We compared vegetation characteristics in the core foraging areas of the three ungulate species with multinomial logistic regression and related our results to previous studies examining their diet composition.

METHODS

Study area

Our study was carried out in the Trupchun valley (46.6° N, 10.08° E) of the SNP, encompassing approximately 22 km^2 close to the Italian border. Elevation in the Trupchun valley ranges from 1775 to 3145 meter above sea level (m a.s.l.), the average annual temperature in the SNP is 0.9±0.5°C (mean±SD) and the mean precipitation is 754±164 mm (2004–2013, recorded at the park's weather station at 1977 m a.s.l.) [71]. The plant' growing season lasts from mid May until mid September. The Trupchun valley is known for its high numbers of co-occurring ungulates; population estimates between 9–10 chamois/km^2, 10–11 ibex/km^2 and 25–31 red deer/km^2 were reported in 2010–2013 [64].

Vegetation data

We collected vegetation data allowing validation of IS data based models in 51 (2010, 2011) to 100 plots (2012, 2013), covering the entire range of exposition, altitude, productivity and plant species composition in the Trupchun valley. These plots were 6 m × 6 m in size, homogenous in vegetation cover and species composition and were grouped into five clusters to enable harvesting within a short time frame after the APEX overflight (approximately four hours; see [41] for details). Georeferencing of the plots was performed using a high-precision GNSS (Global Navigation Satellite System) receiver (Leica 1200+, Leica Geosystems, Heerbrugg, Switzerland) with measurement accuracy < 1 cm. On the day of overflight, 1 m^2 of vegetation was clipped in the centre of each plot and immediately sealed into plastic bags. We weighed the samples the same day to determine fresh weight of plant biomass. Then the samples were dried at 65°C and milled to pass a 0.5 mm screen (Pulverisette 16, Fritsch, Idar-Oberstein, Germany). One third of the vegetation samples were chemically analysed for total plant N and plant NDF (neutral detergent fibre) content using standard laboratory methods (TruSpec CN analyser Leco Corp., St Joseph, MI, USA; Fibre Analyser 200, Ankom Technology, NY, USA). Plant NDF content is a widely used indicator of forage quality and important for ruminal function in ungulates [72]. The reflectance spectra of the vegetation samples were measured using a laboratory near-infrared reflectance spectrometer (NIRS; Multi-purpose near-infrared reflectance spectrometer (NIR-MPA), Bruker Optics, Switzerland) and chemically analysed samples were subsequently used to calibrate models for predicting plant N content and plant NDF content of all vegetation samples. NIRS models achieved predictive accuracies of $R^2=0.93$ for plant N and $R^2=0.81$ for plant NDF content. Since we found strong correlation between plant N content and plant NDF content ($R^2=-0.61$, $p<0.001$), we excluded plant NDF content from further analysis. When comparing the vegetation characteristics in the plots sampled in all four years (n=25), using Wilcoxon rank sum tests for pairwise comparisons, no significant differences regarding plant biomass (all $p \geq 0.19$) and plant N content (all $p \geq 0.35$) were found. This allowed us to combine GPS data of the animals collected in different years (but always within 43 days of the APEX flight) with the corresponding IS data sets.

Imaging spectroscopy data

Imaging spectroscopy (IS) data were collected on June 24, 2010, June 26, 2011, June 29, 2012 and July 12, 2013 using the airborne imaging spectrometer APEX [31,73], mounted on a propeller aircraft (Dornier DO-228) operated by the German Aerospace Centre (DLR). APEX covers the wavelength region between 380 nm and 2500 nm in 334 reconfigurable spectral bands. After removing noisy bands, 285 (2010), 301 (2011), 299 (2012) and 284 (2013) spectral bands remained for analysis. Ground pixel size depended on flight altitude, but was resampled to 2 m x 2 m. APEX IS data were geometrically and atmospherically corrected using the software packages PARGE [74] and ATCOR-4 [75], based on the atmospheric radiative transfer code MODTRAN-5. Geometric mis-registration of the orthorectified data was evaluated using ground-based differential global positioning system (DGPS) measurements and was found to be less than one pixel (± 2 m) in flat terrain [76] and up to two pixels (± 4 m) on steep slopes (A. Damm, personal communication). Generally, IS data collected at different times are not comparable due to differences in sun angle and atmospheric conditions resulting in varying surface anisotropy. Therefore, we used APEX IS data and ground reference vegetation data to model forage quantity and quality for each year separately. Since reference plots measured 6 m × 6 m and APEX pixel size was 2 m × 2 m, a 3 × 3 pixel aggregation scheme was defined to extract the reflectance values from the IS data per plot.

We calculated simple ratios indices (SRI=band i/band j) for all possible band combinations based on the average reflectance of the 9 aggregated pixels per plot and determined the correlation between plant biomass ($g.m^{-2}$) and plant N (%) content and the SRI using Pearson's correlation coefficient (R^2). Next, we used the SRI's with the 100 highest correlations (according to R^2) as input to model plant biomass and plant N content with linear, exponential and second order polynomial functions and validated the models using leave-one-out cross validation (see also [41]). We selected the best model according to Akaike's Information Criterion (AIC) and evaluated model fit with Theil's uncertainty coefficient (Theil's U). Compared to Pearson correlation, Theil's U has the advantage of taking deviations of the slope from its ideal value of 1 and deviations of the intercept from its ideal value of 0 into account [77]. Theil's U normalizes the sum of the squared prediction errors between observed and predicted values to a value between zero and one, with zero indicating perfect agreement [77]. Generally, values of Theil's U<0.2 indicate high, values between 0.2 and 0.4 moderately high predictive power. Due to their frequent use, we also added R^2 values in our text and graphs. Additionally, we determined predictive accuracy by calculating the root mean squared error of prediction (RMSE) and the proportion of samples predicted within less than 20% RMSE. Finally, we applied the best models to predict and map plant biomass and plant N content in all 2 m × 2 m raster cells of the grasslands in the Trupchun valley. Since our models were designed to predict plant biomass and plant N content only in grasslands, we used linear spectral unmixing (LSU) and applied a 50% threshold to exclude areas dominated by forest, rock, snow or water from mapping [78]. IS data were prepared using ENVI (version 4.7, Exelis Visual Information Solutions, Boulder, CO, US). All analyses were conducted in R [79]. For the map layout we used ArcGIS (version 10.1, Environmental Systems Research Institute, Redlands, CA, US).

GPS data collection

To match the temporal scale of IS data collection we used GPS data from five chamois (two in 2011, one in 2012, two in 2013), seven ibex (four in 2010, three in 2013) and two red deer (both in 2013) recorded within three weeks before and three weeks after the APEX IS flights (total of 43 days per year). All animals were caught and handled by SNP rangers experienced in the procedures and regularly supervised by a veterinarian. Chamois and female ibex were caught in box traps and marked without narcosis. Red deer and male ibex were darted and injected with 1 ml to 3 ml Hellabrunner Mischung (125 mg Xylazin + 100 mg Ketamin per ml), dependent on body weight. The animals were released within 30 minutes after an injection of 1 ml to 3 ml Antipamezol, an antagonistic drug. The animals were equipped with GPS PLUS collars (Vectronic Aerospace GmbH, Berlin, Germany). The fix rate was set to either four or two hours, but was resampled to the common interval of four hours during data preparation. We performed a GPS accuracy test, placing two collars at six georeferenced (Leica GNSS 1200+, see above) locations, two in the main valley, two in the forest and two in the grasslands of the Trupchun valley. We placed the collars around wooden frames with heights ranging from 120 cm to 140 cm and rotated the collars between the locations on a weekly basis. The location error of the collars was 11.3±4.7 m (mean±SD; [= SQRT ((SD (x-coordinate)^2)+(SD (y-coordinate)^2))]). If the animal was captured or recaptured during the 43 day time window around the APEX flights, the first and last days of data collection were excluded from analysis. GPS data were screened for unrealistic movement following the method of Bjørneraas et al. [80], with limiting parameters set to $\alpha = 1.5$ km/h and $\cos\theta = -0.97$ (velocity and turning angle defining erroneous turnarounds, i.e. spikes in the data), $\mu = 50$ km (possible distance travelled within 20 h) and $\Delta = 200$ km (distance impossible to travel within 20 h; for details see [80]).

Behavioural maps

We used the T-LoCoH package in R [81], a non-parametric UD method to construct behavioural maps [61] that serve as proxy to delineate the core foraging areas of the three ungulate species. T-LoCoH models space use by constructing local MCP's or hulls around each data point, which are then sorted and progressively merged to form isopleths. Sorting of the hulls can be based on different time-use metrics that serve as proxies for the animals' behaviour, such as duration of stay, directionality of movement and revisitation rates. The time-stamp of each location is incorporated in both, the selection of nearest neighbours for local hull construction and the sorting of the hulls. For hull construction, two points have to be close in time and in space to be considered nearest neighbours. T-LoCoH introduces a distance function that transforms a unit of time into a unit of distance, called time-scaled distance (TSD). The time and space components of TSD are weighted by setting parameter "s". To make comparisons between the animals possible, we used the same process for all individuals and species and set "s" to a consistent proportion of 60% time selected hulls [61,82].

The number of nearest neighbours can be defined by selecting the "k" closest points in space and time ("k-method"), the points within a defined time-scaled radius "r" ("r-method") or by identifying the nearest neighbours up to a cumulative distance "a" in space and time ("a-method"). We decided to use the "a-method" as this method is better suited for studies where both, high and low point densities of GPS locations can be expected [47]. As before, we used the same process to define parameter "a" for all individuals of all species. We set parameter "a" to a cumulative distance that stabilised the isopleths' edge to area ratio [61,82] before creating a jump in the isopleths' area, thus balancing type I (including area that is not used) and type II errors (omitting area that is used) [83]. Since absolute values for the optimal "a" across all individuals of a specific ungulate species were very close, it was reasonable to use the same value for "a" for all three species.

While there are guidelines available for selecting the weight placed on the time-component ("s"-value) and the threshold for nearest neighbour selection ("a"-value) [61], the parameters for hull sorting and the isopleths' threshold have to be based on the aim of the study and the knowledge of the animals' ecology. In our study area, the three ungulate species show distinct diurnal movement patterns. They are known to regularly return to the same areas for foraging, ruminating and resting and have multiple activity peaks per day [67,68].We therefore calculated the revisitation rate for each hull based on an inter-visit gap (IVG, time to pass for an observation to count as a separate visit) of 12 hours, sorted the hulls according to the mean number of separate visits normalised (NNSV) and merged them until 30% of all points were included (creating the 30% isopleths). While it is likely crucial to limit type II errors (omitting areas important for a species) in conservation projects (e.g. the delineation of protected areas), studies of animal behaviour profit from limiting type I errors (including area that is not used by the species) to detect the patterns of interest. Thus we decided to choose a tight threshold, i.e. the 30% isopleths, to delineate the animals' core foraging areas.

Species comparison

We fitted multinomial logistic regression models using the three ungulate species as the response and plant biomass and plant N content in the animals' core foraging areas as predictor variables. We rescaled plant biomass to a level similar to plant N content by dividing all biomass values by 100 (BiomRS=Biomass/100). As candidate models we chose i) the two models containing only one predictor variable (plant biomass or plant N content), ii) the main effects model containing both terms (plant biomass and plant N content), iii) the model including both terms plus their interaction and iv) the intercept-only model. We selected the best model based on differences of AIC (Δ AIC) and confirmed our selection using the likelihood ratio test. To evaluate model fit we calculated the Hosmer-Lemeshow goodness-of-fit statistic and the area under curve (AUC) of the receiver-operating characteristic (ROC) for each of the two logits separately [84]. The ROC is obtained by plotting all sensitivity values (true-positive fraction) on the y-axis against their equivalent 1-specificity values (false-positive fraction) for all thresholds on the x-axis. Thus, this measure of overall accuracy is independent of any threshold [85]. AUC values between 0.7 and 0.8 indicate good, values between 0.8 and 0.9 excellent discriminative ability [86]. We assessed the sensitivity of our results with regard to the size of the animals' core foraging areas by re-running the analysis after adding and subtracting a 6 m buffer to the core foraging areas, respectively, and tested the hypothesis of equality of the model coefficients. For analyses and graphs we used the packages nnet [87], pROC [88] and effects [89] in R [79].

RESULTS

The SRI models of grassland vegetation developed from APEX IS data predicted plant biomass and plant N content with high to moderately high predictive power (Table 1). Generally, the grasslands in the Trupchun valley showed high heterogeneity regarding plant biomass and plant N content (Table 2). The core foraging areas of chamois, ibex and red deer delineated using T-LoCoH's revisitation index (NNSV) were in agreement with the areas where the three ungulate species are frequently observed (SNP, personal communication). Figure 1 shows examples of spatially segregated core foraging areas in the Trupchun valley. We found minor overlaps of the core foraging areas both within and between species, however they occurred at different times of our observation period. The best multinomial logistic regression model included plant biomass, plant N content and their interaction (df=8, Table 3). When comparing the other candidate models to this model (the interaction model), the main effects model reached a Δ AIC of 38.6 (df=6), the model including only plant N content a Δ AIC of 668.0 (df=4), the model including only plant biomass a Δ

AIC of 2013.1 (df=4) and the intercept-only model a Δ AIC of 2538.8 for (df=2). According to likelihood ratio tests, the interaction model performed clearly better (all $p<0.001$) than all other models. Therefore, we chose the interaction model as our best model. The Hosmer-Lemeshow goodness-of-fit statistic for each of the two logits reached a value of $p<0.001$ indicating very good model fit. The best model's AUC was 0.82 for logit 1, indicating excellent ability to discriminate between chamois and ibex coreforaging areas, and 0.75 for logit 2, indicating good ability to discriminate between chamois and red deer coreforaging areas. Sensitivity analysis revealed no significant differences (all $p>0.25$) between the best model›s coefficients for the two logits, neither after increasing, nor after decreasing the animals' core foraging areas by a 6 m buffer.

Table 1: Imaging spectroscopy models predicting fresh weight of plant biomass and plant nitrogen content

		Biomass (g.m $^{-2}$)	Nitrogen (%)
2010	Theil's U	0.19	0.11
	adj. R^2	0.65	0.53
	RMSE	174.37	0.53
	< 20 % RMSE (%)	44.19	62.79
2011	Theil's U	0.15	0.07
	adj. R^2	0.70	0.43
	RSME	155.71	0.28
	< 20 % RMSE (%)	53.57	88.80
2012	Theil's U	0.23	0.07
	adj. R^2	0.49	0.39
	RSME	174.35	0.26
	< 20 % RMSE (%)	36.44	84.48
2013	Theil's U	0.22	0.08
	adj. R^2	0.43	0.36
	RSME	241.30	0.27
	< 20 % RMSE (%)	36.93	81.67

Models were generated separately for each year using data from the imaging spectrometer APEX and in situ vegetation data. Model performance is described using Theil's uncertainty coefficient (Theil's U), adjusted Pearson's correlation coefficient (adj. R^2), root mean squared error of prediction (RMSE) and % of predicted values below 20 % RMSE (< 20 % RMSE (%)).

Table 2: Resources in the Trupchun valley and in core foraging areas of chamois, ibex and red deer

	Trupchun	Chamois	Ibex	Red deer
Mean biomass (g.m $^{-2}$)	295.26	192.28	242.63	276.79
SD biomass (g.m $^{-2}$)	230.10	84.83	119.75	107.15
Biomass min (g.m $^{-2}$)	0.10	32.48	0.10	83.84
Biomass max (g.m $^{-2}$)	2799.10	667.61	680.30	986.63
Mean nitrogen (%)	2.11	1.84	2.37	1.78
SD nitrogen (%)	0.53	0.25	0.64	0.22
Nitrogen min (%)	0.01	1.29	0.97	1.04
Nitrogen max (%)	4.80	2.75	4.75	3.69

Biomass (g.m $^{-2}$)=fresh weight of plant biomass (g.m $^{-2}$), N (%)=plant nitrogen content (%), SD=standard deviation, min=minimum, max=maximum.

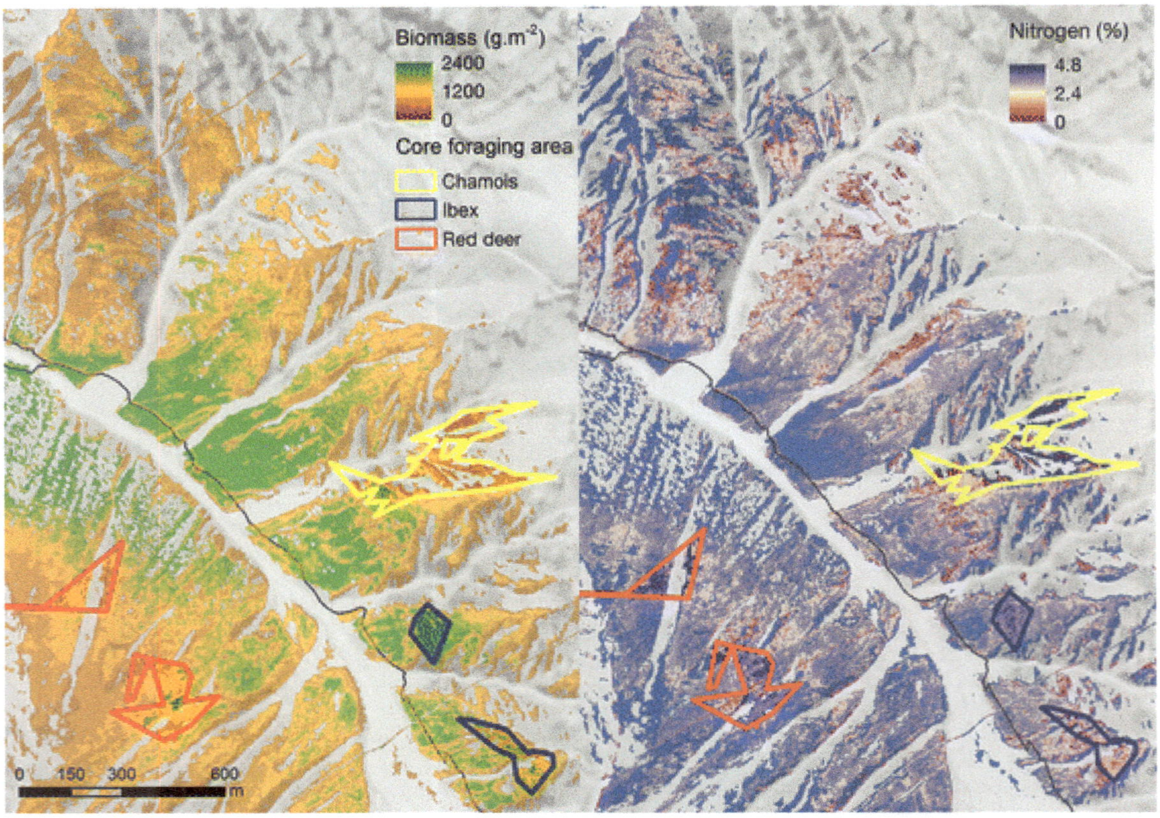

Figure 1: Examples of core foraging areas of chamois, ibex and red deer, respectively. Core foraging areas represent the 30% isopleths of T-LoCoH's revisitation index. The map shows fresh weight of plant biomass ($g.m^{-2}$) (left panel) and plant nitrogen content (%) (right panel). Grey colours represent areas covered by forest, rock, snow or water, identified using linear spectral unmixing (LSU) and subsequently excluded from analysis.

Table 3: Best multinomial logistic regression model comparing chamois, ibex and red deer core foraging areas

	Variable	Coeff	SE	z	p
Logit 1	Intercept	−8.7218	0.5825	−14.9738	0.0000
Chamois vs. ibex	BiomRS	1.0043	0.2658	3.7786	0.0002
	N	3.6048	0.2898	12.4381	0.0000
	BiomRS:N	−0.1314	0.1369	−0.9602	0.3369
Logit 2					
Chamois vs. deer	Intercept	2.2434	0.6461	3.4722	0.0005
	BiomRS	−0.1548	0.2731	−0.5670	0.5707
	N	−2.4917	0.3538	−7.0432	0.0000
	BiomRS:N	0.5535	0.1471	3.7640	0.0002

Logit 1 represents the logistic link function for chamois vs. ibex core foraging areas, logit 2 the logistic link function for chamois vs. red deer core foraging areas. Coefficients of the parameters (Coeff) for plant biomass rescaled (BiomRS = fresh weight of plant biomass/100 ($g.m^{-2}$)), plant nitrogen content (N (%)) and their interaction (BiomRS:N), standard errors (SE), Wald Z-statistic values (z) and corresponding p-values (p) are indicated.

The core foraging areas of chamois were characterised by generally low plant biomass (< 200 $g.m^{-2}$ fresh weight, Figures 2 and 3), and a low but slightly increased level of plant N content (around 2%, unimodal relationship, Figures 3 and 4). In contrast, vegetation in the core foraging areas of ibex was characterised by high plant N content, but variable plant biomass (Figure 2, Figure 4), while vegetation in the core foraging areas of red deer was characterised by high levels of plant biomass, but variable plant N content (Figure 2). Ibex showed a tendency to use areas with the highest plant biomass and plant N content (Figure 3). However, the core foraging areas with the highest plant biomass and highest plant N content had an almost 50:50 modelled chance of being used by either ibex or red deer (Figure 2).

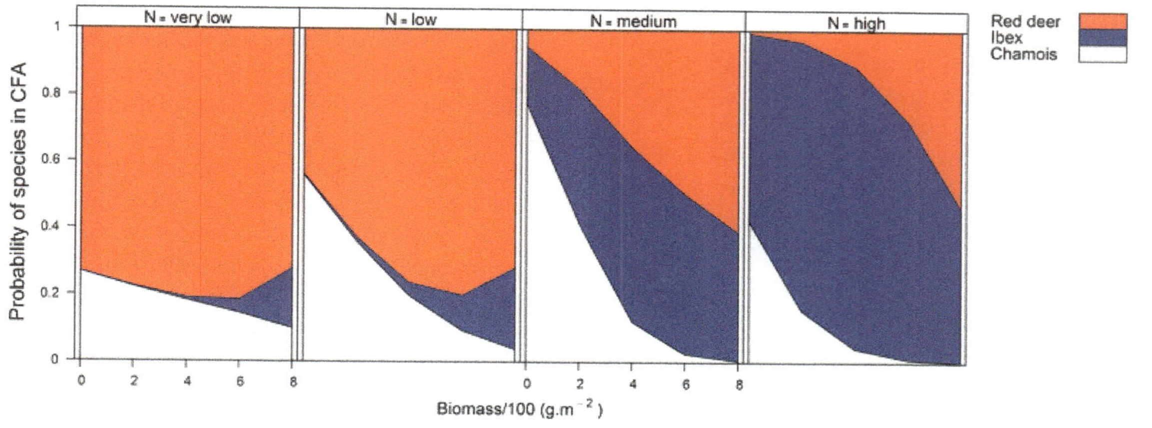

Figure 2: Probabilities for de facto use of core foraging areas by red deer, ibex and chamois. Predicted probabilities for chamois, ibex and red deer using core foraging areas (CFA) depending on plant biomass (Biomass/100 (g.m^{-2}), x-axis) at increasing levels of plant nitrogen content (very low (<1%), low (< 2%), medium (< 2.5%), high (≥ 2.5%)) displayed in the panels from left to right. Probabilities were generated from the model in Table 3.

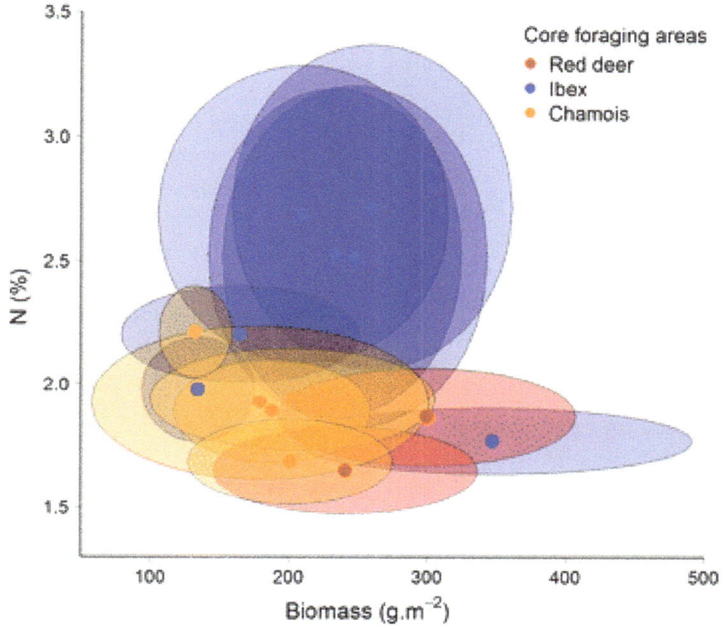

Figure 3: Core foraging areas of red deer, ibex and chamois. Core foraging areas of red deer, ibex and chamois regarding plant biomass and plant nitrogen (N) content. Dots represent mean values and the axes of the ellipses standard deviations (SD) in either direction.

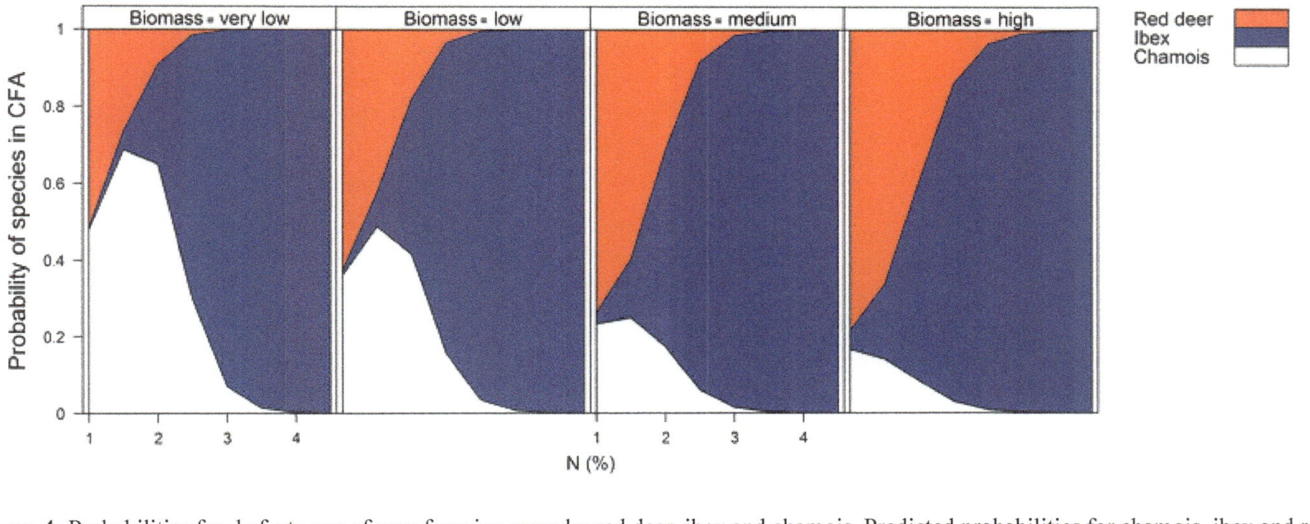

Figure 4: Probabilities for de facto use of core foraging areas by red deer, ibex and chamois. Predicted probabilities for chamois, ibex and red deer using core foraging areas (CFA) depending on plant nitrogen content (x-axis) at increasing levels of plant biomass (very low (< 200 g.m^{-2}), low (< 350 g.m^{-2}), medium (< 450 g.m^{-2}), high (≥ 450 g.m^{-2})) displayed in the panels from left to right. Probabilities were generated from the model in Table 3.

DISCUSSION

Previous studies conducted in the Trupchun valley found no difference in spring and summer forage composition between chamois, ibex and red deer when using micro-histological analysis of plant fragments in faecal pellets [22-24]. Likewise, a large overlap in the diet of chamois and red deer was found when they co-occurred with roe deer (*Capreolus capreolus* L.) in Southern Germany [90], with mouflon (*Ovis ammon musimon* Pallas) in the Western Alps [17] and with re-introduced red deer in the Italian Apennine [20]. While overlap in resource use (of both habitat and forage) is a prerequisite for competition [91], it could also be a sign of coexistence between species with no need of specialisation or segregation [92]. Similarly, low overlap in diet and high specialisation may point towards species living in coexistence [1], but could also be an effect of active competition, with the species trying to relieve competitive pressure [92]. This makes coexistence and competition extremely difficult to demonstrate in the field and without experimental manipulation [91,93,94], and conclusion have to be drawn with care.

The studies mentioned above used micro-histological analyses of faecal pellets and thus identified forage remains at the level of plant functional groups or plant families. However, the ratios of the two main forage components, graminoids and forbs, varied considerably within the ungulate species studied, which suggests that the animals have some flexibility in their dietary choices. Moreover, ungulates are assumed to partition forage resources at levels below the scale of vegetation types [3], and several studies confirm that plant biomass, plant nutrient and mineral content are major drivers for the spatial distribution and forage resource selection in ungulates [35-39]. Therefore, studies investigating forage resource selection in multi-species ungulate communities profit from including forage quantity and quality in their analyses, especially in areas where these vegetation characteristics are expected to vary.

Finding core foraging areas of chamois predominantly where plant biomass was low, red deer core foraging areas where plant biomass was high and ibex in between is in line with traditional feeding type definitions [12]. Regarding body size, chamois as the smallest of our three study species (body weight: 30 – 50 kg) is more limited in terms of forage intake than ibex (body weight: 40 – 150 kg) and red deer (body weight: 60 – 200 kg). However, the differences of plant N content in the core foraging areas of the three species, with chamois foraging in areas with low, ibex in areas with high and red deer in areas with variable plant N content warrant some explanation. Chamois have smaller and less complex rumens resulting in shorter retention time of digesta. This makes them less able to digest fibre, which could indicate that chamois depend on forage with higher plant N content [12]. However, it was found that by comparison to red deer, chamois foraged on lower quality vegetation consisting predominantly of graminoids in areas where food supply was limited [95]. Additionally, chamois have been found to select high quality forage in high quality habitats, while foraging in a more generalist pattern in low quality habitats [14]. Indeed, as mixed feeders [12] chamois can be expected to show high plasticity in forage selection, which was supported by our results.

In contrast to chamois, vegetation in the core foraging areas of ibex was characterised by high plant N content and variable plant biomass. Generally, we expected ibex to forage in rocky terrain with little, but nutrient-rich vegetation. While our results suggested that ibex did indeed forage in areas where plant biomass was low but of high nutritious value (high plant N content), we also found ibex core foraging areas in the highest quality meadows of the Trupchun valley where both plant biomass and plant N content were high. Generally, terrain roughness and slope create a template of risk [96,97], in which herbivores have to trade off between resource acquisition (e.g. foraging in high quality habitats, finding mates) and predator avoidance [98,99]. Ibex are very good climbers that find protection from predators and the possibility to overview large areas in predominantly rocky terrain with steep slopes. Within the SNP predators are absent, hunting is prohibited and visitors are obliged to stay on the marked paths. Thus, ibex might have abandoned part of their anti-predator behaviour in favour of maximising forage resource acquisition. Visual observations (SNP, unpublished observations) confirm that the rather flat, high quality meadows are regularly visited by ibex, where they forage together with red deer and occasionally also chamois.

The core foraging areas of red deer were always located on the rather flat meadows where the animals are expected to be able to cover their forage intake needs as plant biomass is high. Red deer inhabit predominantly open, flat terrain, have good running skills and thus withdraw themselves from predator attacks by using areas with high lateral cover, such as areas with forest or tall-growing shrubs. Similar to ibex, red deer might have abandoned part of their anti-predator behaviour as they can be observed grazing, resting and ruminating on the alpine meadows of the Trupchun valley during daytime [67,68]. However, besides the absence of predators and limited disturbance, finding ungulates foraging in open areas with higher plant biomass could also be an effect of high population densities [100]. When forage availability in habitats with more protection declines with increasing population numbers, the animals might be forced to use more open terrain for foraging. To assess whether the animals choose the high-quality meadows in the Trupchun valley voluntarily (in order to maximise resource acquisition) or if they are forced to use these areas (due to high population numbers) would require a comparison of core foraging areas at variable population densities or in the presence of predators. Wolf, lynx and bear are expected to return to the SNP in the future [101], which could have profound impacts on the abundance, population dynamics and spatial distribution of ungulates [102]. Thus, our results provide an important basis against which to assess future changes.

Due to the fixed dates of the APEX IS data acquisition, the results of our study represent the animals' behaviour during a specific time, i.e. during early summer. However, forage quantity and quality are expected to influence ungulate movement patterns in our study area in particular during this time of the year (peak of the plant growing season), when females have to nourish their offspring, and all individuals have to build up winter reserves. Naturally, the quantity and quality of forage resources will change during the course of the year and therefore also the ungulates' habitat use patterns can be expected to change [103,104]. The habitat use patterns of ungulates are apart from vegetation quantity and quality, also influenced by physical landscape characteristics, such as elevation, aspect and slope, as they can facilitate e.g. effective temperature regulation [104-106], ease of movement and anti-predator behaviour [98,99]. However, vegetation composition and thus the quantity and quality of forage resources depend on microclimate and soil, which are also influenced by elevation, aspect and slope. Disentangling the effects of "pure" physical landscape characteristics and "derived" vegetation properties would be challenging but provide important insights into trade-off mechanisms in habitat choice.

The distribution of vegetation quantity and quality influences the space use of herbivores on several spatial and temporal scales [104,107]. Large ungulates show HR establishment at the regional or landscape scale, they choose suitable feeding areas and vegetation communities at the local scale, select vegetation communities of favourable quantity and quality at the patch scale and certain plant species or plant parts at the bite scale [41,107]. APEX data gathered in this study represents vegetation quantity and quality at the patch scale of 2 m x 2 m. It is therefore possible, that some of the ungulate species, especially the smaller chamois, feed more selectively within these patches [3,11,108]. However, visually observing and exactly locating the animals, sampling browsed plants and determining their nutrient content is difficult in an area where access is limited due to challenging terrain and the high protection status. As advances in GPS and remote sensing technology continue, spatially accurate, high-temporal resolution GPS data (e.g. at minute intervals) that allow following the exact movement paths of animals will become more widely available. Combined with temporally flexible, very high spatial resolution remote sensing instruments such as unmanned aerial vehicles (UAV's, drones; [109]), this would provide opportunities to investigate forage quantity and quality at the individual plant level. Further differentiation might be achieved by not only assessing commonly observable vegetation traits by remote sensing [110], but also by adding advanced retrievals of pigment composition [111].

CONCLUSIONS

We detected significant differences in plant biomass and plant N content in the core foraging areas of sympatric chamois, ibex and red deer when combining resource maps developed from airborne imaging spectroscopy data with behavioural maps developed using the T-LoCoH algorithm: T-LoCoH enables to detect behavioural phases in GPS data by making use of their temporal and spatial autocorrelation. The combination of behavioural and resource maps proved to be valuable for studying a classical issue in resource ecology, resource partitioning between sympatric species. For the future, we expect that the development of remote sensing instruments with increased spatial resolution and temporal flexibility together with highly accurate and short interval GPS systems will continue to deepen our understanding of the foraging ecology of multi-species communities.

DECLARATIONS

Acknowledgements

Radio-collaring was conducted under permit from the Canton of Grisons (no. 2011–07) and the Swiss Federal Office for the Environment FOEN (no. J 074–0727). We thank A. Lyons and W. Getz for discussion on T-LoCoH analyses and two anonymous reviewers for their comments that helped to improve a previous version of the manuscript. We acknowledge the SNP rangers Fadri Bott, Andri Cuonz, Curdin Eichholzer, Domenic Godly, Ueli Nef, Alfons à Porta, Reto Strimer and Not Armon Willy for collaring the animals and Dominik Affolter, Thomas Baumann, Rolf Bösch, Martin Brüllhardt, Antonia Eisenhut, Tamara Estermann, Niculin Geer, Alan Haynes, Miriam Herrmann, Melanie Hodel, Christopher Jenni, Uli Kias, Thomas Niederberger, Maia Rapp, Ursina Raschein, Thomas Rempfler, Stefan Röber, Mirjam von Rütte, Christian Schmid, Crest Simeon and Remo Wild for their help during fieldwork. The contribution of MES is supported by the University of Zurich Research Priority Program on "Global Change and Biodiversity".

Authors' contributions

AKS participated in designing the study, planned and carried out field work, performed the statistical analysis and wrote the manuscript. MS designed the study, planned and carried out field work, provided statistical support and wrote the manuscript. PA provided statistical support and provided comments to the manuscript. MES designed the study and wrote the manuscript. MK designed the study, pre-processed IS data and wrote the manuscript. RH designed the study and provided comments to the manuscript. ACR designed the study, planned and carried out field work, provided statistical support and wrote the manuscript. All authors read and approved the final manuscript.

REFERENCES

1. Hutchinson GE. Homage to Santa Rosalia or why are there so many kinds of animals? American Naturalist 1959:145–159.
2. Van Langevelde F, Prins HHT. Introduction to Resource Ecology. In Resource Ecology. Edited by Prins HHT, Van Langevelde F. Springer Netherlands; 2008: 1–6
3. Hanley TA. A nutritional view of understanding and complexity in the problem of diet selection by deer (Cervidae). Oikos. 1997;79:209–18.
4. McNaughton SJ. Grazing as an Optimization Process: Grass-Ungulate Relationships in the Serengeti. Am Nat. 1979;113:691–703.
5. McNaughton SJ, Banyikwa FF, McNaughton MM. Promotion of the cycling of diet-enhancing nutrients by African grazers. Science. 1997;278:1798–800.
6. Risch AC, Frank DA. Carbon dioxide fluxes in a spatially and temporally heterogeneous temperate grassland. Oecologia. 2006;147:291–302.
7. De Knegt HJ, Groen TA, Van De Vijver CADM, Prins HHT, Van Langevelde F. Herbivores as architects of savannas: Inducing and modifying spatial vegetation patterning. Oikos. 2008;117:543–54.
8. De Jager NR, Pastor J. Declines in moose population density at Isle Royle National Park, MI, USA and accompanied changes in landscape patterns. Landscape Ecol. 2009;24:1389–403.
9. Coughenour MB. Spatial components of plant-herbivore interactions in pastoral, ranching, and native ungulate ecosystems. J Range Manage. 1991;44:530–42.

10. Wardle DA, Bardgett RD, Klironomos JN, Setälä H, Van Der Putten WH, Wall DH. Ecological linkages between aboveground and belowground biota. Science. 2004;304:1629–33.

11. Bailey DW, Gross JE, Laca EA, Rittenhouse LR, Coughenour MB, Swift DM, et al. Mechanisms that result in large herbivore grazing distribution patterns. J Range Manage. 1996;49:386–400.

12. Hofmann RR. Evolutionary Steps of Ecophysiological Adaptation and Diversification of Ruminants: A Comparative View of Their Digestive System. Oecologia. 1989;78:443–57.

13. Gordon IJ, Illius AW. Resource partitioning by ungulates on the Isle of Rhum. Oecologia. 1989;79:383–9.

14. Brambilla P, Bocci A, Ferrari C, Lovari S. Food patch distribution determines home range size of adult male chamois only in rich habitats. Ethology Ecology & Evolution. 2006;18:185–93.

15. Hülber K, Ertl S, Gottfried M, Reiter K, Grabherr G. Gourmets or gourmands?—Diet selection by large ungulates in high-alpine plant communities and possible impacts on plant propagation. Basic and Applied Ecology. 2005;6:1–10.

16. Bagchi S, Goyal SP, Sankar K. Habitat separation among ungulates in dry tropical forests of Ranthambhore national park Rajasthan. Tropical Ecology. 2003;44:175–81.

17. Bertolino S, Di Montezemolo NC, Bassano B. Food-niche relationships within a guild of alpine ungulates including an introduced species. Journal of Zoology. 2009;277:63–9.

18. La Morgia V, Bassano B. Feeding habits, forage selection, and diet overlap in Alpine chamois (Rupicapra rupicapra L.) and domestic sheep. Ecological Research. 2009;24:1043–50.

19. Klansek E, Vavra I, Onderscheka K. Die Äsungszusammensetzung des Alpensteinwildes (Capra i. ibex L.) in Abhängigkeit von Jahreszeit, Alter und Äsungsangebot in Graubünden. Zeitschrift für Jagdwissenschaften. 1995;41:171–81.

20. Lovari S, Ferretti F, Corazza M, Minder I, Troiani N, Ferrari C, et al. Unexpected consequences of reintroductions: competition between increasing red deer and threatened Apennine chamois. Animal Conservation 2014.

21. Pérez-Barbería FJ, Gordon IJ, Nores C. Evolutionary transitions among feeding styles and habitats in ungulates. Evolutionary Ecology Research. 2001;3:221–30.

22. Zingg A. Seasonal variability in the diet composition of Alpine ibex (Capra ibex L.) in the Swiss National Park. University of Zurich, Institute of Evolutionary Biology and Environmental Studies; 2009.

23. Trutmann C. Diet composition of alpine chamois (Rupicapra rupicapra L.): Is there evidence for forage competition to the alpine ibex (Capra ibex L.)? University of Zurich, Institute of Evolutionary Biology and Environmental Studies; 2009.

24. Anderwald P, Haller R, Risch AC, Schütz M, Schweiger AK, Filli F. Resource competition between chamois, alpine ibex and red deer in the Swiss National Park? In 5th Symposium for Research in Protected Areas; Mittersill. 2013

25. Duparc A, Redjadj C, Viard-Crétat F, Lavorel S, Austrheim G, Loison A. Co-variation between plant above-ground biomass and phenology in sub-alpine grasslands. Applied Vegetation Science. 2013;16:305–16.

26. Blix AW, Mysterud A, Loe LE, Austrheim G. Temporal scales of density-dependent habitat selection in a large grazing herbivore. Oikos. 2014;123:933–42.

27. Kerr JT, Ostrovsky M. From space to species: Ecological applications for remote sensing. Trends in Ecology and Evolution. 2003;18:299–305.

28. Aplin P. Remote sensing: Ecology. Progress in Physical Geography. 2005;29:104–13.

29. Kneubühler M, Damm A, Schweiger A, Risch A, Schütz M, Schaepman M. Continuous Fields From Imaging Spectrometer Data for Ecosystem Parameter Mapping and Their Potential for Animal Habitat Assessment in Alpine Regions. IEEE J Selected Topics Appl Earth Observations Remote Sensing. 2014;7:2600–10.

30. Schaepman ME, Ustin SL, Plaza AJ, Painter TH, Verrelst J, Liang S. Earth system science related imaging spectroscopy-An assessment. Remote Sensing of Environment. 2009;113:S123–37.

31. Schaepman ME, Jehle M, Hueni A, D'Odorico P, Damm A, Weyermann J, et al. Advanced radiometry measurements and Earth science applications with the Airborne Prism Experiment (APEX). Remote Sensing of Environment. 2015;158:207–19.

32. Ustin SL, Roberts DA, Gamon JA, Asner GP, Green RO. Using imaging spectroscopy to study ecosystem processes and properties. BioScience. 2004;54:523–34.

33. Wang K, Franklin SE, Guo X, Cattet M. Remote sensing of ecology, biodiversity and conservation: A review from the perspective of remote sensing specialists. Sensors. 2010;10:9647–67.

34. Pottier J, Malenovský Z, Psomas A, Homolová L, Schaepman ME, Choler P, et al. Modelling plant species distribution in alpine grasslands using airborne imaging spectroscopy. Biology letters. 2014;10:20140347.

35. Skidmore AK, Ferwerda JG, Mutanga O, Van Wieren SE, Peel M, Grant RC, et al. Forage quality of savannas - Simultaneously mapping foliar protein and polyphenols for trees and grass using hyperspectral imagery. Remote Sensing of Environment. 2010;114:64–72.

36. McNaughton S. Mineral nutrition and spatial concentrations of African ungulates. Nature. 1988;334:343.

37. Langvatn R, Hanley TA. Feeding-patch choice by red deer in relation to foraging efficiency. Oecologia. 1993;95:164–70.

38. Wilmshurst JF, Fryxell JM. Patch selection by red deer in relation to energy and protein intake: a re-evaluation of Langvatn and Hanley's (1993) results. Oecologia. 1995;104:297–300.

39. Smallegange IM, Brunsting AMH. Food supply and demand, a simulation model of the functional response of grazing ruminants. Ecological Modelling. 2002;149:179–92.

40. Cho MA, Skidmore AK. Hyperspectral predictors for monitoring biomass production in Mediterranean mountain grasslands: Majella National Park, Italy. International Journal of Remote Sensing. 2009;30:499–515.

41. Schweiger AK, Risch AC, Damm A, Kneubühler M, Haller R, Schaepman ME, et al. Using imaging spectroscopy to predict above-ground plant biomass in alpine grasslands grazed by large ungulates. Journal of Vegetation Science. 2015;26:175–90.

42. Mutanga O, Skidmore AK. Integrating imaging spectroscopy and neural networks to map grass quality in the Kruger National Park, South Africa. Remote Sensing of Environment. 2004;90:104–15.

43. Burt WH. Territoriality and home range concepts as applied to mammals. Journal of Mammalogy. 1943;24:346–52.

44. Cagnacci F, Boitani L, Powell RA, Boyce MS. Animal ecology meets GPS-based radiotelemetry: a perfect storm of opportunities and challenges. Philosophical Transactions of the Royal Society B: Biological Sciences. 2010;365:2157–62.

45. Tomkiewicz SM, Fuller MR, Kie JG, Bates KK. Global positioning system and associated technologies in animal behaviour and ecological research. Philosophical Transactions of the Royal Society B: Biological Sciences. 2010;365:2163–76.

46. Burgman MA, Fox JC. Bias in species range estimates from minimum convex polygons: implications for conservation and options for improved planning. Animal Conservation. 2003;6:19–28.

47. Getz WM, Wilmers CC. A local nearest-neighbor convex-hull construction of home ranges and utilization distributions. Ecography. 2004;27:489–505.

48. Otis DL, White GC. Autocorrelation of location estimates and the analysis of radiotracking data. J Wildlife Manage. 1999:1039–1044.

49. McNay RS, Morgan JA, Bunnell FL. Characterizing independence of observations in movements of Columbian black-tailed deer. J Wildlife Manage. 1994:422–429.

50. Fieberg J, Matthiopoulos J, Hebblewhite M, Boyce MS, Frair JL. Correlation and studies of habitat selection: problem, red herring or opportunity? Philosophical Trans Royal Soc B Biol Sci. 2010;365:2233–44.

51. Fieberg J. Kernel density estimators of home range: smoothing and the autocorrelation red herring. Ecology. 2007;88:1059–66.

52. Solla D, Shane R, Bonduriansky R, Brooks RJ. Eliminating autocorrelation reduces biological relevance of home range estimates. J Anim Ecol. 1999;68:221–34.

53. Horne JS, Garton EO, Krone SM, Lewis JS. Analyzing animal movements using Brownian bridges. Ecology. 2007;88:2354–63.

54. Kranstauber B, Kays R, LaPoint SD, Wikelski M, Safi K. A dynamic Brownian bridge movement model to estimate utilization distributions for heterogeneous animal movement. J An Ecol. 2012;81:738–46.

55. Viswanathan GM, Buldyrev SV, Havlin S, da Luz MGE, Raposo EP, Stanley HE. Optimizing the success of random searches. Nature. 1999;401:911–4.

56. Benhamou S, Riotte-Lambert L. Beyond the Utilization Distribution: Identifying home range areas that are intensively exploited or repeatedly visited. Ecological Modelling. 2012;227:112–6.

57. Wall J, Wittemyer G, LeMay V, Douglas-Hamilton I, Klinkenberg B. Elliptical Time-Density model to estimate wildlife utilization distributions. Methods in Ecology and Evolution. 2014;5:780–90.

58. Long JA, Nelson TA. Time geography and wildlife home range delineation. The Journal of Wildlife Management. 2012;76:407–13.

59. Bartumeus F, Levin SA. Fractal reorientation clocks: Linking animal behavior to statistical patterns of search. Proceedings of the National Academy of Sciences. 2008;105:19072–7.

60. Patterson TA, Thomas L, Wilcox C, Ovaskainen O, Matthiopoulos J. State–space models of individual animal movement. Trends in ecology & evolution. 2008;23:87–94.

61. Lyons AJ, Turner WC, Getz WM. Home range plus: A space-time characterization of movement over real landscapes. Movement Ecology. 2013;1:1–14.

62. Brown DR, Sherry TW. Solitary winter roosting of Ovenbirds in core foraging area. The Wilson Journal of Ornithology. 2008;120:455–9.

63. Bontadina F, Schofield H, Naef-Daenzer B. Radio-tracking reveals that lesser horseshoe bats (Rhinolophus hipposideros) forage in woodland. Journal of Zoology. 2002;258:281–90.

64. SchweizerischeEidgenössischeNationalparkkommission (ENPK): Geschäftsbericht 2010–2013 [https://www.parcs.ch/mmds/pdf_public/3941_GB_SNP_2010.pdf, https://www.parcs.ch/mmds/pdf_public/4282_GB_SNP_2011.pdf, http://www.parcs.ch/snp/pdf_public/2013/9746_20130731_124944_Geschaeftsbericht_2012_DEF.pdf, http://www.parcs.ch/snp/pdf_public/2014/30465_20141029_123957_Geschaeftsbericht_2013.pdf]

65. Hudson R, Frank S. Foraging ecology of bison in aspen boreal habitats. Journal of Range Management 1987:71–75.

66. Hamel S, Côté SD. Trade-offs in activity budget in an alpine ungulate: contrasting lactating and nonlactating females. Animal Behaviour. 2008;75:217–27.

67. Hegglin I. Verteilung, Dichte, Aktivitätsmuster und Konkurrenz bei Steinbock (Capra ibex), Gemse (Rupicapra rupicapra) und Rothirsch (Cervus elaphus) in der Val Trupchun, Schweizerischer Nationalpark. Diploma thesis. University of Zurich, 1996.

68. Hofmann A. Das jahreszeitliche Verteilungsmuster und der Aesungsdruck von Alpensteinbock, Gemse,

Rothirsch und Reh in einem begrenzten Gebiet im Oberengadin. Diploma thesis. University of Zurich, 1971.

69. Georgii B, Schröder W. Home range and activity patterns of male red deer (Cervus elaphus L.) in the alps. Oecologia. 1983;58:238–48.

70. Kamler JF, Jędrzejewska B, Jędrzejewski W. Activity Patterns of Red Deer in Białowieża National Park, Poland. 2007.

71. MeteoSwiss: IDA web. The data portal of MeteoSwiss for research and teaching [https://gate.meteoswiss.ch/idaweb/login.do?language=en]

72. Van Soest PJ. Nutritional ecology of the ruminant. Cornell University Press; 1994.

73. Jehle M, Hueni A, Damm A, D›Odorico P, Weyermann J, Kneubühler M, Schläpfer D, et al. APEX-Current status, performance and validation concept. In Sensors IEEE 2010 Waikoloa, HI, US; 2010: 533–537

74. Schläpfer D, Richter R. Geo-atmospheric processing of airborne imaging spectrometry data. Part 1: Parametric orthorectification. International Journal of Remote Sensing. 2002;23:2609–30.

75. Richter R, Schläpfer D. Geo-atmospheric processing of airborne imaging spectrometry data. Part 2: Atmospheric/topographic correction. International Journal of Remote Sensing. 2002;23:2631–49.

76. Damm A, Kneubuhler M, Schaepman ME, Rascher U, editors. Evaluation of gross primary production (GPP) variability over several ecosystems in Switzerland using sun-induced chlorophyll fluorescence derived from APEX data. Munich, Germany: International Geoscience and Remote Sensing Symposium (IGARSS); 2012.

77. Smith EP, Rose KA. Model goodness-of-fit analysis using regression and related techniques. Ecological Modelling. 1995;77:49–64.

78. Roberts DA, Smith MO, Adams JB. Green vegetation, nonphotosynthetic vegetation, and soils in AVIRIS data. Remote Sensing of Environment. 1993;44:255–69.

79. R Core Team. R: A Language and Environment for Statistical Computing. Vienna, Austria: R Foundation for Statistical Computing; 2013. http://www.R-project.org/.

80. Bjørneraas K, Moorter B, Rolandsen CM, Herfindal I. Screening global positioning system location data for errors using animal movement characteristics. The Journal of Wildlife Management. 2010;74:1361–6.

81. Lyons AJ, Getz WM, R Development Core Team. T-LoCoH: Time Local Convex Hull Homerange and Time Use Analysis. R package version 1.16. 2014.

82. Dürr S, Ward MP. Roaming behaviour and home range estimation of domestic dogs in Aboriginal and Torres Strait Islander communities in northern Australia using four different methods. Preventive veterinary medicine. 2014;117:340–57.

83. T-LoCoH for R - Tutorial and Users Guide [http://tlocoh.r-forge.r-project.org/tlocoh_tutorial_2014-08-17.pdf]

84. Hosmer DW. Lemeshow S. Applied logistic regression. John Wiley & Sons: Sturdivant RX; 2013.

85. Fielding AH, Bell JF. A review of methods for the assessment of prediction errors in conservation presence/absence models. Environmental Conservation. 1997;24:38–49.

86. Pearce J, Ferrier S. Evaluating the predictive performance of habitat models developed using logistic regression. Ecological Modelling. 2000;133:225–45.

87. Venables WN, Ripley BD. Modern applied statistics with S. Statistics and computing. New York: Springer; 2002.

88. Robin X, Turck N, Hainard A, Tiberti N, Lisacek F, Sanchez J-C, et al. pROC: an open-source package for R and S+ to analyze and compare ROC curves. BMC bioinformatics. 2011;12:77.

89. Fox J, Hong J. Effect displays in R for multinomial and proportional-odds logit models: Extensions to the effects package. Journal of Statistical Software. 2009;32:1–24.

90. Schröder J, Schröder W. Niche breadth and overlap in red deer Cervus elaphus, roe deer Capreolus capreolus and chamois Rupicapra rupicapra. Acta Zoologica Fennica. 1984;172:85–6.

91. Putman RJ. Competition and resource partitioning in temperate ungulate assemblies. London: Chapman & Hall; 1996.

92. De Boer WF, Prins HHT. Large Herbivores That Strive Mightily but Eat and Drink as Friends. Oecologia. 1990;82:264–74.

93. Chase JM, Leibold MA. Ecological niches: linking classical and contemporary approaches. Chicago: London: University of Chicago Press; 2003.

94. Connell JH. Diversity and the Coevolution of Competitors, or the Ghost of Competition Past. Oikos. 1980;35:131–8.

95. Kamler J. Feeding strategy of wild herbivores in habitats of limited food resources. Wildlife Biology in Practice. 2011;7:46–55.

96. Ripple WJ, Beschta RL. Wolves and the ecology of fear: Can predation risk structure ecosystems? BioScience. 2004;54:755–66.

97. Laundré JW, Hernández L, Altendorf KB. Wolves, elk, and bison: reestablishing the" landscape of fear" in Yellowstone National Park, USA. Canadian Journal of Zoology. 2001;79:1401–9.

98. Anderson TM, Hopcraft JGC, Stephanie E, Ritchie M, Grace JB, Olff H. Landscape-scale analyses suggest both nutrient and antipredator advantages to Serengeti herbivore hotspots. Ecology. 2010;91:1519–29.

99. Wang G, Hobbs NT, Twombly S, Boone RB, Illius AW, Gordon IJ, et al. Density dependence in northern ungulates: Interactions with predation and resources. Population Ecology. 2009;51:123–32.

100. Coulombe ML, Huot J, Massé A, Côté SD. Influence of forage biomass and cover on deer space use at a fine scale: A controlled-density experiment. Ecoscience. 2011;18:262–72.

101. Filli F. Ungulate research in the Swiss National Park: development, current issues and future challenges. In: Filli F, Suter W, editors. Huftierforschung im Schweizerischen Nationalpark. Nationalpark-Forschung in der Schweiz. 2006. p. 9–29.

102. Andersen R, Linnell J, Solberg E. The future role of large carnivores in terrestrial trophic interactions: the northern temperate view. In: Danell K, Bergström R, Duncan P, Pastor J, editors. Large herbivore ecology, ecosystem dynamics and conservation, vol. Volume 11. Cambridge: Cambridge Univ. Press; 2006. p. 413–48.

103. Hebblewhite M, Merrill E, McDermid G. A Multi-Scale Test Of The Forage Maturation Hypothesis In A Partially Migratory Ungulate Population. Ecological Monographs. 2008;78:141–66.

104. Van Beest FM, Rivrud IM, Loe LE, Milner JM, Mysterud A. What determines variation in home range size across spatiotemporal scales in a large browsing herbivore? Journal of Animal Ecology. 2011;80:771–85.

105. Aublet J-F, Festa-Bianchet M, Bergero D, Bassano B. Temperature constraints on foraging behaviour of male Alpine ibex (Capra ibex) in summer. Oecologia. 2009;159:237–47.

106. Signer C, Ruf T, Arnold W. Hypometabolism and basking: The strategies of Alpine ibex to endure harsh over-wintering conditions. Functional Ecology. 2011;25:537–47.

107. Senft RL, Coughenour MB, Bailey DW, Rittenhouse LR, Sala OE, Swift DM. Large Herbivore Foraging and Ecological Hierarchies. BioScience. 1987;37:789–99.

108. Brivio F, Grignolio S, Brambilla A, Apollonio M. Intra-sexual variability in feeding behaviour of a mountain ungulate: size matters. Behavioral Ecology and Sociobiology. 2014;68:1649–60.

109. Anderson K, Gaston KJ. Lightweight unmanned aerial vehicles will revolutionize spatial ecology. Frontiers in Ecology and the Environment. 2013;11:138–46.

110. Homolová L, Malenovský Z, Clevers JG, García-Santos G, Schaepman ME. Review of optical-based remote sensing for plant trait mapping. Ecological Complexity. 2013;15:1–16.

111. Ustin SL, Gitelson AA, Jacquemoud S, Schaepman M, Asner GP, Gamon JA, et al. Retrieval of foliar information about plant pigment systems from high resolution spectroscopy. Remote Sensing of Environment. 2009;113:S67–77.

CHAPTER 4

IMPROVING LANDSAT AND IRS IMAGE CLASSIFICATION

Krishna Bahadur K.C.

Keywords: Landsat, IRS, mountain shadow, land use land cover classification, band ratios, ancillary data, Nepal

INTRODUCTION

Satellite remote sensing has become an important tool for monitoring and management of natural resources and the environment. Remotely sensed data are widely used in land use/land cover classification. Several techniques have been reported to improve classification results in terms of land use discrimination and accuracy of resulting classes while processing remotely sensed data [1]. However there are some problems in getting highly accurate land use data, particularly in the mountainous landscapes of tropical humid climates. One of the main problems when generating land use maps from digital images is the confusion of spectral responses from different features. Discrimination of land use land cover types, including vegetation types, through the use of remote sensing techniques in the mountainous areas of Nepal is a very difficult task because of the complex structure and composition of vegetation communities. In many cases, the size of old secondary growth areas is too small to be detected by Landsat and IRS. In addition, the mountain topography leads to a significant shadowing effect, which becomes a particular problem in the digital image processing [1,2,3,4,5].

One of the techniques to achieve improvement in digital classification is the incorporation of ancillary data, such as a digital elevation model (DEM) [1,6]. There are several approaches for incorporating ancillary data such as DEM, slope, aspects etc. in the image classification process, and earlier studies have demonstrated the use of a DEM for a number of purposes [1]. DEM integration in image classification has helped increase the classification accuracy of digital data [7,8,9,10,11,12]. A DEM has also been used to describe the distribution of terrain components which contribute to spectral response [13], identify sites for fieldwork [14], geographically stratify training areas or homogeneous regions [17], and provide topographic normalization of Landsat TM digital imagery [16].

Eiumnoh and Shrestha [1] mentioned in their paper that in areas of rugged terrain, integration of other ancillary information together with elevation data has improved classification accuracy. For example, Franklin [15] reported that classification accuracy using Landsat MSS increased from 58 to 79 percent with the addition of elevation data. Furthermore, the addition of four other geomorphometric variables (relief, convexity, slope, and incidence) yielded 87 percent classification accuracy. Cibula and Nyquist [18] used elevation along with slope and aspect variables to improve MSS classification. A DEM has also been used with microwave data for surface pattern discrimination [19], forestry studies [20], and snow monitoring [21], among others.

Agricultural applications of remote sensing are time critical, and, hence, careful selections of image dates are important because land-use/land-cover class separations are usually related to plant phenology [1,18]. Experience shows that, in the tropics, high-resolution visible and infrared satellite sensors cannot always provide the desired information due to constraints related to cloud cover and revisit schedules [22]. Langford and Bell [23] have also identified the problem of frequent cloud cover, as well as topographic effects and mixed pixels, in the tropical environment of Colombia. Cloud-free data are generally acquired during the dry season when most of the agricultural fields are fallow. Such a situation makes the process of discriminating land use/land cover a difficult task.

Franklin [17] mentioned that topography exerts a powerful influence on the spectral response observed by polar orbiting

Citation: Krishna Bahadur K.C., Improving Landsat and IRS Image Classification, doi:10.3390/rs1041257.

satellites, the lack of studies on digital image classification using band ratios and ancillary information in Nepal provided the main impetus for this work. The objectives of the study were to evaluate the use of band ratios and elevation data in Landsat and IRS image classification, and to produce a thematic map of land use of the Galaudu Watershed of Nepal. In this study, DEM was prepared from the elevation data using the topographic contours of 1:25,000 scales. The DEM was using with standard digital image processing operations as a component band during image classification process. Improvement of the classification of different land use classes were explored using supervised and unsupervised classification techniques for several feature sets of Landsat and IRS data.

METHODS

Study Area

The study area constitutes a mountainous watershed named Galaudu/Pokhare Khola sub-watershed (hereafter refer as Galaudu watershed) situated in Dhading district of Nepal (Figure 1).

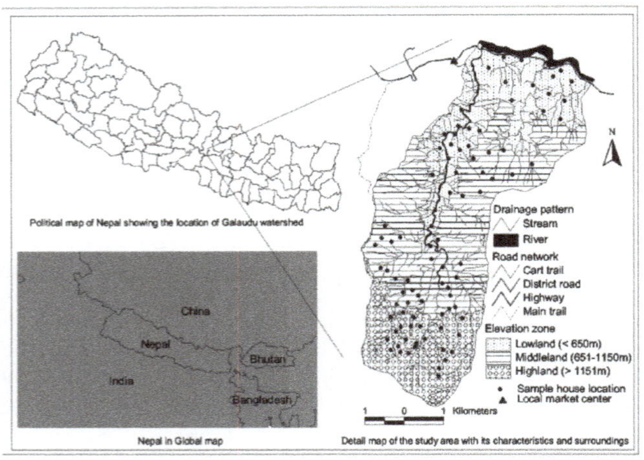

Figure 1: Location of the study area.

Topography of the watershed is mountainous, with an average slope exceeding 30 percent and showing features which are often found in mountain zones of Asia. Most of the watershed is a mountainous region under hill forest and upland cultivation. The soils of the watershed are loam, sandy loam, clay loam, silt loam and sandy clay loam. The area has a sub-tropical climate, with a mean annual rainfall of 1,404 mm. The elevations of the highest and lowest point are 1,960 m and 217 m a.s.l., respectively. The watershed can be divided into fertile, relatively flat valleys along the rivers and surrounding uplands with medium to steep slopes (as shown in Figure 1). Agricultural lands in the valleys are under intensive management with multiple cropping systems and are mostly irrigated. Paddy, potato, wheat and vegetables are the major crops cultivated in the valley.

Data and Materials

Spatial data

The basic data used to prepare the land use map of the study area were Landsat multispectral (MSS) and Thematic Mapper (TM) digital data, Indian Remote Sensing (IRS) digital data. Black-and-white aerial photographs of 1:50,000 scales from 1978 and 1992 were used for "ground-truth" information required for classification and accuracy estimation of classified MSS and TM images respectively. Topographic maps of 1:25,000 scales published by the Nepal Department of Survey and digital topographic data with contour interval of 20 m produced by the same agency were also used. Extensive ground truthing was carried out to collect the training samples for TM image for 2000 and IRS image 2002. The spatial database consisted of Landsat-MSS data acquired on 10th October 1976 (Path 152, Row 41) (Figure 2a) and TM data acquired on 4th February 1990 and 13th March 2000 (path/row-141/41) (Figure 2b and c) in digital format. Similarly IRS data was acquired on 16th February 2002 (Path Row 104/052) (Figure 2d) in digital format. RS data acquired in the dry season has less spectral response from the vegetation as most of the agricultural area are in harvested condition. This situation makes it difficult to interpret such data. Use of DEM or elevation data is one of the techniques, which has been used successfully in the temperate and alpine regions to interpret the remote sensing data.

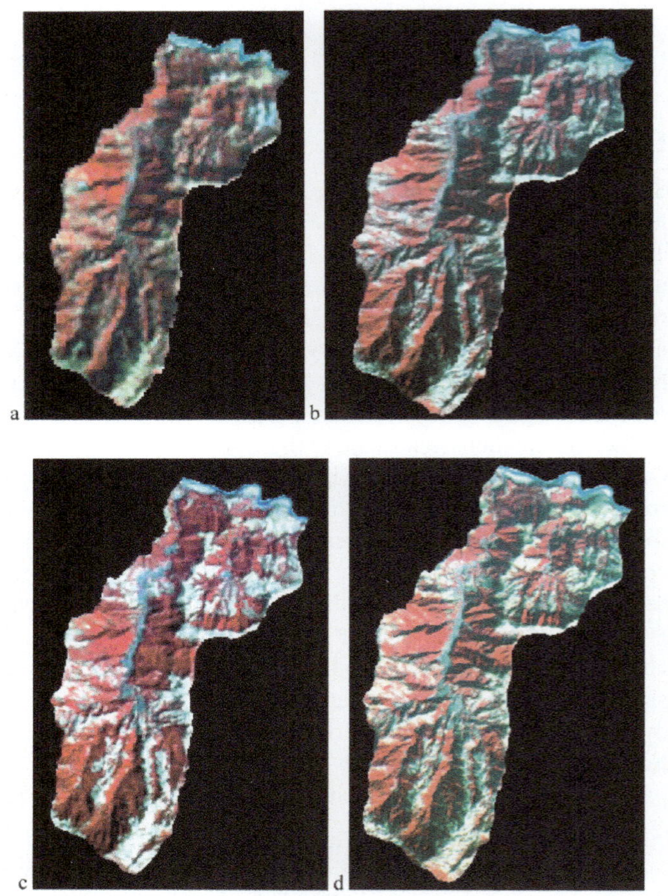

Figure 2: Satellite images over Galaudu watershed (a) Landsat MSS 1976 (b) Landsat TM 1990 (c) Landsat TM 2000 (d) IRS LISS III 2002.

Image processing of the satellite data

Digital image processing was carried out to obtain the land use and land cover maps from the remote sensing (RS) data. Digital image processing of RS data for generating land use map involved several steps as shown in Figure 3. RS data pre-processing was designed to remove any undesirable image characteristics produced by the sensor to calibrate the image radiometry, remove noise and correct geometric distortions [24]. The obtained image, which was geometrically corrected at smaller scale, was further corrected by registering the images to the 1:25,000 scale topographic map sheets by selecting ground control points (GCPs). The root mean square (RMS) error of less than 1 pixel at the first order and the nearest neighborhood transformation was accepted. The study area was clipped with a vector boundary layer. Classification of remote sensing data was performed by extracting different feature sets using Band ratios; normalized difference vegetation index (NDVI) and principal component analysis (PCA) is among the standard image processing techniques [25]. Basically, these techniques are the data reduction techniques, which are carried out to enhance an image, as the quantity of information carried by satellite data is not necessarily same as the amount of data. Six different feature set for MSS, twelve different feature sets for TM and six different features sets for IRS (Table 1) containing original bands, derived bands and DEM were decided before running the classification.

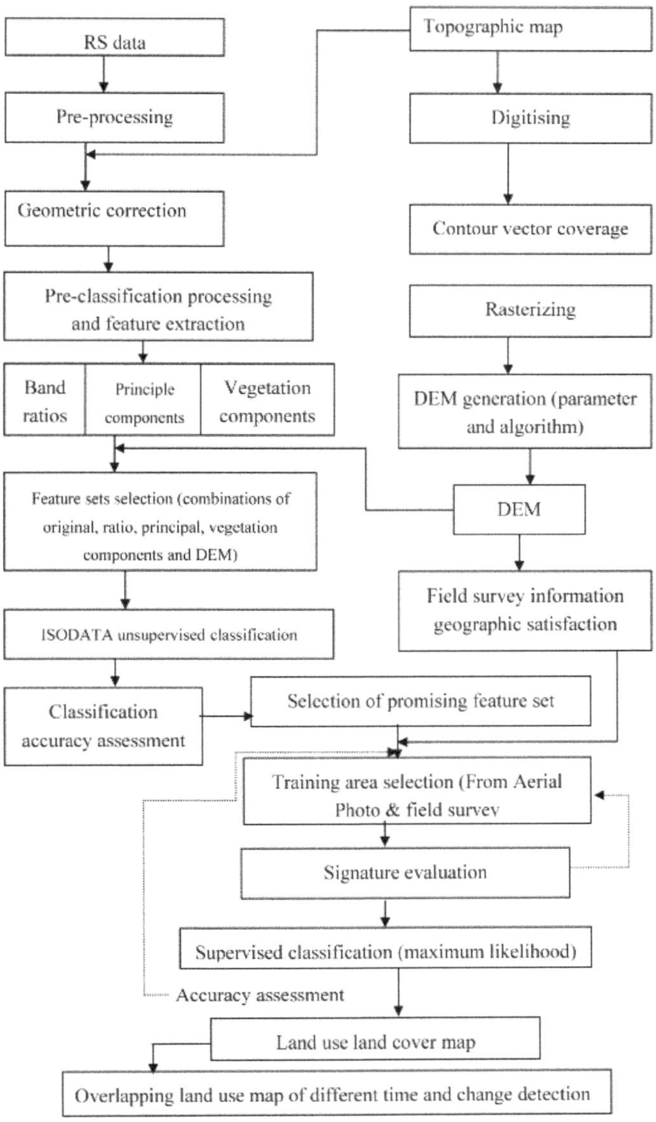

Figure 3: Steps of digital image processing to produce land use land cover map.

Table 1: Feature sets prepared and tested

Feature set	MSS bands	TM bands	IRS bands
1	B1, B2, B3	B2, B3, B4	B2, B3, B4
2	B2, B3, B4	B1, B4, B5	B2, B4, B5
3	B2, B4, NDVI	B2, B4, NDVI	B2, B4, NDVI
4	B2, NDVI, DEM	B4, B7/B5, NDVI	B2, NDVI, DEM
5	B4,NDVI, DEM	B2/B1, B7/B5, NDVI	B4,NDVI, DEM
6	PC1,NDVI ,DEM	B2, NDVI, DEM	PC1,NDVI ,DEM
7		B4,NDVI, DEM	
8		PC1,NDVI ,DEM	
9		B7/B5,NDVI,DEM	
10		B5/B2,B5/B4,B5/B7	
11		B4/B3,B5/B2,B5/B4	
12		B4/B3,B5/B4,B5/B7	

Notes: B = Band DEM = Digital elevation model NDVI = (Band 4 − Band 3) / (Band 4 + Band 3)
PC1 = Principal component 1

Of the unsupervised classification techniques, the Iterative Self-Organizing Data analysis technique (ISODATA) was used. This technique repeatedly performs an entire classification and recalculates statistics with minimum user inputs for locating clusters and also is relatively simple and has considerable intuitive appeal. However, the output of this technique could be affected by the choice of initial parameters and their interactions with each other [26]. Parameters assigned for each classification scheme were kept same including maximum number of clusters (40 clusters) to be formed. The clusters formed were regrouped with Ward's method of hierarchical clustering technique, which is designed to optimize the minimum variance with in clusters [27]. This technique calculates means for each variable within each cluster and squared Euclidean distance to the cluster means for each case [23]. Distances are summed for all of the cases and at each step, the two clusters merge are those that result in the smallest increase in the overall sum of the squared within-cluster distances. Resulting classes were identified on the basis of knowledge drawn from the field survey and previous land use map. Results of each classification scheme were compared creating the error matrices and the overall classification accuracy.

Finally, only one classification scheme that showed superior result during unsupervised classification was selected. Supervised classification was performed on the selected classification scheme employing Bayesian maximum likelihood classifier (MLC). MLC, a parametric decision rule, is a well-developed method from statistical decision theory that has been applied to problem of classifying image data [28,29]. At first, training signatures for identifiable classes were established with the field knowledge, which were evaluated for possible discrimination of individual class. After obtaining a suitable indication for satisfactory discrimination between the classes during signature evaluation, final classification was run to produce land use map. Training areas corresponding to each classification item (hereafter, land use class), in case of the IRS image, were chosen from among the training samples collected from the field and in case of the MSS and TM images they were generated from the interpretation of aerial photographs of 1978 and 1992 of the study area. Although the dates of the aerial photographs used as reference information in these classifications do not exactly match with the dates of the satellite images, they were used in the research with the assumption that land use in the watershed was not substantially changed between the time of aerial photography and satellite observation dates. Moreover, this was the best feasible option that could be used in this research.

For producing land use maps for 1976, 1990, 2000 and 2002 and to investigate changes that occurred between these periods, the following five land use land cover (hereafter, land use) classes were considered in image classification: forestland, scrubland, lowland agriculture, and upland agriculture. Choice of these classes was guided by: i) the objective of the research, ii) expected certain degree of accuracy in image classification, and iii) the easiness of identifying classes on aerial photographs. A brief description of each of the land use classes is given in Table 2.

Table 2: Land use classes considered in image classification and changes detection

Land use classes	General description
Forest	Forest areas with estimated 75 percent or more of the existing crown covered by trees.
Scrublands	Land covered by shrubs, bushes and young regeneration. Degraded forest areas with estimated < 10% tree crown cover are also included.
Lowland agriculture	Irrigated, level-terraced agricultural lands in river valleys, used for multiple cropping including winter crops. Wheat and potato are two major winter crops cultivated in these lands after the harvest of paddy rice in November-December.
Upland agriculture	Non-irrigated agricultural lands with or without slopping terraces, barren lands, settlements, roads, construction sites and other built-up areas.
Vegetables	Irrigated agricultural land under vegetables during winter season

Among all the land use classes, "upland agriculture" is the most complex class. In fact, it includes all other combinations of land uses, which are not included in the remaining classes. During winter, uplands in the study area, like most of the Middle Hills, are mostly barren and have spectral values similar to those of barren lands such as non-vegetative hills and riverbeds [30]. Moreover, during the time the satellite imageries were taken (particularly the IRS image), many upland terraces had exposed soil due to fresh ploughing by farmers as a preparation for the next summer crop. This condition of the cultivated uplands made it impossible to distinguish them from rough roads, new construction sites and other built-up areas. This justifies combining settlements, barren lands and built-up areas with upland agricultural lands in this study, which may not be acceptable at any other time of the year.

Post classification was performed after selectively combining classes; classified images were sieved, clumped and filtered before producing final output. Sieving removes isolated classified pixels using blob grouping, while clumping helps maintain spatial coherency by removing unclassified black pixels (speckle or holes) in classified images. Finally a 3 × 3 median filter was applied to smoothen the classified images. All activities related to image processing were performed in ERDAS Imagine version 8.7 [31]. Classified images were then exported to Arc View GIS Version 3.2 [32] from ERDAS and rest of the analyses was performed in GIS environments. The classified images were first converted to grid and then to shape format in Arc View. The polygon themes so generated, were exported to Arc Info GIS Version 3.5.1 [32] and polygons of <0.5 ha in size were "eliminated" in Arc Info. This elimination was necessary to minimize the effects of classification errors arising from resolution differences among the three satellite images while at the same time without significantly altering the area under each land use class. The resultant polygon themes were used in further analyses.

Detection of land use changes

The land use polygon themes for 1976, 1990 and 2000 obtained from the digital classification of satellite data and subsequent GIS analyses were overlaid two at a time in Arc View GIS and the area converted from each of the classes to any of the other classes was computed.

RESULTS AND DISCUSSION

Land Use/Land Cover Assessment through Image Classification

The mean spectral values for the different land cover classes show that for classification scheme containing only original bands does not attain higher accuracy, as only bare soil and forests were distinguishable. A mixed class situation was observed between forest and lowland agriculture. Especially Band 2 of TM and IRS were not very useful with the similar response for all classes. Band 3 responded higher for bare soil area. Band 4 of TM, which is the near infrared band, responded the highest for vegetation but did not discriminate due to similar response patterns of these classes in other bands. Features set containing NDVI gave the more accuracy in regards to vegetation because of the NDVI responded the highest for vegetation and the lowest for bare soil and burnt field. Some of the bare soil and burnt field was clearly discriminated but at the same times others were found mixed with forest area. One unique class was discriminated upon the addition of DEM together with NDVI. This was the scrubland with relatively scattered and no big trees. The forest types were easily distinguishable, unlike in the earlier mixed situation.

The former mixed class situation between scrub and forest was improved due to the ratio band of 5 and 2, 5 and 4 and 5 and 7. Thus the major land cover classes were differentiated, including forest types from the classification schemes with these kinds of ratio bands. Vegetation classes obtained from these combinations gave the highest accuracy. Overall classification accuracy ranged from 45 to 68 percent for unsupervised techniques. Among the twelve classification schemes for TM, and six for MSS and IRS inclusion of DEM as component band in the classification schemes showed very promising results as it helped to differentiate the vegetation types such as forest types which otherwise not possible to differentiate with NDVI due to similar signatures. Vegetation types, bare soil and upland cultivation areas were discriminated by using the combination of three different ratio bands (4/3, 5/4, 5/7) that gave the highest accuracy. This is due to the ratio bands substantially reducing the effect of shadow, which is a particular problem for satellite image processing and interpretation in the mountain areas.

Combinations of ratioed band schemes was selected for the supervised classification because this scheme gave the higher accuracy in the unsupervised classification. Supervised classification was run on the three bands containing ratio of band 4 and 3, ratio of band 5 and 4 and ratio of band 5 and 7. Trisurat [33], in his study conducted in the Khao Yai National Park of Thailand having a similar bio-climate as the Galaudu watershed, reported that the ratio 5/7 was useful for discriminating the forest classes.

The maximum number of classes yielded by this set than any other sets during unsupervised classification was the basis of testing it in supervised classification. Mixed deciduous forest, which was misclassifying either with scrub or other vegetation in unsupervised classification, could be discriminated on a knowledge base selection of training areas using elevation as a criterion. The spectral signatures of the training area for individual classes showed quite satisfactory indication for separability of the classes selected in either of ratio 4 and 3, ratio 5 and 4 or ratio 5 and 7 bands. An overall map accuracy (OCA) of 94.08 percent was obtained through error matrix of training signatures. As the post-classification accuracy evaluation, an OCA of 82.86 percent and kappa coefficient of agreement (KHAT) statistics of 0.74 (Table 3) were obtained for the classified image.

Table 3: Error Matrix of the Supervised Classification map of Galaudu Watershed, Nepal

Land use classes	Reference data				User Accuracy (%)
	Forest	Scrub	Upland agriculture	Lowland agriculture	
Forest	3				100.0
Scrub		4	1		80.0
Upland agriculture		1	12	2	80
Lowland agriculture			2	10	83.3
Producer's acuracy	100	80	80.0	83.3	82.86

Overall Map Accuracy = 82.86 %; OM = Omission error; Co = commission error; KHAT=0.74

Land Use/ Land Cover Change Detection

The land use maps for 1976, 1990, 2000 and 2002 are presented in Figure 4 and the area under the four land use classes during the three periods is shown in Table 4. The results show that forest area decreased while agriculture area increased continuously over the study period. Scrublands (degraded forest land) were assessed at 1976 but they were already converted into the good condition of forest at 1990, while lowland agricultural area was expanded very much during the first period whereas upland agriculture area were expanded very much at latter period (Table 4 and Table 5).

Table 4: Land uses of Galaudu watershed, Nepal, 1976, 1990 and 2000

Land use classes	1976		1990		2000	
	Area (ha)	%	Area (ha)	%	Area (ha)	%
Upland agriculture	430.1	15.9	414.7	15.3	671.2	24.8
Lowland agriculture	428.1	15.8	809.8	29.9	839.4	31.0
Forest	1,597.2	59.1	1,475.9	54.8	1,189.7	44.2
Scrubland	242.5	9.2	0.0	0.0	0.0	0.0

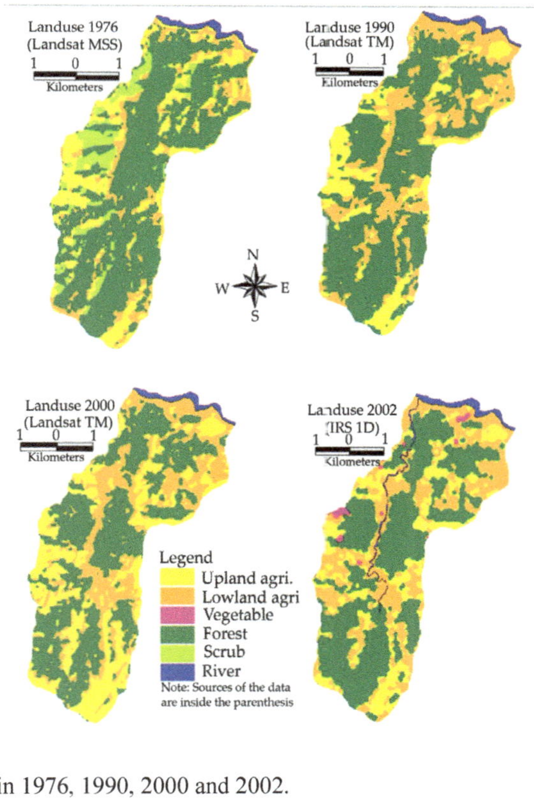

Figure 4: Land use in Galaudu watershed in 1976, 1990, 2000 and 2002.

Among the major land use groups, around 65 percent of upland agriculture, 52 percent of lowland agriculture and 45 percent of the forest area in 1976 remained unchanged until 2000. Forest lands shrunk by about 55 percent of 1976 area in between 1976 and 2000 (Table 5). The observed trends of decreasing forest and increasing agricultural areas in the watershed could be explained by the following three main reasons: first, a substantial proportion of the agricultural lands in the study area are in higher inclinations areas where slope stability and soil erosion is of critical concern [34,35]. Those steep agricultural fields suffer from rapid soil erosion and nutrient depletion, which forces farmers to find new land to meet their growing food needs, so farmers could have moved towards the forest area for their agricultural activities. Some other studies in the mountains regions of Southeast Asia have found that many households are practicing shifting cultivation [36,37,38]. There is also evidence from the hills of Thailand [39] and Honduras [40] that declining soil productivity and increased weed competition leads to the farmers to find new land for the fulfillment of their basic needs.

Table 5: Percent change in land use during the three periods, Galaudu watershed, Nepal

Land use classes	1976-1990	1990-2000	1976-2000
Upland agriculture	−3.71	+38.21	+35.92
Lowland agriculture	+47.13	+3.52	+48.99
Forestland (forest + scrub)	−25.15	−24.05	−55.26

Second, most of the settlements in the upland area do not have good quality land (less productive, out facing sloping terraces, no irrigation facilities) compared to lower elevation areas because of which a higher level of human-forest interaction can be expected in these areas, thereby bringing about more pronounced forest losses compared to the lower elevation areas. Third, most of the community forestry activities that are expected to have positive influences on the balance of forest cover were concentrated in lower elevations. A lower amount of forest loss in lower elevation zones suggests that forest conservation efforts by the local communities and concerned agencies played important roles by bringing positive outcomes in the balance of forestry land use in the watershed. The same could not happen at higher elevations (highland) because of the inability of community-based forest management programs to cover those areas and virtually non-existent forest monitoring by the forest department thereby leading to an open access condition of the high altitude forests.

The existing model of community forestry systems was unable to bring high elevation forests under management. This was probably because of difficulties in identifying users and use patterns arising from a continuous extensive forest accessed by widespread settlements. Although there was a net loss in forest area, a substantial proportion of degraded forests (scrubland) at lowland were converted into good condition of forest especially in the first period (1976–1990), which might have been due to success of community forestry program at lower elevation and near to road and market center because accessibility of those areas. A continuous loss in forest area over time despite different forestry development program challenges the efforts of forest conservation and development, the forest department and the donor agency. A combined investment from multiple actors at various levels is indeed one of the important conditions for successful outcomes from collective actions at a local level [41].

The expansion of lowland agricultural area during the first period at the expense of upland agriculture indicated increased agricultural intensification and diversification during the period. From conversations with the local farmers it was revealed that there was indeed a big shift in the use pattern of lowlands during this period because of farmers' attractions towards winter cropping of mainly wheat and potato on irrigated lands. More recently potato cultivation for commercial purposes has gained momentum on the lowlands due mainly to improved access to local markets and higher profitability compared to wheat and other cereal crops.

CONCLUSIONS

Satellite remote sensing has become an important tool for monitoring and management of natural resources and environment. One of the main problems when generating land covers maps from digital images is the confusion of spectral responses from different features. Discrimination of land use land cover types including vegetation types through the use of remote sensing techniques in mountainous areas of Nepal is a very difficult task because of the complex structure and composition of vegetation communities. In many cases, the size of old secondary growth areas is too small to be detected by Landsat and IRS. In addition, the mountain topography leads to a significant shadowing effect, which becomes a particular problem in the digital image processing. Various techniques of improving classification accuracy in digital image classification are found in practice. Geo-spatial data also has great potential in enhancing the digital image classification apart of their multi-purpose use. In this study, DEM was prepared from the elevation data from the topographic contours of 1:25,000 scales. DEM was incorporated with standard digital image processing operations as a component band during image classification process which helped to improve the classification result.

In this study, improvement of the classification of different vegetation types including the upland agriculture areas in the Galaudu watershed, Nepal, was explored using supervised and unsupervised classification techniques for several feature sets of Landsat and IRS data. Overall, the supervised classification method produced higher accuracy than the unsupervised approach. The result from the combination of bands R4/3, R5/4 and R5/7 ranked the highest in terms of accuracy (82.86%), while the combination of bands 2, 3 and 4 ranked the lowest (45.29%). This result may be influenced by the shadowing effects, which were substantially reduced in the ratio band combinations. It can be inferred that the digital classification can help to improve vegetation mapping including upland agriculture areas compared to visual classification, because digital classification allows the interpretation of smaller mapping units, offering greater detail compared to the visual technique, where the mapping units are generalized.

ACKNOWLEDGEMENTS

I would like to thank to DAAD (German academic exchange service) and University of Hohenheim, Germany for the financial support for this research. I am especially grateful to Werner Doppler for his helpful discussions, encouragements and his valuable criticism and constructive comments on the draft paper. The Author is grateful to editor and anonymous reviewers whose valuable comments and suggestions helped to consolidate and strengthen this article.

REFERENCES AND NOTES

1. Eiumnoh, A.; Shrestha, R.P. Application of DEM data to landsat image classification: Evaluation in a tropical wet-dry landscape of Thailand. *Photogramm. Eng. Remote Sens.* 2000, *66*, 297–1304.
2. Gautam, A.P.; Webb, E.L.; Eiumnoh, A. GIS assessment of land use-land cover changes associated with community forestry implementation in the Middle Hills of Nepal. *Mt. Res. Dev.* 2002, *22*, 63–69.
3. Gautam, A.P.; Shivakoti, G.P.; Webb, E.L. Forest cover change, physiography, local economy, and institution in a mountain watershed in Nepal. *Environ. Manage.* 2004, *33*, 48–61.
4. Semwal, R.L.; Nautiyal, S.; Sen, K.K.; Rana, R.K.; Maikuri, R.K.; Rao, K.S.; Saxena, K.G. Patterns and ecological implications of agricultural land use changes: a case study from central Himalaya, India. *Agri. Ecosys. Environ.* 2004, *102*, 81–92.
5. Rao, K.S.; Pant, R. Land use dynamics and landscape change pattern in a typical micro watershed in the mid elevation zone of central Himalaya, India. *Agri. Ecosys. Environ.* 2001, *86*, 113–123.
6. Hulchinson, C.K. Techniques for combining landsat and ancillary data for digital classification improvement. *Photogramm. Eng. Remote Sens.* 1982, *48*, 123–130.
7. Bonner, W.J.; Rohde, W.G.; Miller, W.A. *Mapping Wildland resources with Digital Landsat and Terrain Data, Remote Sensing and Resource Management*; Johannesen, C.J., Sanders, J.L., Eds.; Soil Conservation Society of America: Iowa, IW, USA, 1982; pp. 73–80.
8. Jones, A.R.; Settle, J.J.; Wyatt, B.K. Use of digital terrain data in the interpretation of SPOT-1 HRV multispectral imagery. *Int. J. Remote Sens.* 1988, *9*, 669–682.
9. Frank, T.D. Mapping dominant vegetation communities in the colorado rocky mountain front range with LANDSAT thematic mapper and digital terrain data. *Photograpmm. Eng. Remote Sens.* 1988, *54*, 727–734.
10. Janssen, L.F.; Jaarsma, J.; Van der Linder, E. Integrating topographic data with remote sensing for land-cover classification. *Photogramm. Eng. Remote Sens.* 1990, *56*, 1503–1506.
11. Gastellu-Etchegorry, J.P.; Estreguil, C.; Mougin, E. A GIS based methodology for small scale monitoring of tropical forests: a case study in Sumatra. *Int. J. Remote Sens.* 1993, *14*, 2349–2368.
12. Palacio-Prieto, J.L.; Luna-Gonzalez, L. Improving spectral results in a GIS context. *Int. J. Remote Sens.* 1996, *17*, 2201–2209.
13. Strahler, A.H.; Logan, T.L.; Bryant, N.A. Improving forest cover classification accuracy form landsat by incorporating topographic information. In Proceedings of 12th International Symposium on Remote Sensing of the Environment, Manila, Philippines, 1978; pp. 927–942.
14. Walsh, S. Variability of Landsat MSS spectral responses of forests in relation to stand and site characteristics. *Int. J. Remote Sens.* 1987, *8*, 1289–1299.
15. Franklin, S.E. Terrain analysis from digital patterns in geomorphometry and landsat MSS spectral response. *Photogram. Eng. Remote Sens.* 1987, *53*, 59–65.
16. Franklin, S.E. Topographic context of satellite spectral response. *Comput. Geosci.* 1990, *16*, 1003–1010.
17. Civco, D.L. Topographic normalization of landsat thematic mapper digital imagery. *Photogramm. Eng. Remote Sens.* 1989, *55*, 1303–1309.
18. Cibula, W.G.; Nyquist, M.O. Use of topographic and climatological models in a geographical data base to improve landsat MSS classification for olympic national park. *Photogramm. Eng. Remote Sens.* 1987, *53*, 67–75.
19. Peddle, D.R.; Franklin, S.E. Image texture processing and data integration for surface pattern discrimination. *Photogramm. Eng. Remote Sens.* 1991, *57*, 413–420.
20. Teillet, P.M.; Guindon, B.; Menuier, J.F.; Goodenough, D.G. Slope aspect effects in synthetic aperture radar imagery. *Can. J. Remote Sens.* 1985, *11*, 39–49.
21. Guneriussen, T.; Johnsen, H. DEM corrected ERS-1 SAR data for snow monitoring. *Int. J. Remote Sens.* 1996, *17*, 181–195.
22. Aldenderfer, M.S.; Blashfield, R.K. *Cluster Analysis*; Sage Press: Beverly Hills, CA, USA, 1984; p. 87.
23. Langford, M.; Bell, W. Land cover mapping in a tropical hillsides environment: a case study in the cauca region of colombia. *Int. J. Remote Sens.* 1997, *18*, 1289–1306.
24. Schowengerdt, R.A. *Techniques for Image Processing and Classification in Remote Sensing*; Academic Press: London, UK, 1983.
25. Matheson, W.; Ringgrose, S. The development of image processing techniques to assess change in green vegetation cover along a climatic gradient through northern territory, Australia. *Int. J. Remote Sens.* 1994, *15*, 17–47.
26. Vanderzee, D.; Ehrichlich, D. Sensitivity of ISODATA to change in sampling procedures and processing

parameters when applied to AVHRR time-series NDVI Data. *Int. J. Remote Sens.* 1995, *16*, 673–686.

27. Ward, J. Hierarchical grouping to optimise an objectives function. *J. Amer. Statist. Assn.* 1963, *58*, 236–244.
28. Niblack, W. *An Introduction to Digital Image Processing*; Strandberg Publications Co.: Birkeroed, Denmark, 1985.
29. Settle, J.J.; Briggs, S.S. Fast maximum likelihood classification of remotely sensed imagery. *Int. J. Remote Sens.* 1987, *8*, 723–734.
30. Tokola, T.; Sarkeala, J.; Van der Linden, M. Use of topographic correction in landsat TM-based forest interpretation in Nepal. *Int. J. Remote Sens.* 2001, *22*, 551–563.
31. ERDAS. *Earth Resources Data Analysis System (ERDAS)*; ERDAS Inc.: Atlanta, GA, USA, 1997.
32. ESRI. *Understanding GIS the ARC/INFO Method*; John Wiley & Sons Inc.: California, CA, USA, 1997.
33. Trisurat, Y. Spatial-Ecological Impact Assessment of the Proposed Dam Construction in Khao Yai National Park, Thailand. Ph.D. Dissertation, Asian Institute of Technology, Bangkok, Thailand, 1997.
34. *Kabhrepalanchok District, Vol II Part I: Assessment of Current Conditions*; International Centre for Integrated Mountain Development (ICIMOD): Kathmandu, Nepal, 1993.
35. *Application of GIS in Rural Development Planning in Nepal*; International Centre for Integrated Mountain Development (ICIMOD): Kathmandu, Nepal, 1994.
36. KC, K.B. Combining socioeconomic and spatial methodologies in rural resources and livelihood development: a case from mountains of Nepal. In *Farming and Rural System Economics*; Doppler, W., Bauer, S., Eds.; Margraf Publishers: Wuerzburg, Germany, 2005; Volume 69, p. 233.
37. KC, K.B. Mapping soil erosion susceptibility using remote sensing and GIS. a case of the upper Nam Wa Watershed, Nan Province, Thailand. *Environ. Geology* 2009, *57*, 695–705.
38. Fox, J.; Krummel, J.; Yarnasarn, S.; Ekasingh, M.; Podger, N. Land use and landscape dynamics in northern Thailand: assessing change in three upland watersheds. *Ambio* 1995, *24*, 328–334.
39. Kammerbauer, J.; Ardon, C. Land use dynamics and landscape change pattern in a typical watershed in the hillside region of central Honduras. *Agr. Ecosys. Environ.* 1999, *75*, 93–100.
40. Ostrom, E. *Governing the Commons: the Evolution of Institutions for Collective Action*; Cambridge University Press: Cambridge, London, UK, 1990.

CHAPTER 5

INSAR-BASED MAPPING OF TIDAL INUNDATION EXTENT AND AMPLITUDE

Talib Oliver-Cabrera and Shimon Wdowinski

Keywords: InSAR, wetlands, tidal inundation, remote sensing

INTRODUCTION

Louisiana has more coastal wetlands than any other state in the US, covering an area of roughly 20,000 km^2. These coastal wetlands provide important ecosystem services including habitat for a wide variety of wildlife, storm protection, flood control, water storage and others. Louisiana coastal wetlands also have high economical values, including habitat for fishing industry, recreational activities and location for oil and gas exploitation. The excessive use of the ecosystem services, climatic drivers, and human sprawl have threaten these valuable wetlands [1]. Coastal Louisiana has undergone a decrease in land area of approximately 4747 km^2 in the period between 1932 and 2010, reaching a rate of 42.9 km^2 of yearly wetland loss between 1985 to 2010 [2]. To protect these prolific resources, it is important to monitor and quantify the circumstances that provoke wetland loss in coastal Louisiana.

Monitoring coastal wetlands is a difficult task due to their large extent and hostile environment, which limit ground-based efforts. Remote sensing observations provide a suitable solution for analyzing large areas in a safe and economic manner. Several remote sensing studies were conducted to analyze and monitor Louisiana's coastal wetlands. Penland *et al.* [3] used aerial photography to describe land loss of the Mississippi River delta plain and explained the loss by three main processes: erosion, submergence and direct removal. Steyer *et al.* [4] monitored vegetation changes by generating vegetation indices from MODIS and Landsat imagery, using Normalized Difference Vegetation Index (NDVI) to quantify the impacts of hurricanes Katrina and Rita (2005) on the coastal marshes of Louisiana. Barras *et al.* [5] used aerial photography and Landsat images to estimate land loss of coastal Louisiana; they also projected the land loss to 2050. Ramsey *et al.* [6] analyzed L-band and C-band Synthetic Aperture Radar (SAR) amplitude combined with Landsat imagery to study flooding in coastal marshes of Louisiana. They created marsh inundation maps by comparing SAR backscatter from marshes during flooding and non-flooding conditions. A detailed description of the use of the different remote sensing techniques for wetland observations was recently summarized by Gallant [1]. The above studies provide a good assessment of vegetation or land changes. However, they have limited capability to detect water level changes in tidal zone, where active land loss occurs. Coastal wetlands and water levels are interrelated, as water level changes may induce shifts that produce disturbances or environmental changes [7].

A different remote sensing method that can detect water level changes in wetlands is Interferometric Synthetic Aperture Radar (InSAR). Studies have shown that InSAR observations can detect water level change with 3–5 cm accuracy level over wide wetland areas (e.g., [8,9,10,11]). InSAR measures phase difference between two SAR acquisitions from roughly the same location in space. The result generated from the phase-difference operation is a phase change map, called an interferogram, which is sensitive to changes of the Earth's surface [12]. Interferograms are widely used to measure surface displacements after an earthquake, urban subsidence due to ground water withdrawal, magmatic-induce deformation and, as in the case of this research, water-level changes in wetlands.

Most wetland InSAR studies were applied to inland freshwater wetlands, where hydrological changes occur within days or weeks. Several studies showed that the technique works well in woody wetlands [11,13]. Only few studies applied the technique to study water level changes in coastal areas. Ramsey *et al.* [14] analyzed four tandem ERS-1/2 interferograms and noticed phase changes between freshwater and saltwater vegetation. Wdowinski *et al.* [8] also noticed

Citation: Talib Oliver-Cabrera and Shimon Wdowinski, InSAR-Based Mapping of Tidal Inundation Extent and Amplitude, doi:10.3390/rs8050393.

phase change occurring along the transition between fresh and salt-water vegetation in the western Everglades (Florida). Recently, Wdowinski *et al.* [15] observed tide inundation in the coastal wetlands of the Florida Everglades, showing the feasibility of InSAR for detecting tidal inundation propagation and mapping the tidal zone.

In this study we use Synthetic Aperture Radar (SAR) data acquired by the ALOS and Radarsat-1 satellites as well as tide gauge data to map the tidal inundation extent and water level changes throughout the coastal wetlands of Louisiana. We were able to detect water level changes up to 30 cm in some areas. Our observations also show disruptions of water inlet into the wetlands due to man-made structures.

LOUISIANA COASTAL WETLANDS

Louisiana's coastal wetlands are located in the southern part of the state, along the entire state shoreline. Sedimentation processes induced by the Mississippi river dominate the morphology of the coast. The shift of the river's deposition plume formed two main features, the Chenier Plain and the Mississippi Deltaic Plain [16] (Figure 1). The Chenier Plain is located along the western part of the state, where the coastline's morphology consists of wooded beach ridges termed Cheniers and mudflats covered by swamps and marshes [17]. The Mississippi Deltaic Plain is located along the eastern shoreline, where the coastal morphology is a product of sediment deposition. It is also the present location of the Mississippi river delta.

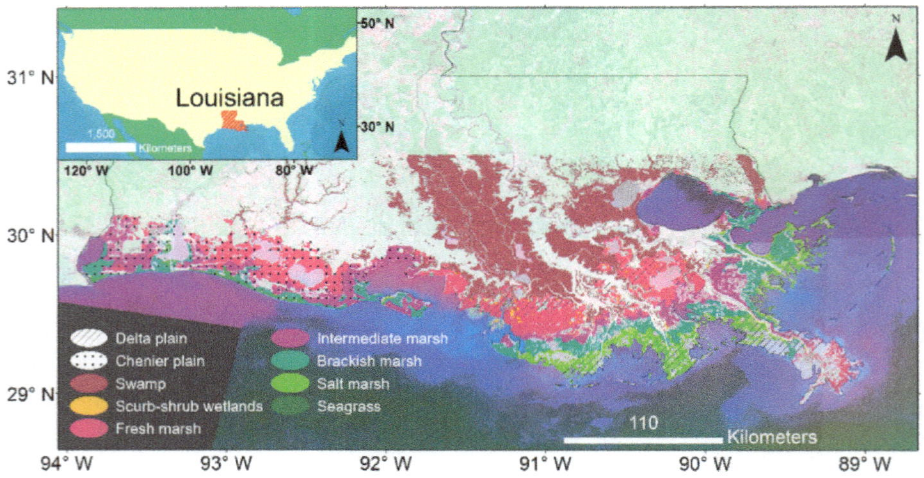

Figure 1: Location map and wetland classification generated using Gulf-Wide Information System (G-WIS) [19], Landsat mosaic [20] as background and Natural Earth data used for the location map. The wetlands with dot pattern correspond to the Chenier Plain and the hatch pattern to the Mississippi River Delta plain.

Wetland vegetation varies along the coast depending mostly on water conditions due to mixed inputs of fresh and salt-water within the coastal zone. Two main types of wetland vegetation are found along the Louisiana coast, forested wetlands, termed swamps, and herbaceous wetlands, termed marshes [18]. Swamps are mostly found in freshwater zones whereas most marshes are located in brackish or saline water.

The map in Figure 1 shows a vegetation transition along Louisiana's coast. The vegetation type works roughly as a marker of the extents of salt-water inputs, for example, the presence of fresh-water marshes in areas that are not affected by ocean tide. The areas on the map with color green represent vegetation that is better adapted to the salt-water intrusions. Meanwhile the purple and orange represent vegetation that requires fresh-water inputs to live. The transition zone between fresh and salt water is marked by the presence of brackish marsh (aqua green color) and intermediate marsh (purple color).

Louisiana coastal wetlands have been threatened by both natural (e.g., sea level rise) and anthropogenic (e.g., infrastructure development) processes causing an average loss of 43 km^2 of wetlands per year [2]. Boesch *et al.* [16] define the loss of wetlands as the change from vegetated wetlands to uplands or drained areas and to nonvegetated wetlands or submerged habitats. Between 1956 and 2004 the state lost around 3000 km^2 of wetlands that were transformed to open water [21]. According to Penland *et al.* [3] between 1932 and 1990, 26% of the coastal land loss was due to natural erosion processes, 36% due to oil and gas extraction, and 23% was due to altered hydrology product of multiple activities as channel construction for navigation or levees. Furthermore the construction of flood control levees has interrupted the normal sediment deposition from the Mississippi River, which created a deficit of deposited material, and resulted in land loss [22]. The vegetation in the submerged areas dies or has a limited growth due to hypoxia, transforming the vegetated areas into tidal mudflats [23].

DATA AND DATA PROCESSING

This study is based on data acquired by two SAR satellites, Advanced Land Observing Satellite (ALOS) and Radarsat-1. ALOS was launched by the Japanese Space Agency (JAXA) and operated from 2006 to 2011. It was equipped with L-band (23.62 cm wavelength) radar and had a repeated orbit of 46 days. Radarsat-1 was launched by the Canadian Space Agency in 1995. It was equipped with C-band (5.66 cm wavelength) radar and had a repeat orbit resolution of 24 days (Table 1). We used all available data acquired by both satellites over the entire length of coastal Louisiana (Figure 2) that are archived in the User Remote Sensing Access (URSA) at the Alaska Satellite Facility (ASF) data pool [24]. We used a total of 246 ALOS scenes and 154 Radarsat-1 scenes (Table 1).

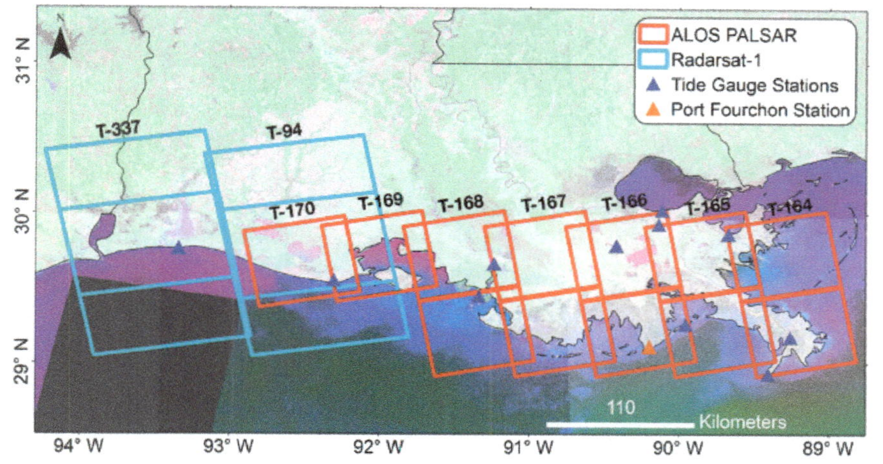

Figure 2: Location map of the Synthetic Aperture Radar (SAR) scenes and the available National Oceanic and Atmospheric Administration (NOAA) tide gauge stations for Louisiana (triangles). The orange triangle marks the location of the tide gauge used in this study.

Table 1: Scenes available and satellite information

Satellite	Path	Frames	Orbit	Number of Scenes	Date
ALOS	164	570 and 580	Ascending	36	2006–2011
ALOS	165	570 and 580	Ascending	40	2007–2011
ALOS	166	570 and 580	Ascending	40	2007–2010
ALOS	167	570 and 580	Ascending	38	2007–2011
ALOS	168	570 and 580	Ascending	50	2007–2011
ALOS	169	580	Ascending	18	2006–2011
ALOS	170	580	Ascending	24	2006–2011
Radarsat-1	94	74 and 75	Ascending	72	2002–2007
Radarsat-1	337	74 and 75	Ascending	82	2002–2008

InSAR data processing aims at detecting phase difference between two acquisitions to generate interferograms, reflecting surface changes between the two acquisitions. InSAR is regularly used to measure displacements of solid surfaces. However, in the case of wetlands, coherent interferometric signal occurs due to double bounce scattering, from the water surface and vegetation, to measure water level changes [25,26]. Data was processed using the ROI_PAC package [27], with 4–8 looks in order to co-register pairs of images. Co-registration searches similarities in morphology between two acquisitions and the looks are an average of pixels with the similar coherence. A lower threshold of looks provides higher resolution results. However, due to the lack of topography in the area, the use of a lower threshold of looks does not permit proper co-registering of the SAR images. Unlike most InSAR applications, we evaluated both filtered and unfiltered results and we didn't unwrap the interferograms. As wetland InSAR phase changes tend to be localized, we avoided these commonly used processing steps, which can affect localized signal in order to increase signal to noise ratio.

Fringes represent phase changed due to water level changes in the interferograms, in which each fringe corresponds to a distance change of $\lambda/2$, where λ is the wavelength of the SAR sensor. Because phase changes in wetlands predominantly reflect vertical movement of the surface, the line of sight (LOS) measurement can be translated to vertical movement, by accounting for the acquisition geometry. The translation from LOS to vertical movement is inversely proportional to the cosine of the acquisition's incidence angle [28]. Based on interferogram interpretation, a vector line representing the tidal inundation extents along the coast of Louisiana was digitized. The InSAR detected extents follow the observed phase changes and disruptions along the coast.

Previous studies indicate that the coherence of wetland interferograms is very sensitive to the temporal and perpendicular baselines (e.g., [11,13]). However, the temporal baseline contribution varies depending on the sensor and wetland types. Kim *et al.* [11] showed that fresh water wetlands with woody vegetation keep better coherence than herbaceous wetlands, mainly because tree trunks tend to be stable scatterers over time, whereas soft-stemmed vegetation are altered by vegetation growth and wind. In order to obtain high coherence interferograms, we generate interferograms with short temporal baselines, 46–92 days for ALOS data and 24–48 days for Radarsat-1 data.

In addition to the satellite data, we also used tide gauge records from Port Fourchon station (Figure 2) installed and operated by the National Oceanic and Atmospheric Administration (NOAA). The tide gauge data is sampled every 6 min and is available online [29].

RESULTS

We analyzed a total of seven ALOS and two Radarsat-1 tracks and generated an average of 10 interferograms per track with temporal baseline ranging between 24–48 days for Radarsat-1 and 46–92 for ALOS. We present here eight representative interferograms, some were combined into a mosaic map that covers the extent of Louisiana's coastline (Figure 3). The rest of the interferograms are presented in the supplementary materials. The mosaic map shows fringe patterns over both inland and coastal areas. The inland fringe patterns are mostly a result of water level changes in wetlands or agriculture activities, like the ones on the central area of tracks 168 and 167. Long wavelength fringes (20–40 km) in the north part of the interferogram from track 165 and 168 are most likely phase changes due to atmospheric effects [10]. Fringe discontinuities occur along swath edges, because the selected interferograms are from different time periods and, hence, represent distinct signal response according to phase changes of each interferogram. We processed ALOS data of the entire coastline. However, the produced interferograms were unable to detect details from the Chenier plain coast. Thus we present Radarsat-1, which has a shorter wavelength interferograms, over the Chenier plain and ALOS interferograms over the Mississippi River delta.

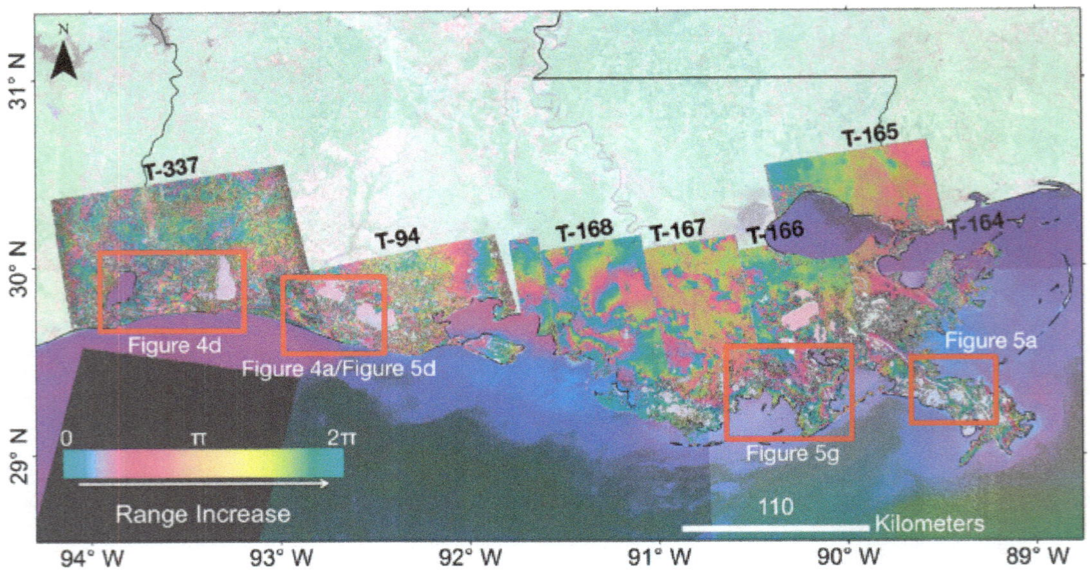

Figure 3: Mosaic of Radarsat-1 (**left side**) and ALOS (**right side**) interferograms for coastal Louisiana. The red frames show the location of enlarged interferograms presented in Figure 4 and Figure 5.

The L-band ALOS interferograms show better coherence than those obtained using C-band Radarsat-1. This is a result of the interaction between the different wavelength of each satellite and the vegetation [11]. Radarsat-1 produces less coherent results because its shorter wavelength (C-band 5.6 cm) interacts with leaves and soft stem vegetation, which change rapidly with time. The L-band ALOS (24 cm) signal penetrates through tree canopies and interacts with branches and tree trunks, which have fewer changes over time.

Coastal wetlands phase changes are characterized by short wavelength patterns and require zooming into some specific areas (Figure 4 and Figure 5). Interferograms from the western most part of coastal Louisiana, the Chenier Plane, show a group of fringe discontinuities aligned with the coastline and relate to the coastal geomorphology in that area (Figure 4a,d). The Chenier Plane is a zone of ridges aligned parallel to the coast; the ridges are 20 to 30 km long and 2 to 6 m high with large mudflats covered with wetland vegetation [30]. InSAR results also show that due to disruptions of the tidal flow along the west Chenier Plain, the extent of tidal inundation ranges between 800 m to 2 km. In the center and east of the Chenier Plain, both ALOS and Radarsat interferograms show noticeable phase discontinuities on different locations along the coastline. The mapped tidal zone in this area extends between 2–4 km (Figure 4a,b and Figure 5d). For the Mississippi River delta, ALOS results show fringe changes due to both natural tidal inundation and anthropogenic activity (Figure 5). Fringe changes due to tidal inundation are most apparent west of the Mississippi delta, where the interferogram shows two fringe cycles, 5 km to the west side and 15 km to the east side, of tidal inundation change over the coast (Figure 5g). The observed discontinuities for this area coincide with man-made structures that where present along the Mississippi River delta (Figure 5a,b). Our results indicate that from the total 743 km length of the Louisiana coast, tidal inundation shows to be disrupted by man-made canals in roughly 45% of the coast and levees disruptions are observed in 15% of the coast length.

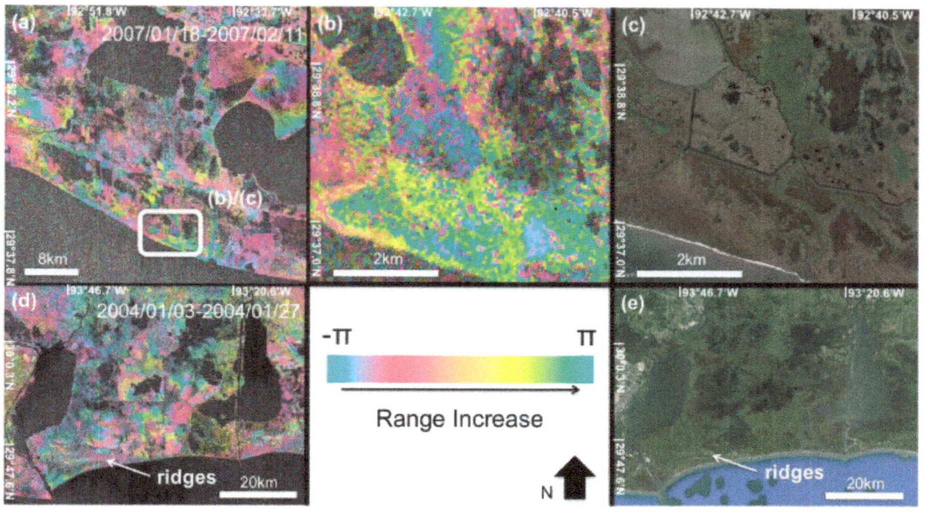

Figure 4: (**a**) Radarsat-1 interferograms of the central part of the Chenier Plain spanning 24 days (18 January 2007–11 February 2007). The white square shows the location of figures (**b,c**); (**b**) Zoom in showing fringe discontinuities sub-parallel to the coastline that coincides with man-made canals observed in the Google Earth image (**c**); (**c**) Google earth image of the same area; (**d**) Radarsat-1 interferogram of the western Chenier Plain spanning 24-day period (3 January 2004–27 January 2004). In the central part of the interferogram fringe patterns show changes in inland water bodies. The white square on the left side of the interferogram shows fringe patterns due to the morphology of the Chenier reflecting water level changes in the mudflats in between the ridges of the Chenier's; (**e**) Google earth image of the same study area.

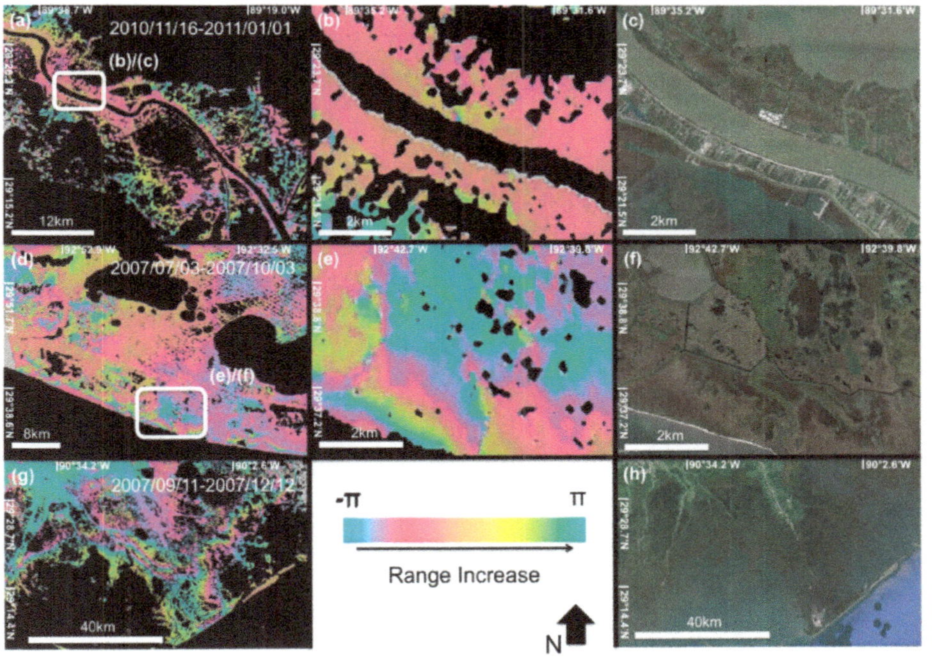

Figure 5: (**a**) ALOS interferograms of the Mississippi Delta showing phase change due to water level changes, spanning 46 days between acquisitions (16 November 2010–1 January 2011); (**b**) Zoom in area showing phase change due to tide inundation disruption caused by the presence of levees along the river; (**c**) Google earth image of the same study area; (**d**) ALOS interferogram of the central Chenier Plain showing phase changes due to water level changes, spanning 92 days (3 July 2007–3 October 2007); (**e**) Elongated fringe patterns sub-parallel to the coastline coinciding with man-made canals; (**f**) Google Earth image of the same study area; (**g**) ALOS interferogram of the west of the Mississippi River delta, spanning 92-day period (11 September 2007–12 December 2007). The interferogram shows fringe patterns due to tidal induced water level changes; (**h**) Google earth image of the same study area.

DISCUSSION

The use of L-band and C-band sensors allowed us to detect different type of tide inundation disruptions along coastal Louisiana due to the difference in wavelength and repeat pass time. We focus our analysis on short wavelength changes, because the interaction between the tide inundation and land or man-made structures happens in localized areas over a few kilometer distance. The Chenier Plane is characterized by phase changes caused by three main causes, which are ridge morphology, inland water bodies, and man-made structures (Figure 4). For this area we found that both ALOS (L-band) and Radarsat-1 (C-band) results show phase disruptions that correlate with numerous man-made canals, dredged mainly for oil recovery or navigation purposes (Figure 4b and Figure 5e). However, Radarsat-1 results also show very short wavelength fringe patterns parallel to the coast due to the geomorphology of the Cheniers (Figure 4d). The geomorphology of the Chenier has been affected by anthropogenic activities including sand mining, road constructions and colonization as described in the Cheniers and Natural Ridges Study of the State of Louisiana (2009). The short wavelength of Radarsat-1 (5.66 cm) enables the satellite to detect phase changes, where ALOS failed to detect. The subtle phase changes induced by the Chenier geomorphology are too small for the 24 cm wavelength of L-band to be detected. Two main types of phase changes, tide inundation disruptions and water level changes through the tidal flush zone characterize the area of the Mississippi River delta. ALOS interferograms were able to detect short wavelength tidal flow disruptions, similar to the ones in Radarsat-1 results, mainly due to levee construction for flood protection (Figure 5a). Interferograms of the western side of the Mississippi River delta show a group of elongated fringes that map the tidal flush zone in that region (Figure 5g, Figure 6 and Figure 7a). Sea level changes in the tidal flush zone are detected because it is an emerged vegetation environment with changing water levels. Although tide-induced water level changes occur also further south in the open water, the lack of vegetation prevents a coherent interferometric signal in such environment (Figure 6). Further north the tidal flush zone is delimited by a man-made canal that interrupts the tidal inundation, obstructing the path to its natural limit in higher lands.

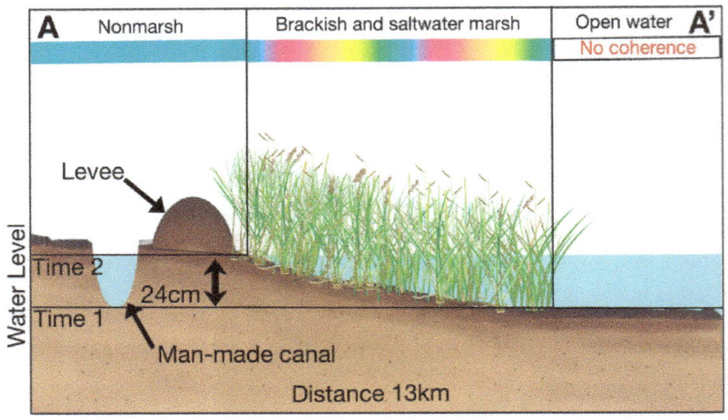

Figure 6: Schematic diagram showing the topography, vegetation and tidal induced level change along transect A-A'.

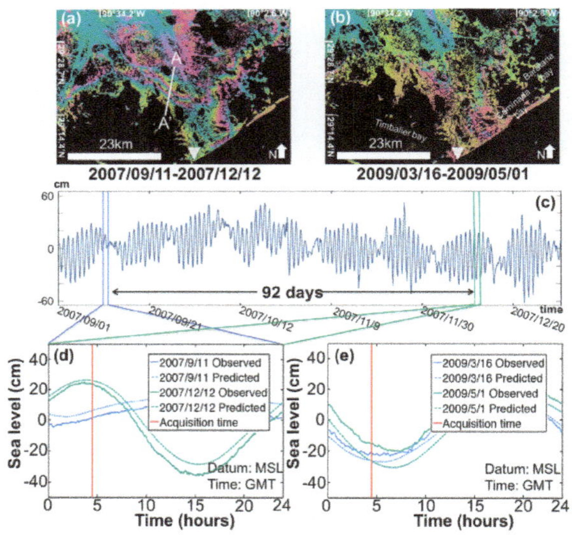

Figure 7: (**a**) ALOS interferogram of track 166 showing two fringe cycles across the tidal zone mapping the tide inundation extent. The white triangle marks the location of the Port Fourchon gauge station. The white line shows the location of the section described on Figure 7; (**b**) ALOS 16 March 2009–1 May 2009 interferogram of the same track showing no fringes; (**c**) Port Fourchon tide gauge data from 1 September 2007 to 31 December 2007. The blue and green vertical lines show the 24-h period recorded by the tide gauge for the two acquisition dates of interferogram (**a**); (**d**) Port Fourchon tide gauge predicted (solid) and observed (dashed) records plotted for 24-hour periods of the two acquisition dates (11 September 2007 in blue and 12 December 2007 in green). The red vertical line marks the ALOS acquisition time 4:32 GMT, which shows roughly 25 cm water level differences between the two acquisitions; (**e**) A similar tide gauge analysis for the two acquisition dates of the interferogram (**b**) showing only 5 cm of water level change between the two acquisition times.

The highest fringe gradient was found along the coastal wetlands in Timbalier bay, Caminada bay and Barataria bay, west of the Mississippi River delta (Figure 5g,h and Figure 7a). However, not all interferograms of the same area show the same amount of fringe changes. For example, the 46-day interferogram of 16 March 2009–1 May 2009 shows no phase change between the acquisitions (Figure 7b). In contrast the 11 September 2007–12 December 2007 interferogram (Figure 7a) shows a set of two fringe cycles, which represent a change of 30 cm of vertical motion occurring between the open water of the Gulf and the wetlands further inland. In order to understand this inconsistent fringe pattern, we compared the InSAR results with the tide gauge information located closest to the area of interest. Due to the tide gauge station distribution and the few places where natural tidal inundation can be observed, the comparison between tide gauges and InSAR results was conducted only for the results of ALOS track 166 and data acquired by the Port Fourchon station (Figure 7a,b). We use a 24-h period of tide information from each acquisition, using Greenwich Mean Time (GMT) and 6 min sampling. We mark the acquisition time of the satellite on the plots (red vertical line in Figure 7d,e) and calculate the water level elevation difference between the two acquisition times, which we compare with the InSAR derived water level change. We also use the NOAA's predicted data for the station, to verify that no extreme events occurred during the InSAR acquisition time that may have affected the InSAR results. The tide gauge records show a water level change of 24 cm between 11 September 2007–12 December 2007 and 5 cm between 16 March 2009 and 1 May 2009, which are consistent with differences in the fringe pattern of the interferograms in Figure 7.

The time difference between the ALOS satellite repeat orbit (46 days) and tidal cycle enables InSAR detection of tide-induced water level variations (Figure 7d). However, if the satellite acquisitions and the tidal changes were synchronized, InSAR observations will not be able to detect tide-induced water level changes occurring at the same tidal amplitude. The tide has both a near daily (23 h) and a two-week cycle, whereas the satellites are sun-synchronous and acquire data exactly at the same time of the day. The phase shift between satellite acquisitions and tide condition results in satellite observations that are collected at different amplitude positions of the tide cycle. The 46-day interferograms have a small amplitude shift (3 days) producing a minor difference between acquisitions and, thus, showing no noticeable tide induced changes in the interferogram. The 92-days interferogram observations are located almost in opposite sides of the tide amplitude cycle, allowing InSAR to better detect differences in the tide inundation level. With the 3–5 cm accuracy of L-band observations [8,10], the patterns inferred from the fringes, which are in the range of 7–30 cm, bearing the method uncertainty level and hence conclusions made are solid.

The mapped tidal inundation extents provide a good perspective of tidal inundation through Louisiana coastal wetlands (Figure 8). The detected tidal inundation extent in the central part of the coast shows an area of approximately 65 km long and 15 km wide area of fresh marsh that is being flushed by tidal currents (Figure 8 and Figure 9c). Detailed observations of the tidal inundation extent along the coast of Louisiana confirmed that the majority of the tidal inundation is being disrupted by man-made canals and other structures along most of the coast (Figure 9).

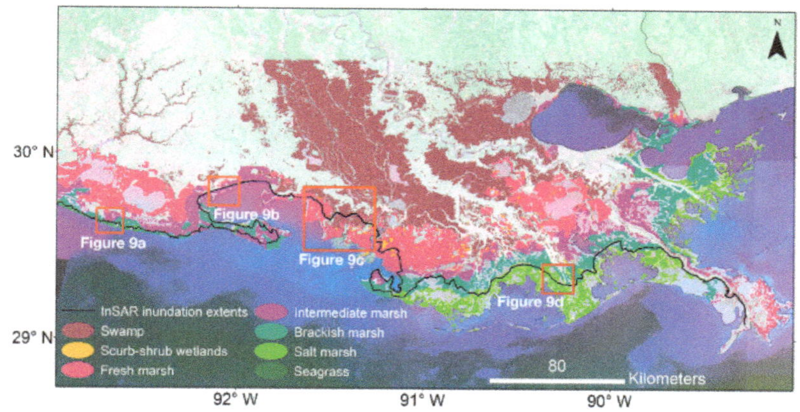

Figure 8: Wetland classification map from Gulf-Wide Information System (G-WIS) [19] and InSAR detected tidal inundation extents through Louisiana coastal wetlands (black line). The red frames show the location of detailed observations presented in Figure 9.

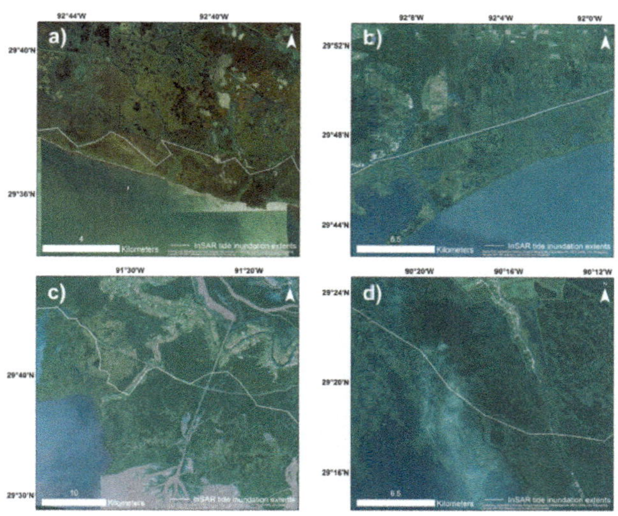

Figure 9: Zoom-in maps showing geographical features along the InSAR detected boundary of tidal inundation extents: (**a,b**) Chenier Plane; (**c**) central coast of Louisiana and (**d**) Timbalier bay, west of the Mississippi River delta. Satellite images source: Esri ArcGIS® world imagery basemap.

The tidal zone along coastal Louisiana shows to be delimited mainly by dredged canals and levees, leaving narrow areas of uninterrupted tidal inundation. Some areas show disruptions by both levees and man-made canals (Figure 9c) while others are disrupted by canals before reaching the populated areas where most of the levees are located (Figure 9d). Our results show that at present approximately 40% of the Louisiana coast is subjected to uninterrupted tidal inundation. However, with rising sea levels and continuous coastal subsidence the length of the shoreline with uninterrupted tidal inundation will decrease due to the existence of man-made structures at short distances (1–6 km) inland of the current tidal inundation zone.

Disruptions along the Louisiana coast are major factors of wetland loss. Anthropogenic intervention is transforming coastal Louisiana, changing the natural tide interaction with the coast and changing the regular sediment deposition with it. The dredged canals that interrupt the water flow work as barriers that divide and separate the coastal wetlands. The dissected wetlands become more sensitive to sea level rise of 3.2 ± 0.4 mm/year [31] and ultimately dying due to land subsidence and increasing sea level.

This study mainly focuses on the analysis of the interaction between tidal inundation and the coastal wetlands of Louisiana by observing changes in the water level along the coast. However it is important to mention that coastal wetlands are strongly affected and driven by temperature changes, rainfall regimes and sea-level rise [32], and they play a permanent role in the synergy between the coastal wetlands and the tidal flow. Nevertheless, our study shows that man-made structures along the Louisiana coast have changed the regular patterns of tidal inundation through coastal wetlands and, hence, have become a dominant element in determining coastal wetland distribution.

CONCLUSIONS

The analysis of four Radarsat-1 and twelve ALOS swaths over the Louisiana coast reveals that tidal inundation is restricted to narrow areas along the coast, where the tide can flow regularly. InSAR results show that the Chenier Plain has the narrowest tidal zone, reaching a maximum of 2 km in some parts, and the Mississippi River delta has wider zones reaching up to 15 km. InSAR technique showed to be an effective tool to map and measure the extents and amplitude of the tide inundation zone through Louisiana coastal wetlands, as well as to detect areas of tidal flow disruptions along the coast. Detailed observations of the produced interferograms indicate that man-made canals and levees are the main sources of tide disruptions along the entire coast.

A comparison between InSAR detected water level changes and tide gauge records indicate that interferometric observations in coastal wetlands highly depend on the tide-induced water level changes observed between satellite acquisitions. In the case of the coast of Louisiana, the 92-day ALOS interferograms observe larger tide-induced water level changes than the 46-day ones, providing clearer information of the tidal inundation extents. Although our study is limited to the Louisiana coast, the use of this technique can be expanded to detect tidal inundation interactions in similar coastal zones. The implementation of InSAR as a monitoring tool to detect tidal inundation propagation can provide useful constrains for coastal tidal inundation models, and can be used to detect and track sources of tidal flow disruptions.

ACKNOWLEDGMENTS

Authors thank the three anonymous reviewers for their very helpful comments on the manuscript. We would like to thank Sang-Hoon Hong and Heresh Fattahi for their help processing the InSAR data. We also acknowledge help of the Japan Aerospace Exploration Agency (JAXA), the Canadian Space Agency (CSA), the Alaska Satellite Facility (ASF) and the National Oceanic and Atmospheric Administration (NOAA) for providing access to the satellite images and tide gauge station information. We thank the Fulbright and CONACYT grants for their support of the first author. The research was supported by the National Science Foundation through Research program under Grant No. DEB-1237517.

AUTHOR CONTRIBUTIONS

Talib Oliver was responsible for InSAR and tide-gauge data processing. Both authors contributed to the data interpretation and the writing of the manuscript.

REFERENCES

1. Gallant, A.L. The challenges of remote monitoring of wetlands. *Remote Sens.* 2015, *7*, 10938–10950.
2. Couvillion, B.R. *Land Area Change in Coastal Louisiana (1932 to 2010)*; US Department of the Interior, US Geological Survey: Reston, VA, USA, 2011.
3. Penland, S. Geomorphic classification of coastal land loss between 1932 and 1990 in the Mississippi River Delta Plain, Southeastern Louisiana. *USGS Open-File Report 00-417*; US Geological Survey Woods Hole: Barnstable County, MA, USA, 2000.
4. Steyer, G.D.; Couvillion, B.R.; Barras, J.A. Monitoring vegetation response to episodic disturbance events by using multitemporal vegetation indices. *J. Coast. Res.* 2013, *63*, 118–130.
5. Barras, J. *Historical and Projected Coastal Louisiana Land Changes: 1978–2050*; United States Geological Survey: Reston, VA, USA, 2003.
6. Ramsey, E.; Rangoonwala, A.; Bannister, T. Coastal flood inundation monitoring with satellite C-band and L-band synthetic aperture radar data. *JAWRA J. Am. Water Resour. Assoc.* 2013, *49*, 1239–1260.
7. Moffett, K.B. Multiple stable states and catastrophic shifts in coastal wetlands: Progress, challenges, and opportunities in validating theory using remote sensing and other methods. *Remote Sens.* 2015, *7*, 10184–10226.
8. Wdowinski, S. Space-based detection of wetlands' surface water level changes from L-band SAR interferometry. *Remote. Sens. Environ.* 2008, *112*, 681–696.
9. Kim, J.-W. Integrated analysis of PALSAR/ Radarsat-1 InSAR and ENVISAT altimeter data for mapping of absolute water level changes in Louisiana wetlands. *Remote Sens. Environ.* 2009, *113*, 2356–2365.
10. Sang-Hoon, H.; Wdowinski, S.; Sang-Wan, K. Evaluation of TerraSAR-X Observations for Wetland

InSAR Application. *IEEE Trans. Geosci. Remote Sens.* 2010, *48*, 864–873.

11. Kim, S.-W. Interferometric Coherence Analysis of the Everglades Wetlands, South Florida. *IEEE Trans. Geosci. Remote Sens.* 2013, *51*, 5210–5224.
12. Pritchard, M.E. InSAR, a tool for measuring Earth's surface deformation. *Phys. Today* 2006, *59*, 68.
13. Lu, Z.; Kwoun, O.-I. Radarsat-1 and ERS InSAR analysis over southeastern coastal Louisiana: Implications for mapping water-level changes beneath swamp forests. *IEEE Trans. Geosci. Remote Sens.* 2008, *46*, 2167–2184.
14. Elijah, R., III; Lu, Z.; Rangoonwala, A.; Rykhus, R. Multiple baseline radar interferometry applied to coastal land cover classification and change analyses. *GISci. Remote Sens.* 2006, *43*, 283–309.
15. Wdowinski, S. Remote-sensing monitoring of tide propagation through coastal wetlands. *Oceanography* 2013, *26*, 64.
16. Boesch, D.F.; Josselyn, M.N.; Mehta, A.J.; Morris, J.T.; Nuttle, W.K.; Simenstad, C.A.; Swift, D.J. Scientific assessment of coastal wetland loss, restoration and management in Louisiana. *J. Coast. Res.* 1994, *20*, i-103.
17. Owen, D.E. Geology of the Chenier Plain of Cameron Parish, southwestern Louisiana. *Field Guides* 2008, *14*, 27–38.
18. Sasser, C.E. Vegetation types in coastal Louisiana in 2013. *U.S. Geological Survey Scientific Investigations Map 3290*; US Department of the Interior, US Geological Survey: Reston, VA, USA, 2013.
19. Minerals Management Service. *G-WIS Environmental Sensitivity Index Datasets*; Minerals Management Service: New Orleans, LA, USA, 2002.
20. MDA Federal. Landsat Geocover 2000/ETM+ Edition Mosaics. Available online: http://glcf.umd.edu/data/mosaic/(accesed on 5 July 2015).
21. Bernier, J. Trends and Causes of Historical Wetland Loss in Coastal Louisiana. Available online: http://pubs.usgs.gov/fs/2013/3017/ (accessed on 4 May 2016).
22. Dixon, T.H. Space geodesy: Subsidence and flooding in New Orleans. *Nature* 2006, *441*, 587–588.
23. Morris, J.T. Responses of coastal wetlands to rising sea level. *Ecology* 2002, *83*, 2869–2877.
24. University of Alaska Fairbanks. The National Aeronautics and Space Administration. Available online: https://ursa.asfdaac.alaska.edu/ (accessed on 5 February 2015).
25. Richards, J.; Woodgate, P.; Skidmore, A. An explanation of enhanced radar backscattering from flooded forests. *Int. J. Remote Sens.* 1987, *8*, 1093–1100.
26. Wdowinski, S.; Hong, S.-H. 7 Wetland InSAR. In *Remote Sensing of Wetlands: Applications and Advances*; CRC Press: Boca Raton, FL, USA, 2015.
27. Rosen, P.; Fielding, E. Repeat Orbit Interferometry Package (ROI_PAC). Available online: http://roipac.org/ (accessed on 6 May 2016).
28. Hanssen, R.F. *Radar Interferometry: Data Interpretation and Error Analysis*; Springer Science & Business Media: Berlin, Germany, 2001.
29. The National Oceanic and Atmospheric Administration. Center for Operational Oceanographic Products and Services. Available online: http://tidesandcurrents.noaa.gov/ (accessed on 5 May 2016).
30. Penland, S.; Suter, J.R. The geomorphology of the Mississippi River Chenier plain. *Mar. Geol.* 1989, *90*, 231–258.
31. Church, J.A.; White, N.J. Sea-level rise from the late 19th to the early 21st century. *Surv. Geophys.* 2011, *32*, 585–602.
32. Osland, M.J. Beyond just sea-level rise: Considering macroclimatic drivers within coastal wetland vulnerability assessments to climate change. *Glob. Chang. Biol.* 2016, *22*, 1–11.

CHAPTER 6

LANDSAT ETM+ AND SRTM DATA PROVIDE NEAR REAL-TIME MONITORING OF CHIMPANZEE (PAN TROGLODYTES) HABITATS

Samuel M. Jantz, Lilian Pintea, Janet Nackoney and Matthew C. Hansen

Keywords: Chimpanzee, conservation, habitat suitability monitoring, MaxEnt, random forests, species distribution model

INTRODUCTION

The dramatic decline in biodiversity has become so great that it is considered an important facet of global change in its own right [1]. Biodiversity loss has multiple causes, with habitat destruction via land-cover and land-use change as the predominant driver [2,3,4]. The majority of future land-use change will likely occur in tropical regions [5], which are home to almost half of all described species [6], including humans' closest living relatives, the chimpanzee (*Pan troglodytes*). It has been estimated that more than 70% of chimpanzee tropical forest habitats are now threatened by infrastructure development and land use change [7]. All four chimpanzee sub-species (Western (*P.t. versus*), Nigeria-Cameroon (*P.t. ellioti*), Central (*P.t. troglodytes*), and Eastern (*P.t. schweinfurthii*)) have been classified as endangered on the International Union for the Conservation of Nature (IUCN) Red List. In fact, it has been estimated that their cumulative population has declined by more than 66% over the past 40 years [8] and there is a decreasing trend in population for all four sub-species [9,10,11,12]. Major threats to current populations include habitat loss from resource extraction activities and land conversion, hunting, disease and the illegal pet trade. Chimpanzees occupy a wide variety of forest and woodland habitats and serve as an umbrella species [13]; thus, protecting chimpanzee habitat could help conserve habitat for many other species.

The Open Standards for the Practice of Conservation (OSPC) provides a framework to guide the development and implementation of conservation projects [14]. OSPC is a collaborative and adaptive management process that strategically focuses conservation decisions on clearly defined objectives and prioritized threats and measures success in a manner that enables adaptation and learning over time. A key component outlined in the OSPC is an overall assessment of the viability or health of the conservation target and the establishment of a monitoring plan that requires indicators that are measurable, precise, consistent and sensitive [14]. Habitat suitability models (HSMs), also referred to as species distribution or ecological niche models, are useful tools that can fit easily within this framework. HSMs use statistical and machine learning techniques to find empirical relationships between observed species' occurrences and environmental descriptors, such as resource, biotic and climatic factors, to estimate conditions that are suitable for population viability [15], or to predict the likelihood of occurrence [16,17] in geographic space. Results from HSMs are most convincing when they are fit with both presence and absence data. However, due to the difficulty of acquiring ecologically meaningful absence data, HSMs are often constructed with so called "presence only" data [18]. When models are fit with presence only data, the output can be interpreted as a relative measure of environmental suitability rather than the probability of presence [19,20] typically with a range between 0 (unsuitable) and 1 (highly suitable).

The use of HSMs to model and map chimpanzee habitat suitability is a recent endeavor and has been used to answer questions applicable to chimpanzee conservation and ecology. Vegetation layers derived from Landsat ETM+ data were used in [21] to model potential habitat suitability for chimpanzees in Western Tanzania at 90 m resolution. Landsat TM derived variables were used to generate habitat distribution maps for three time periods between 1986 and 2003 for a small area in Guinea-Bissau, within the range of *P.t. verus*, where a marked decrease in habitat area and connectivity was found [22]. Habitat suitability was mapped for the *P.t. schweinfurthii* range

Citation: Samuel M. Jantz, Lilian Pintea, Janet Nackoney and Matthew C. Hansen, Landsat ETM+ and SRTM Data Provide Near Real-Time Monitoring of Chimpanzee (Pan troglodytes) Habitats, doi:10.3390/rs8050427.

at 10 km resolution using climate, human impact and vegetation variables to aid in the delineation of areas for new population survey efforts [23]. HSMs were used for *P.t. elliotii*in Nigeria and Cameroon and *P.t. troglodytes* in Cameroon using climate, topographic, human population density and canopy cover variables at 1-km resolution to investigate the role of environmental variation in the genetic dissimilarity between the two sub-species and to assess potential future impacts of climate change on their habitat suitability [24]. The first range-wide study was performed by [25], who used climate, human impact and vegetation variables as model inputs to map habitat suitability for all four chimpanzee sub-species, and all African Great Apes, at 5-km resolution for both the 1990s and 2000s, and found a decrease in suitable habitat over time. Field researchers from several conservation organizations contributed data and were involved in the publication; thus, their results represent the most currently accepted and published state of chimpanzee habitat suitability across the field. We therefore rely on the results from Reference [25] for this study.

The objectives of this study are to map chimpanzee habitat suitability at 30-m resolution and provide information on changes in suitability through time to demonstrate that remotely sensed datasets that are updated on a continual basis can be used in near real-time conservation monitoring plans. Previous studies have attempted to determine the environmental variables that govern the observed distribution of chimpanzees at local, regional and range-wide scales. In this study, we determine which regularly updated remotely sensed variables are most useful indicators for effective and continuous monitoring of chimpanzee habitats. We utilized a modified version of habitat suitability modeling where, instead of using presence data to calibrate our model, we rescaled the coarse resolution habitat suitability map cited in Reference [25] to 30-m resolution using variables derived from remotely sensed datasets as input into a Random Forests regression model. Our results inform field monitoring efforts that support conservation planning and measure success as they portray finer scale habitat variation over several time periods.

DATA AND METHODS

Study Area

The study area covers the ranges of the four chimpanzee sub-species that span central and western Sub-Saharan Africa over an approximately 2.3 × 10^6 km^2 area [26] (Figure 1), which represents approximately 7.5% of the African continental area. Chimpanzees occupy a wide variety of forest habitats across their range, including humid evergreen forests, mosaic woodlands, deciduous forest, and dry savanna woodlands [27]. Their habitats also cover a large elevation range, from sea-level in West Africa up to 2600 m in East Africa [26]. For our study, we used geographic range extents for each of the sub-species from the IUCN Red List. We buffered each range by 100 km to cover any suitable habitat that may fall outside the delineated range.

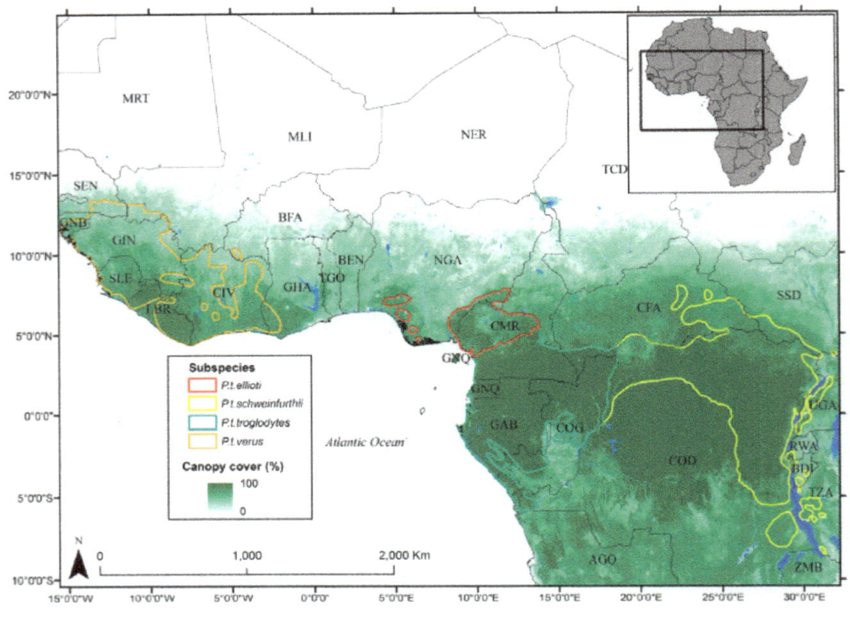

Figure 1: Overview of study area with the International Union for the Conservation of Nature (IUCN) Red List ranges of the four chimpanzee sub-species overlaid on top of data showing percent canopy cover from [28].

Habitat Data

Data on the spatial distribution of habitat suitability for most of the study area are from Reference [25], who used location data from the IUCN/Species Survival Commission A.P.E.S. (Ape Populations, Environments and Surveys) database (http://apes.eva.mpg.de), which serves as a central repository for all ape related survey data. Resulting estimates of habitat suitability were foundational for this study.

Logistic regression modelling was used in Reference [25] to develop HSMs. Logistic regression requires both a positive and negative outcome. In this case, known chimpanzee locations were used as the positive outcome and the negative outcome was composed of "pseudo-absences". Pseudo-absences were selected based on the output from MaxEnt, a software package widely used to generate habitat suitability models where presence only data are available [29]. Grid-cells with low predicted suitability from the MaxEnt model were more likely to be used as the negative outcome in the logistic regression models. Predictor variables used in Reference [25] included bioclimatic variables from the WorldClim dataset, forest cover products derived from the Moderate Resolution Imaging Spectrometer (MODIS) and the Advanced Very High Resolution Radiometer (AVHRR) sensors, human population density, proximity to roads and proximity to navigable rivers.

Location data covering the dry forests and miombo woodlands of western Tanzania were not included in the APES database; therefore these important chimpanzee habitats were not reflected in the output from Reference [25]. In order to include these habitats in our model, we incorporated a habitat suitability model created by the authors of [21] specifically for this region based on nest locations (Figure 2B). Mahalanobis distance was used to relate chimpanzee location data, collected as part of biodiversity surveys conducted by the authors of [30], to three environmental predictors: distance to the nearest forest patch, distance to steep slopes (*i.e.*, slopes greater than 20°) and elevation [30]. In the context of habitat suitability modeling, the Mahalanobis distance is the squared difference (in feature space) between a vector of values for the environmental predictors at un-sampled locations and the vector containing the mean values for the same predictors at sampled locations [31]. Mahalanobis distances are approximately distributed as Chi-square with n degrees of freedom where n is the number predictor variables, therefore a location that has a vector identical to the mean vector will have a typicality of 1.0 while larger distances will approach a typicality of 0 [32]. In order for the prediction from this model to be on the same scale as [25], Mahalanobis distances were converted to typicality probabilities by assuming a Chi-square distribution with three degrees of freedom. The suitability map was then coarsened from its native 90-m resolution to approximately 5-km resolution and combined with the 2000s prediction produced by Reference [25].

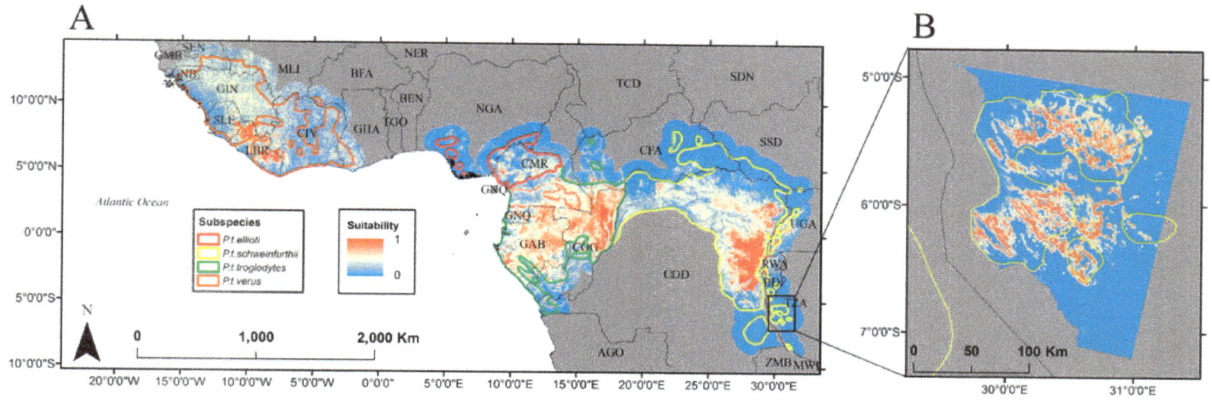

Figure 2: (**A**) Habitat suitability data used for Random Forest calibration [25] and (**B**) habitat suitability data for the miombo woodlands of western Tanzania [21].

Satellite Data

Spectral reflectance, forest structure and forest loss data were from a recent processing and synthesis of the Landsat ETM+ data archive performed by the authors of [28]. This dataset consists of global annual composites of percent canopy cover, canopy height, and top of canopy reflectance for Landsat ETM+ bands 3, 4, 5 and 7, as well as annual data of forest loss from 2000 to 2012. The global products are in a geographic coordinate system and were disaggregated into 10° × 10° tiles with a resolution of 0.000278 decimal degrees (~30 m). To reduce the effects of noise and to fill data gaps due to persistent cloud cover, we divided the annual composites into four time periods by applying a three year median to the canopy cover, canopy height, and spectral band layers, resulting in datasets representing 2000–2003, 2004–2006, 2007–2009, and 2010–2012 time periods. The method for generating Landsat based canopy cover and forest loss data can be found in Reference [28], however the canopy height data were not included as part of the initial data release. Global tree height maps were generated on an annual time scale using a regression tree model that related height measurements from the Geoscience Laser Altimeter System (GLAS), part of the ICESat satellite mission, to Landsat time-series metrics. These height maps were used in Reference [33] to estimate carbon emissions from natural and managed forests; model error was quantified using a random 90% subset of the available GLAS data to train the model while the remaining 10% was used for testing model performance. Root mean square error and mean absolute error for the tropics was 8.1 m and 5.9 m, respectively. For additional information on the method used to generate the tree height maps from Landsat data, see Reference [34].

Topographic data for the study region were from the Shuttle Radar Topography Mission (SRTM) 3 arc-second resolution dataset and locations of rivers were from the SRTM Waterbodies Dataset (SWBD). The SWBD was created as a by-product of the finished SRTM DEM and includes water bodies detectable at approximately 30-m resolution at the time of the SRTM shuttle flight, circa February 2000 [35]. The SWBD is distributed in vector format, which we rasterized to ~30-m resolution to match the resolution of the Hansen *et al.* 2013 [28] dataset. We also resampled the elevation and slope layers to ~30 m to match the resolution of the dataset in Reference [28].

Derivation of Predictor Variables and Calibration Data

A full list of predictor variables used to construct the habitat suitability models is shown in Table 1. We identified the set of predictor variables based on the variables used to generate the HSMs in Reference [25] and variables used in other studies on chimpanzee habitat suitability. Canopy cover and canopy height were included in the set of predictor variables because measurements of forest structure (e.g., percent canopy cover) were found to be important in the determination of chimpanzee habitat suitability in several previous studies [23,24,25]. Spectral metrics, namely NDVI, were found to be an important predictor in Reference [22], which they argued served as a proxy for ecosystem functioning. We chose to include all Landsat ETM+ bands and band derivatives from the dataset in Reference [28] for this reason and because several studies have found an association between spectral data and forest structure metrics [36,37,38].

Table 1: List of predictor variables used as input to Random Forest regression models

Variable	Units	Source	Abbreviation
Landsat ETM+ band 3 (0.63–0.69 μM)	% reflectance	Hansen *et al.* 2013 [28]	B3
Landsat ETM+ band 4 (0.77–0.90 μM)	% reflectance	Hansen *et al.* 2013 [28]	B4
Landsat ETM+ band 5 (1.55–1.75 μM)	% reflectance	Hansen *et al.* 2013 [28]	B5 *
Landsat ETM+ band 7 (2.09–2.35 μM)	% reflectance	Hansen *et al.* 2013 [28]	B7
Normalized difference band4/band3	Unitless	Hansen *et al.* 2013 [28]	NormB4/B3
Normalized difference band4/band5	Unitless	Hansen *et al.* 2013 [28]	NormB4/B5
Normalized difference band4/band7	Unitless	Hansen *et al.* 2013 [28]	NormB4/B7 *
band3/band5	Unitless	Hansen *et al.* 2013 [28]	B3/B5 *
band3/band7	Unitless	Hansen *et al.* 2013 [28]	B3/B7
band5/band7	Unitless	Hansen *et al.* 2013 [28]	B5/B7
Canopy cover	Percent	Hansen *et al.* 2013 [28]	CC *
Canopy height	M	Hansen *et al.* 2013 [28]	HT *
Distance to forest	M	Hansen *et al.* 2013 [28]	DF
Distance to forest loss	M	Hansen *et al.* 2013 [28]	DL
Forest loss in 1 km buffer	Proportion	Hansen *et al.* 2013 [28]	L1K
Forest loss in 25 km buffer	Proportion	Hansen *et al.* 2013 [28]	L25K *
Distance to forest edge	M	Hansen *et al.* 2013 [28]	DE *
Forest edge in 1 km buffer	Proportion	Hansen *et al.* 2013 [28]	E1K
Forest edge in 25 km buffer	Proportion	Hansen *et al.* 2013 [28]	E25K *
Distance to rivers	M	SWBD [35]	DR *
Distance to steep slopes	M	SRTM [35]	DS *
Elevation	M	SRTM [35]	EL *
Slope	Degrees	SRTM [35]	SLP *

* indicates variable was included in final model. See Section 2.4.1 for details on variable selection.

Distance to the nearest forest patch was used as a predictor variable both in the habitat suitability models for Western Tanzania [21] and in Reference [22], and was therefore included in the set of predictor variables. In this study, we defined forest as 40% canopy cover with a minimum patch size of 0.5 ha, which is a combination of definitions used by the United Nations Environment Programme (UNEP) and Food and Agriculture Organization (FAO) for closed forests [39,40]. We applied a threshold of 40% or greater to the canopy cover layers and removed patches smaller than 5 Landsat pixels or approximately 0.5 ha. We used these layers to calculate the minimum per pixel distance to the nearest forest patch for all four time periods.

Human impacts on chimpanzee habitat were found to be important in References [24,25], and Reference [23] that used some form of gridded human population as a predictor variable. Deforestation and forest fragmentation are the direct result of human activity and are known to adversely affect biodiversity [41]. In this study, we used forest loss and fragmentation metrics as proxies for human impact on chimpanzee habitat. We created forest edge layers for each time period by calculating the per pixel distance from the interior of each forest patch to the nearest forest edge. In addition, we calculated the proportion of forest edge within a buffer of 1 km and 25 km for each pixel. We chose a buffer of 1 km to capture the influence of local scale fragmentation while the 25 km diameter represents the upper limit of a chimpanzee's home range [42,43,44].

We used the forest loss data from Reference [28] to create four binary loss/no-loss layers for 2000–2003, 2004–2007, 2008–2009, and 2010–2012 time periods. To clean the data, we removed pixels that were labeled as forest-loss in non-forest pixels and used the same minimum size to define forest patches (*i.e.*, 0.5 ha). We used these layers to calculate the per-pixel distance to the nearest forest loss patch and the proportion of loss in 1 km and 25 km in diameter, respectively.

We used topographic information including elevation, slope and distance to steep slopes as predictor variables in our model because previous studies have found they factor into the prediction of chimpanzee habitat suitability. Chimpanzee nests were generally found to be absent in forests and woodlands in elevations greater than approximately 1900 m [21,23]. In addition, chimpanzees have been found to potentially prefer to nest on steep slopes, with a grade greater than 20°. Slope was found to be a prominent factor in the prediction of habitat suitability in Reference [24], particularly for the *P.t. elliotis*ub-species. Finally, we used the rasterized SWBD dataset to calculate the per pixel distance to rivers because the authors of [25] found proximity to rivers to be a significant predictor of habitat suitability in all of their models. We decided to use the SWBD dataset to map the range-wide river network rather than the Digital Chart of the World data used in Reference [25], as it is a more current depiction of the river network.

In order to match the spatial resolution of the calibration dataset, all predictor variables for the 2000–2003 time period were resampled from 30 m to 5 km by averaging all 30 m pixels within a 5 km grid-cell. Values for the habitat suitability layer and predictor layers were extracted for each of the four sub-species ranges resulting in 18,000, 71,500, 45,000, and 48,000 samples (*i.e.*, 5-km grid-cells) for *P.t. ellioti*, *P.t. schweinfurthii*, *P.t. troglodtyes* and *P.t. verus*, respectively. We generated fifty boot-strapped training and testing datasets for each sub-species by randomly sampling 75% of each dataset for model calibration and 25% for model validation. We created an additional fifty training and testing datasets that represented the entire chimpanzee range by combining training and testing datasets from each sub-species. We then predicted habitat suitability from the range-wide, 5-km resolution calibration map using a suite of remotely sensed derived metrics (Table 1).

Habitat Suitability Estimation

We used Random Forests regression to estimate range-wide chimpanzee habitat suitability with the Sci-kit Learn module in the python software environment [45]. Random Forests is an ensemble technique in the Classification and Regression Trees (CART) family of machine learning methods [46]. Random Forests works by constructing several regression trees based on bootstrapped samples of the input dataset and each node is split using the best from a random subset of predictors and each tree is grown to full size. Overfitting in Random Forests is avoided by averaging the predictions produced by individual trees. Random Forests have the attractive properties of being insensitive to non-normally distributed variables, non-linear relationships, and co-linear predictor variables [47]. In addition, only two primary parameters are required for Random Forest models; the number of trees constructed in each forest and the number of variables to consider at each node split.

We calibrated Random Forests regression models using satellite data from 2000–2003 to have a close temporal match with Reference [25]. The climate, demographic and topographic variable data used in Reference [25] are not updated regularly and could presumably be used to predict habitat suitability over a fairly wide temporal range. Canopy cover metrics, on the other hand, were from AVHRR and MODIS continuous field estimates representing forest in the first few years of the 2000 decade.

While the authors of [25] created a separate model for each sub-species, we investigated if a general, range-wide model could reasonably predict habitat suitability for each of the sub-species. An initial Random Forests regression model was constructed for each bootstrapped training sample for each sub-species as well as a general, range-wide model from the pooled sub-species data. Predictions were made from each model on the test data, and Root Mean Square Error (RMSE) was computed for each bootstrapped sample. A preliminary analysis indicated that using a range-wide model to predict suitability for each of the sub-species ranges did not result in a substantial increase in error for any of the sub-species (Figure 3). We therefore focused on the development of a general, range-wide model for the remainder of this manuscript.

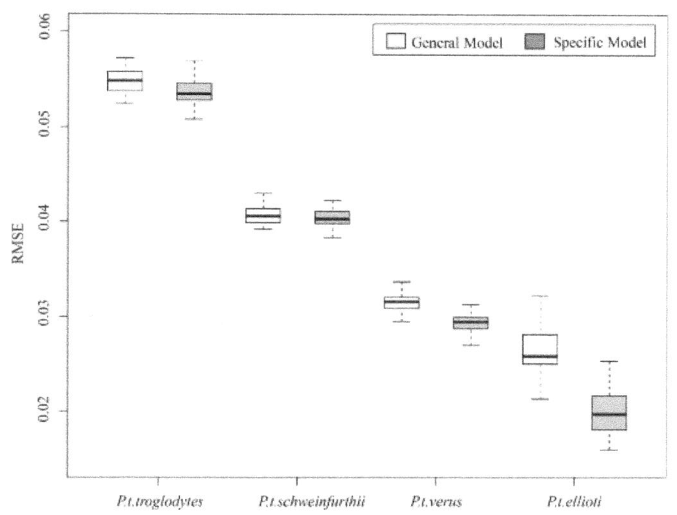

Figure 3: Boxplots of 50-fold cross-validated RMSE of a general, range-wide model (white) and four models trained on data for each of the individual sub-species (grey).

Selection of Predictor Variables

To create the most parsimonious model possible, we first removed highly correlated variables in addition to the variables that contributed little to the predictive capability of the model. We created a correlation matrix for all predictor variables to determine the coefficient of determination for each combination of variables. For each pair of highly correlated variables ($r^2 >$ 0.9), we calculated the coefficient of determination between each variable and habitat suitability and retained the predictor variables with the highest coefficient of determination. We found canopy height and percent canopy cover to have a correlation coefficient greater than 0.9; however, we decided to retain both variables because of their recognizable ecological importance to chimpanzee habitat. Next, we used recursive feature elimination to remove variables that had little influence on the model's predictive power. Recursive feature elimination operates by first ranking all variables by a measure of importance. Here, we ranked variables by the mean increase in RMSE when the values for a particular variable were randomly permuted [48]. The variable that resulted in the smallest increase in RMSE was dropped and a new model was created, and new ranks were then calculated for each variable. We continued this process until RMSE reached a minimum [49]. The lowest RMSE occurred when a subset of 12 variables was used to construct Random Forest models. Mean RMSE for models created using a subset of variables (0.041 ± 0.002) was slightly lower, but not significantly different than the mean RMSE (0.042 ± 0.003) from models created using the full set of variables.

Habitat Suitability Prediction

We used all samples from the calibration dataset to train the final Random Forests model of habitat suitability. We found that out of bag (*i.e.*, the sample withheld during the construction of trees) error stabilized when a forest was created with 250 trees and 7 of the variables considered at each node split. The final Random Forest model was used to create range-wide habitat suitability maps for the 2000–2003, 2004–2006, 2007–2009 and 2010–2012 time periods at the 30-m Landsat pixel scale.

RESULTS

We found similar spatial patterns of habitat suitability between the calibration data (Figure 4A) and the predictions from our model (Figure 4B). Moreover, we were able to model chimpanzee habitat suitability reasonably well compared to the calibration data, explaining 82% (±0.2%) of the variance for the entire chimpanzee range (Table 2). However, there was considerable variation in the model's predictive capability among the four sub-species. Our model was able to explain 35% (±1.95%), 89% (±0.18%), 66% (±0.47%), and 73% (±0.44%) of the variance for *P.t. ellioti*, *P.t. schweinfurthii*, *P.t. troglodytes*, and *P.t. verus*, respectively (Table 2). Compared to the calibration data, our model tended to underestimate habitat suitability (Figure 5A,B). In general, areas where suitability was underestimated occurred at low elevations with sparse canopy cover, while the opposite was true for overestimated areas. Linear error patterns could be seen within the range of *P.t. troglodtyes*(Figure 5A), which resulted from a spatial mismatch of rivers between our study and Reference [25] due to differing river network datasets.

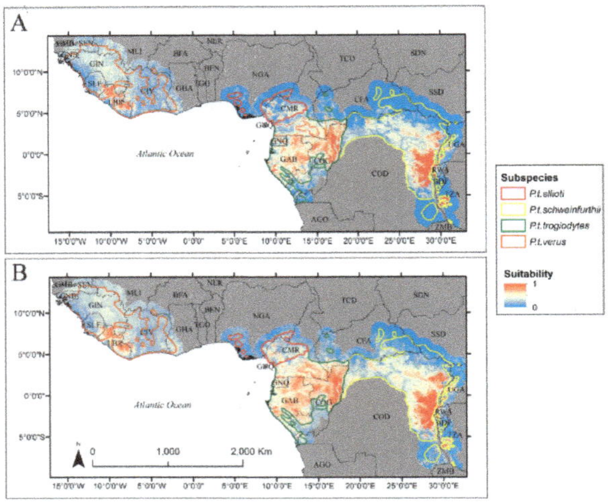

Figure 4: Spatial distribution of suitability values for 5-km resolution map (**A**), and Random Forests prediction at 30-m resolution (**B**).

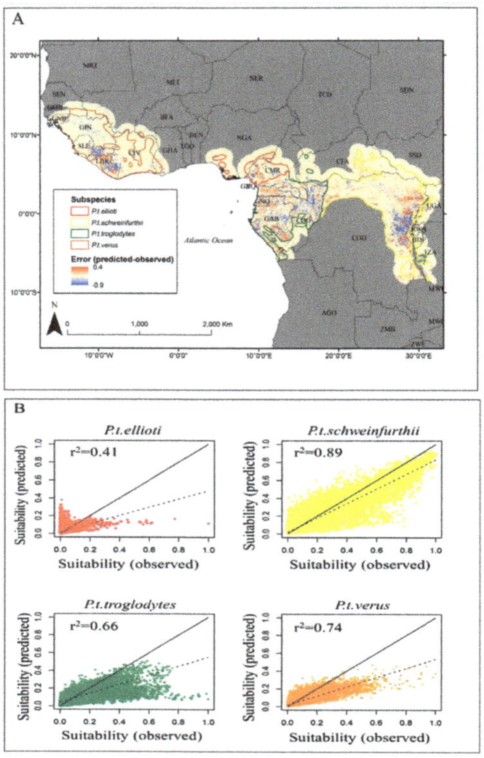

Figure 5: Spatial distribution of differences between the 5-km resolution map used for model calibration and the 30-m map after coarsening to 5-km resolution (**A**), distribution of suitability values of 5-km calibration map (x-axis) and coarsened 30-m map (y-axis) for each sub-species (**B**). The one-to-one line is shown in solid black and results of regressing values from the coarsened 30-m map onto values from the 5-km map for each sub-species are shown by the dashed black line.

Table 2: Cross-validated mean correlation coefficients with standard errors ($N = 50$ cross validations) reported for the entire chimpanzee range, as well as for each sub-species

Range	Mean r^2	Standard Error
Range-wide	0.82	0.001
P.t. ellioti	0.35	0.01
P.t. schweinfurthii	0.89	0.001
P.t. troglodytes	0.66	0.002
P.t. verus	0.73	0.002

Elevation, Landsat ETM+ band 5 and Landsat derived canopy cover were the strongest predictor variables in our model (Figure 6). A significant association between habitat suitability and bioclimatic variables was found in the suitability models for all four sub-species in Reference [25]. The bioclimatic variables used in Reference [25] were taken from the WorldClim set of global climate layers, which were created by interpolating weather station data using latitude, longitude and elevation as independent variables [50]. It is therefore probable that elevation acted as a proxy for suitable climate conditions in our model. Shortwave infrared reflectance is highly correlated with forest/non-forest, where forests have lower reflectance due to absorption properties of live vegetation and light extinguishing effects of tall forest canopies. The value of band 5 in discriminating Congo Basin forest from non-forest was illustrated in Reference [51]. Trees in the CART family have been shown to exhibit a bias toward variables that have many cut-points (*i.e.*, variables with many unique values) [48]. Band 5 is tightly correlated with canopy cover ($r^2 = 0.83$) and canopy height ($r^2 = 0.82$) in the calibration data. Canopy cover values were all integers between 0 and 100, while canopy height values were all integers between 0 and 23. On the other hand, band 5 values were continuous on the interval 0 to 0.7, which provided the algorithm with many more possible cut-points relative to canopy cover and canopy height. We believe it may therefore likely be driving the model as a surrogate for forest structure.

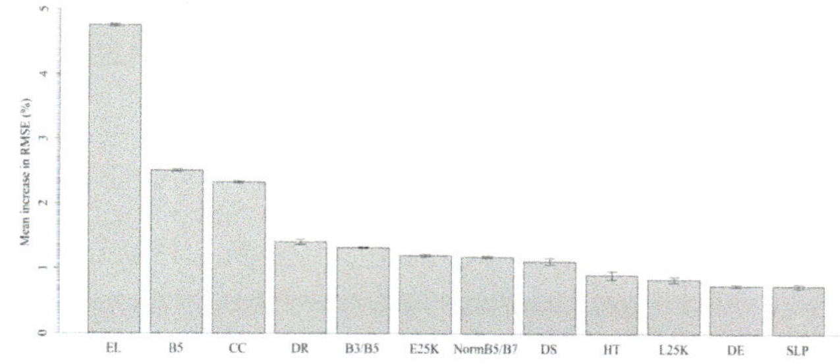

Figure 6: Relative importance of 12 variables selected for final Random Forest model after recursive feature elimination. Confidence intervals (95%) were calculated using bootstrap sampling ($N = 50$). Variable importance was measured by the mean increase in RMSE % when values for a variable were randomly permuted.

DISCUSSION

This study represents the first attempt to map temporal dynamics of range-wide chimpanzee habitat suitability at a finer spatial scale. Our model demonstrated sensitivity to environmental change, which is a key feature for monitoring applications. The area depicted in Figure 7 is situated in the Republic of Congo, which historically has not experienced any substantial human disturbance but recently has seen the most rapid increase in the establishment of logging roads among all Central African nations [51]. Our model was able to detect the fine-scale changes in habitat suitability due to human activities in the area (Figure 7B–E), whereas the coarse scale model was unable to resolve any fine-scale variation (Figure 7F). These results highlight the value of integrating continuously updated variables derived from satellite remote sensing into habitat suitability models for near real-time monitoring and decision support. The Landsat suite of satellites has provided the longest continuing record of standardized data of Earth's land cover at 30-m resolution; metrics derived from these data will be crucial for the development of standardized habitat suitability models that can be repeated and integrated into long-term monitoring plans.

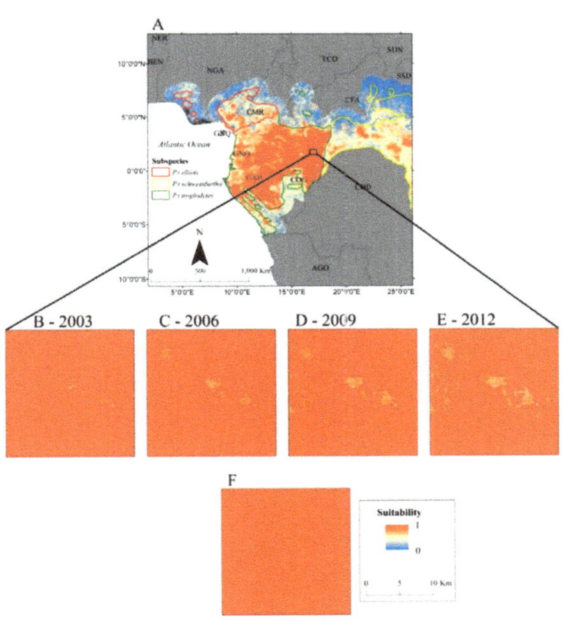

Figure 7: Suitability values predicted by a Random Forests model for Central Africa (**A**), an area experiencing selective logging circa 2003 (**B**), circa 2006 (**C**), circa 2009 (**D**), circa 2012 (**E**) and the area as depicted in the 5-km resolution calibration data (**F**).

Previous studies have reported successful efforts to model chimpanzee habitat suitability, including the two sources from which our model was calibrated. Direct comparison between studies is difficult because, so far, there are only a few published studies and they generally do not have overlapping study areas or represent habitat at substantially different spatial scales. All studies except for Reference [30] utilized the MaxEnt software to generate their models, which reports model accuracy based on the Area Under the Curve of the Receiver Operating Characteristic (AUC). AUC characterizes the performance of a model at all possible thresholds by a single number that represents the probability that a known presence location will be ranked higher than a random background location [29]. Typical values for AUC are between 1.0–0.5 where a value of 1.0 represents a perfect fit while 0.5 indicates the model is no better than a random expectation [29]. The first study to use Landsat data to model chimpanzee habitat suitability was Reference [22] and reported an average AUC value greater than 0.9 indicating a good model fit. In addition, they were also able to detect changes in habitat suitability between two time periods. While their study area was much smaller than ours, a 2700 km^2 region in Guinea-Bissau, it lends support to the approach of using Landsat data to detect changes in chimpanzee habitat suitability. Similar AUC values were reported in Reference [25], the primary calibration data source for our model, which ranged from 0.86 to 0.93. The authors of [25] were able to compare their habitat suitability model for *P.t. schweinfurthii* to the results from Reference [23] and found large differences in the spatial distribution of habitat suitability for this sub-species. However, the maximum AUC reported in Reference [23] was 0.67, which suggests the results from Reference [25] are a better representation of habitat for this sub-species. Due to a lack of calibration data, important miombo woodland habitat in Western Tanzania was predicted as unsuitable in Reference [25], which is the reason we included model output from [30] in our calibration data. While the authors of [30] did not provide AUC values for their model, they did assess model predictions with a 25% excluded portion of the nest locations that was not used to create the Mahalanobis distance model and an independent dataset from historical surveys. They found 92% of nests that were not used in model calibration and 84% of nests from historical surveys fell into the moderate to high suitability range. This indicates that the model was able to predict suitable nesting sites, but unfortunately does not provide any information about the model's commission error.

While our model performed well across the chimpanzee range, it did show variation in its predictive capability. Compared to the calibration data, our model underestimated regions of high suitability in the *P.t. ellioti* and *P.t. troglodytes*ranges, which is likely due to missing predictor variables. We found dense canopy cover to be an important predictor of habitat suitability in our model, however; the authors of [25] did not find canopy cover to be a significant predictor variable in their models for these two sub-species. Both of these sub-species are found in central Africa, which is thought to be the last bastion of contiguous high quality chimpanzee habitat. While these large tracts of undisturbed forest are likely suitable places for chimpanzees to live, recent research has found a catastrophic decline in populations due to intense bush-meat hunting and disease [52]. High value tree species are selectively logged in the region, which leaves forests largely intact; however, the creation of logging roads opens up previously isolated forests, increasing the frequency of human contact and allowing easier access for hunters [52,53]. An absence of chimpanzees in these intact forests is likely the reason why habitat suitability is low in these areas and incorporating information on the proximity to logging roads and camps would benefit model accuracy.

It is important to note that inaccuracies in our model could also arise from errors in the input datasets to our model as well as errors in the datasets used to calibrate our model. Tree canopy cover was an important predictor variable in our model and was derived from Landsat ETM+ spectral data [28]. Canopy cover estimates from satellite data can be influenced by the composition of understory vegetation, satellite signal saturation and phenological noise [54,55]. Moreover, tree canopy cover from Reference [28] does not discriminate between forest types which could lead to predicting suitable habitat in intensely managed forests. Habitat suitability models in Reference [25] were generated using bioclimatic variables from the WorldClim dataset and considerable uncertainty surrounds the estimated climate conditions in Africa due to an extremely sparse network of weather stations found there [53]. Since elevation likely served as a proxy for climate in our model, errors in Reference [25] associated with poorly estimated climate conditions will be present in our model.

An additional source of uncertainty in our model could come from the assumption that habitat suitability is insensitive to changes in spatial resolution. However, our model used only continuous variables, and the scaling properties of continuous data have been known for a long time [56]. Reflectance and canopy cover, which figure prominently in our estimates of habitat suitability, have been demonstrated to be insensitive to changes in resolution [54,57]. In addition, the authors of [58] showed that coarsening the grain size of predictive variables in a species distribution model did not significantly impact model performance, "Change in grain size does not have a substantial effect on species distribution models. The trend is overall weakly significant towards degradation of model performance, but improvement can also be observed for some species". We further evaluated model sensitivity to changes in spatial resolution by averaging all pixels of the 30-m suitability model output within the extent of the 5-km grid-cells from the map used to calibrate the model. Figure 5A portrays the difference between averaged predictions from our model at 5-km resolution and the map used for calibration (observed), where areas of underestimation are seen in blue while overestimates are seen in red. Figure 5B depicts the relationship of suitability values from the calibration data (x-axis) and average predicted suitability from the Random Forests model in 5-km grid-cells for each sub-species. The correlation coefficients between the estimated habitat suitability and suitability from the calibration map were nearly identical to the relationships found from the cross-validation error assessment (Table 1), which suggests that error scales with spatial resolution.

CONCLUSIONS

In this study, we extended previous efforts to create maps of range-wide chimpanzee habitat suitability by using Random Forest regression to relate predictor variables derived from higher temporal and spatial resolution remotely sensed datasets to a coarse resolution chimpanzee habitat suitability map. We were able to map habitat suitability reasonably well, explaining 82% (±0.2%) of the variance in the calibration data based on 50-fold cross-validation; however, there was considerable variation in predictive ability among the four different chimpanzee sub-species. Our model accounted for 35% (±1.95%), 89% (±0.18%), 66% (±0.47%) and 73% (±0.44%) of the variance for *P.t. ellioti* (Nigeria-Cameroon Chimpanzee), *P.t. schweinfurthii* (Eastern Chimpanzee), *P.t. troglodytes* (Central Chimpanzee) and *P.t. verus* (Western Chimpanzee), respectively. We found that highly suitable areas were associated with dense tree canopy cover for all but the Nigeria-Cameroon and Central Chimpanzee sub-species. A recent proliferation of logging roads within the ranges of these sub-species has resulted in increased hunting pressure and disease transmission, causing severe declines in chimpanzee populations; therefore, seemingly undisturbed tracts of forest are devoid of chimpanzees. Incorporating information on the proximity to logging roads and camps may therefore benefit future modeling efforts. An important aspect of our approach that lends itself to conservation planning is the value of integrating continuously updated variables derived from remote sensing into habitat suitability models in order to detect and map change in suitability through time. With recent increases in data processing capabilities, developing such abilities to dynamically model and map changes in habitat suitability will be crucial for near real-time monitoring applications and the development of management plans that consider long-term trends in human-induced land cover change.

ACKNOWLEDGMENTS

We would express our gratitude to Hjalmar Kuehl for providing the habitat suitability layer from the Junker *et al.*, 2012, study. This work was supported by a grant from NASA's Ecological Forecasting for Conservation and Natural Resource Management program (NNX14AB96G) and the United States Agency for International Development (USAID) Central Africa Regional Program for the Environment (CARPE) program (NNX14AR46G). We would also like to recognize the support of our other partners in achieving the objectives of our work, the Arcus Foundation (DR Congo), Esri, DigitalGlobe and Google Earth Outreach, the American people through USAID and the African Wildlife Foundation. This paper does not reflect the official views of any of these organizations or the United States Government.

AUTHOR CONTRIBUTIONS

S.J. performed the analyses and wrote the manuscript. S.J., L.P., J.N. and M.H. contributed to the discussion of ideas and manuscript revisions.

REFERENCES

1. Sala, O.E. Global Change and Ecosystem Complexity. In *Global Change and Terrestrial Ecosystems*; Walker, B., Steffen, W., Eds.; Press Syndicate of the University of Cambridge: New York, NY, USA, 1996.
2. Purvis, A.; Jones, K.E.; Mace, G.M. Extinction. *Bioessays* 2000, *22*, 1123–1133.
3. Baillie, J.E.M.; Hilton-taylor, C.; Stuart, S.N. *2004 IUCN Red List of Threatened Species: A Global Species Assessment*; IUCN: Gland, Switzerland; Cambridge, UK, 2004.
4. Pimm, S.L.; Jenkins, C.N.; Abell, R.; Brooks, T.M.; Gittleman, J.L.; Joppa, L.N.; Raven, P.H.; Roberts, C.M.; Sexton, J.O. The biodiversity of species and their rates of extinction, distribution, and protection. *Science* 2014, *344*, 987–997.
5. Sala, O.E.; Chapin, F.S.; Armesto, J.J.; Berlow, E.; Bloomfield, J.; Dirzo, R.; Huber-Sanwald, E.; Huenneke, L.F.; Jackson, R.B.; Kinzig, A.; *et al.* Global biodiversity scenarios for the year 2100. *Science* 2000, *287*, 1770–1774.
6. Wright, S.J. Tropical forests in a changing environment. *Trends Ecol. Evol.* 2005, *20*, 553–560.
7. Nellemann, C.; Newton, A. The Great Apes: The Road Ahead: A Globio Perspective on the Impacts of Infrastructural Development on The Great Apes. 2008. Available online: http://www.globio.info/downloads/249/Great+Apes+-+The+Road+Ahead.pdf (accessed on 17 May 2016).
8. Kormos, R.; Boesch, C.; Bakarr, M.I.; Butynski, T.M. *West African Chimpanzees. Status Survey and Conservation Action Plan*; IUCN: Gland, Switzerland; Cambridge, UK, 2003.
9. Oates, J.F.; Dunn, A.; Greengrass, E.; Morgan, B.J. *Pan troglodytes* ssp. *ellioti*. The IUCN Red List of Threatened Species 2008. Available online: http://dx.doi.org/10.2305/IUCN.UK.2008.RLTS.T40014A10301774.en (accessed on 17 May 2016).
10. Wilson, M.L.; Balmforth, Z.; Cox, D.; Davenport, T.; Hart, J.; Hicks, C.; Hunt, K.D.; Kamenya, S.; Mitani, J.C.; Moore, J.; *et al. Pan troglodytes* ssp. *schweinfurthii*. The IUCN Red List of Threatened Species 2008. Available online: http://dx.doi.org/10.2305/IUCN.UK.2008.RLTS.T15937A5324021.en (accessed on 17 May 2016).
11. Tutin, C.E.G.; Baillie, J.E.M.; Dupain, J.; Gatti, S.; Maisels, F.; Stokes, E.J.; Morgan, D.B.; Walsh, P.D. *Pan troglodytes* ssp. *troglodytes*. The IUCN Red List of Threatened Species 2008. Available online: http://dx.doi.org/10.2305/IUCN.UK.2008.RLTS.T15936A5323646.en (accessed on 17 May 2016).
12. Humle, T.; Boesch, C.; Duvall, C.; Ellis, C.M.; Farmer, K.H.; Herbinger, I.; Blom, A.; Oates, J.F. *Pan troglodytes* ssp. *verus*. The IUCN Red List of Threatened Species 2008. Available online: http://dx.doi.org/10.2305/IUCN.UK.2008.RLTS.T15935A5323101.en (accessed on 17 May 2016).
13. Wrangham, R.W.; Hagel, G.; Leighton, M.; Marshall, A.J.; Waldau, P.; Nishida, T. The Great Ape World Heritage Species Project. In *Conservation in the 21st Century: Gorillas as A Case Study*; Stoinski, T.S., Steklis, H.D., Mehlman, P.T., Eds.; Springer Science and Business Media: New York, NY, USA, 2008; pp. 282–296.
14. *Open Standards for the Practice of Conservation*; Version 3.0; The Conservation Measures Partnership, 2013; Available online: http://cmp-openstandards.org/

wp-content/uploads/2014/03/CMP-OS-V3-0-Final.pdf (Accessed on 17 May 2016).
15. Araújo, M.; Peterson, A. Uses and misuses of bioclimatic envelope modeling. *Ecology* 2012, *93*, 1527–1539.
16. Franklin, J. Predictive vegetation mapping: Geographic modelling of biospatial patterns in relation to environmental gradients. *Prog. Phys. Geogr.* 1995, *19*, 474–499.
17. Guisan, A.; Zimmerman, N.E. Predictive habitat distribution models in ecology. *Ecol. Modell.* 2011, *135*, 147–186.
18. Hastie, T.; Fithian, W. Inference from presence-only data; the ongoing controversy. *Ecography* 2013, *36*, 864–867.
19. Fitzpatrick, M.C.; Gotelli, N.J.; Ellison, A.M. MaxEnt versus MaxLike: Empirical comparisons with ant species distributions. *Ecosphere* 2013, *4*, 1–15.
20. Merow, C.; Smith, M.J.; Silander, J.A. A practical guide to MaxEnt for modeling species' distributions: What it does, and why inputs and settings matter. *Ecography* 2013, *36*, 1058–1069.
21. Pintea, L.; Plumptre, A.J. Prediction of suitable habitat for chimpanzees using remote sensing and GIS. In *Surveys of Chimpanzees and Other Biodiversity in Western Tanzania*; Moyer, D., Plumptre, A.J., Pintea, L., Hernandez-Aguilar, A., Moore, J., Stewart, F., Davenport, T.R.B., Piel, A., Kamenya, S., Mugabe, H., *et al.*, Eds.; Wildlife Conservation Society, the Jane Goodall Institute, US Fish and Wildlife Service; 2006; pp. 47–54. Available online: http://pages.ucsd.edu/~jmoore/courses/methprimconsweb08/MoyerEtAl2006WCSTanzania.pdf (accessed on 17 May 2016).
22. Torres, J.; Brito, J.C.; Vasconcelos, M.J.; Catarino, L.; Gonçalves, J.; Honrado, J. Ensemble models of habitat suitability relate chimpanzee (*Pan troglodytes*) conservation to forest and landscape dynamics in Western Africa. *Biol. Conserv.*2010, *143*, 416–425.
23. Plumptre, A.; Rose, R.; Nangendo, G.; Williamson, E.; Didier, K.; Hart, J.; Mulindahabi, F.; Hicks, C.; Griffin, B.; Ogawa, H.; *et al. Eastern Chimpanzee (Pan troglodytes Schweinfurthii): Status Survey and Conservation Action Plan, 2010–2020*; IUCN: Gland, Switzerland, 2010.
24. Clee, P.R.S.; Abwe, E.E.; Ambahe, R.D.; Anthony, N.M.; Fotso, R.; Locatelli, S.; Maisels, F.; Mitchell, M.W.; Morgan, B.J.; Pokempner, A.A.; *et al.* Chimpanzee population structure in Cameroon and Nigeria is associated with habitat variation that may be lost under climate change. *BMC Evol. Biol.* 2015, *15*, 1–13.
25. Junker, J.; Blake, S.; Boesch, C.; Campbell, G.; Toit, L.D.; Duvall, C.; Ekobo, A.; Etoga, G.; Galat-Luong, A.; Gamys, J.; *et al.* Recent decline in suitable environmental conditions for African great apes. *Divers. Distrib.* 2012, *18*, 1077–1091.
26. Caldecott, J.; Miles, L. *World Atlas of the Great Apes and Their Conservation*; UNEP World Conservation Monitoring Centre: Cambridge, UK, 2005.
27. Kortlandt, A. Marginal habitats of chimpanzees. *J. Hum. Evol.* 1983, *12*, 231–278.
28. Hansen, M.C.; Potapov, P.V.; Moore, R.; Hancher, M.; Turubanova, S.A.; Tyukavina, A.; Thau, D.; Stehman, S.V.; Goetz, S.J.; Loveland, T.R.; *et al.* High-resolution global maps of 21st-century forest cover change. *Science* 2013, *342*, 850–853.
29. Phillips, S.J.; Anderson, R.P.; Schapire, R.E. Maximum entropy modeling of species geographic distributions. *Ecol. Modell.* 2006, *190*, 231–259.
30. Moyer, D.; Plumptre, A.J.; Pintea, L.; Hernandez-Aguilar, A.; Moore, J.; Stewart, F.A.; Davenport, T.R.B.; Piel, A.K.; Kamenya, S.; Mugabe, H.; *et al.* Surveys of Chimpanzees and other Biodiversity in Western Tanzania. 2006. Available online: http://pages.ucsd.edu/~jmoore/courses/methprimconsweb08/MoyerEtAl2006WCSTanzania.pdf (accessed on 17 May 2016).
31. Farber, O.; Kadmon, R. Assessment of alternative approaches for bioclimatic modeling with special emphasis on the Mahalanobis distance. *Ecol. Modell.* 2003, *160*, 115–130.
32. Clark, J.D.; Dunn, J.E.; Smith, K.G.A. Multivariate Model of Female Black Bear Habitat Use for a Geographic Information System. *J. Wildl. Manag.* 1993, *57*, 519–526.
33. Tyukavina, A.; Baccini, A.; Hansen, M.C.; Potapov, P.V.; Stehman, S.V.; Houghton, R.A.; Krylov, A.M.; Turubanova, S.; Goetz, S.J. Aboveground carbon loss in natural and managed tropical forests from 2000 to 2012. *Environ. Res. Lett.*2015, *10*, 074002.
34. Hansen, M.C.; Potapov, P.; Goetz, S.J.; Turubanova, S.; Tyukavina, A.; Krylov, A.; Kommareddy, A.; Egorov, A. Mapping tree height distributions in Sub-Saharan Africa using Landsat 7 and 8 data. *Remote Sens. Environ.* 2016, 2–13.
35. United States Geological Survey. Documentation for the Shuttle Radar Topography Mission (SRTM) Water Body Data Files. 2003. Available online: https://dds.cr.usgs.gov/srtm/version2_1/SWBD/SWBD_Documentation/Readme_SRTM_Water_Body_Data.pdf (accessed on 17 May 2016).
36. Helmer, E.H.; Ruzycki, T.S.; Wunderle, J.M.; Vogesser, S.; Ruefenacht, B.; Kwit, C.; Brandeis, T.J.; Ewert, D.N. Mapping tropical dry forest height, foliage height profiles and disturbance type and age with a time series of cloud-cleared Landsat and ALI image mosaics to characterize avian habitat. *Remote Sens. Environ.* 2010, *114*, 2457–2473.
37. Wilkes, P.; Jones, S.D.; Suarez, L.; Mellor, A.; Woodgate, W.; Soto-Berelov, M.; Haywood, A.; Skidmore, A.K. Mapping forest canopy height across large areas by upscaling als estimates with freely available satellite data. *Remote Sens.*2015, *7*, 2–25.
38. Baccini, A.; Goetz, S.J.; Walker, W.S.; Laporte, N.T.; Sun, M.; Sulla-Menashe, D.; Hackler, J.; Beck, P.S.A.; Dubayah, R.; Friedl, M.A.; *et al.* Estimated carbon dioxide emissions from tropical deforestation improved by carbon-density maps. *Nat. Clim. Chang.* 2012, *2*, 182–185.

39. Food and Agriculture Organization of the United Nations. *Global Forest Resources Assessment 2005, Main Report. Progress Towards Sustainable Forest Management*; FAO: Rome, Italy, 2006.

40. Singh, A.; Shi, H.; Zhu, Z.; Foresman, T. *An Assessment of the Status of the World's Remaining Closed Forests*; UNEP: Nairobi, Kenya, 2001.

41. Vitousek, P.M. Beyond Global Warming: Ecology and Global Change. *Ecology* 1994, *75*, 1861–1876.

42. Goodall, J. *The Chimpanzees of Gombe: Patterns of Behavior*; The Belknap Press of Harvard University Press: Cambridge, MA, USA, 1986.

43. Nowak, R. *Walkers Mammals of the World*, 6th ed.; Johns Hopkins University Press: Baltimore, MD, USA; London, UK, 1999.

44. Jones, C.; Jones, C.; Jones, J.; Wilson, D. Pan Troglodytes. *Mamm. Species* 1996, *529*, 1–9.

45. Pedregosa, F.; Varoquaux, G. Scikit-learn: Machine Learning in Python. *J. Mach. Learn. Res.* 2011, *12*, 2825–2830.

46. Breiman, L.; Friedman, J.; Stone, C.J.; Olsen, R.A. *Classification and Regression Trees*; Chapman and Hall/CRC: Washington, WA, USA, 1984.

47. Breiman, L. Random forests. *Mach. Learn.* 2001, *45*, 5–32.

48. Strobl, C.; Boulesteix, A.-L.; Zeileis, A.; Hothorn, T. Bias in random forest variable importance measures: Illustrations, sources and a solution. *BMC Bioinform.* 2007, *8*, 25.

49. Granitto, P.M.; Furlanello, C.; Biasioli, F.; Gasperi, F. Recursive feature elimination with random forest for PTR-MS analysis of agroindustrial products. *Chemom. Intell. Lab. Syst.* 2006, *83*, 83–90.

50. Hijmans, R.J.; Cameron, S.E.; Parra, J.L.; Jones, P.G.; Jarvis, A. Very high resolution interpolated climate surfaces for global land areas. *Int. J. Climatol.* 2005, *25*, 1965–1978.

51. Hansen, M.C.; Roy, D.P.; Lindquist, E.; Adusei, B.; Justice, C.O.; Altstatt, A. A method for integrating MODIS and Landsat data for systematic monitoring of forest cover and change in the Congo Basin. *Remote Sens. Environ.* 2008, *112*, 2495–2513.

52. Walsh, P.; Abernethy, K.; Bermejo, M. Catastrophic ape decline in western equatorial Africa. *Nature* 2003, *422*, 611–614.

53. Laporte, N.T.; Stabach, J.A.; Grosch, R.; Lin, T.S.; Goetz, S.J. Expansion of industrial logging in Central Africa. *Science*2007, *316*, 1451.

54. Sexton, J.O.; Song, X.-P.; Feng, M.; Noojipady, P.; Anand, A.; Huang, C.; Kim, D.-H.; Collins, K.M.; Channan, S.; Dimiceli, C.; *et al.* Global, 30-m resolution continuous fields of tree cover: Landsat-based rescaling of MODIS Vegetation Continuous Fields with lidar-based estimates of error. *Int. J. Digit. Earth* 2013.

55. Ozdemir, I. Linear transformation to minimize the effects of variability in understory to estimate percent tree canopy cover using RapidEye data. *GISci. Remote Sens.* 2014, *1603*, 37–41.

56. Stevens, S.S. On the Theory of Scales of Measurement. *Science* 1946, *103*, 677–680.

57. Gao, F.; Masek, J.; Schwaller, M.; Hall, F. On the Blending of the MODIS and Landsat ETM+ Surface Reflectance. *IEEE Trans. Geosci. Remote Sens.* 2006, *44*, 2207–2218.

58. Guisan, A.; Graham, C.H.; Elith, J.; Huettmann, F.; Dudik, M.; Ferrier, S.; Hijmans, R.; Lehmann, A.; Li, J.; Lohmann, L.G.; *et al.* Sensitivity of predictive species distribution models to change in grain size. *Divers. Distrib.* 2007, *13*, 332–340.

CHAPTER 7

ECOSYSTEM SPECIFIC DYNAMICS, AND DETAILED CONTEMPORARY MAPPING OF DISTINCT ECOSYSTEMS

Trevor G. Jones, Leah Glass, Samir Gandhi, Lalao Ravaoarinorotsihoarana, Aude Carro, Lisa Benson, Harifidy Rakoto Ratsimba, Chandra Giri, Dannick Randriamanatena and Garth Cripps

Keywords: Madagascar, mangrove, Landsat, dynamics, coastal

INTRODUCTION

Found in over 120 countries and territories between 30°N and 30°S latitude, mangrove ecosystems support high floral and faunal biodiversity and provide a diverse range of goods (e.g., food, fuel, building materials) and services (e.g., water filtration; storm protection) to coastal populations. Furthermore, a growing number of studies report that mangroves possess similar or greater above- and exceptionally larger below-ground carbon stocks as compared with their terrestrial peers [1,2,3,4,5,6,7,8,9,10,11,12,13,14,15,16,17,18,19,20,21,22,23, 24,25], meaning that intact mangrove ecosystems contribute to global climate change mitigation through sequestering significant amounts of CO_2. Despite their tremendous and diverse value, throughout the world mangrove ecosystems are rapidly being degraded, deforested and converted for other uses. Over the past several decades, annual global mangrove loss is estimated at 1%–2%, exceeding rates in many inland tropical forests [6,26,27,28,29,30]. The primary drivers of loss include conversion for agri- and aquaculture, coastal development, over-extraction of woody materials, and the ripple effects of upstream terrestrial agriculture and deforestation (*i.e.*, erosion, sedimentation and siltation) [31,32,33,34,35,36,37,38,39,40]. While anthropogenic activities are the primary drivers of loss, natural phenomena such as tropical storms and rising ocean temperatures and sea-levels also play a significant role; the impacts of which are expected to continue to increase based on current climate change projections [6,11,30,31,37,40,41,42,43 ,44,45]. Many of the world's enduring mangrove ecosystems have already been degraded, and may functionally collapse within this century without intervention [27,46,47].

To quantify dynamics and initiate effective management and decision-making, contemporary information on the extent and status of mangrove ecosystems is required. Remotely sensed data can be employed to map and monitor mangrove ecosystems, providing managers and decision makers with detailed, accurate, timely and updateable information [15,48]. In particular, the Landsat archive, which extends back >40 years, is well tested for identifying and quantifying mangrove distribution and dynamics, and for mapping and monitoring specific mangrove ecosystems at the level of thematic detail required for management strategies [25,49,50,51,52,53,54,55, 56,57,58,59,60,61,62,63,64,65]. As described in [66], there is a growing need to use remotely sensed data such as Landsat to monitor shifts in stable states in coastal wetlands, such as mangroves. New techniques which go beyond single-date mapping and multi-date change detection are needed to create evolutionary models of mangroves; pushing the boundaries of Landsat-like remotely sensed data [66]. Along these lines, a recent study in the Mekong delta by [67] demonstrates how to study the spatial and temporal evolution of mangroves using a time series of Landsat data augmented by Shuttle Radar Topography Mission (SRTM) data. Factoring in the need for and benefit of monitoring shifts in stable states, detailed, contemporary single-date maps are still required for many management tasks and are unavailable for much of the world's mangrove ecosystems [25,61].

Citation: Trevor G. Jones, Leah Glass, Samir Gandhi, Lalao Ravaoarinorotsihoarana, Aude Carro, Lisa Benson, Harifidy Rakoto Ratsimba, Chandra Giri, Dannick Randriamanatena and Garth Cripps, Ecosystem Specific Dynamics, and Detailed Contemporary Mapping of Distinct Ecosystems, doi:10.3390/rs8020106.

The island nation of Madagascar contains extensive mangrove ecosystems, which as of 2005 totaled approximately 280,000 hectares (ha); Africa's fourth largest amount and 2% of the global distribution [10,57,68,69]. Many of Madagascar's mangrove ecosystems have undergone or are starting to show signs of wide-spread degradation and deforestation. To date, numerous studies have employed remotely sensed data to produce single- or multi-date nation–wide mangrove maps (*i.e.*, [68,69,70,71,72]). However, national-level maps, while critical for country-wide overviews, do not provide the thematic detail required for many localized (e.g., ecosystem-specific) applications. Prior to 2010, localized mapping was undertaken for several specific mangrove ecosystems (*i.e.*, Bombetoka Bay, Betsiboka Estuary, Mahajamba Bay (MHJ), and Mangoky Delta) (*i.e.*, [73,74,75,76,77]). However, these maps are either out-of-date or lack the thematic detail required for certain management initiatives (e.g., carbon stock inventories); and for many of Madagascar's mangrove ecosystems, are simply non-existent.

Here, we inventory and compare existing national-level mangrove maps for Madagascar. Using the most comprehensive and contemporary available data, we quantify nation-wide mangrove dynamics from 1990 to 2010. To use these data to examine dynamics at a finer scale, we partition Madagascar's mangroves in to non-contiguous, primary (*i.e.*, >1000 ha) mangrove ecosystems; with focus on four distinct mangrove ecosystems representing areas of interest (AOIs): Ambaro-Ambanja Bays (AAB), MHJ, Tsiribihina Manambolo Delta (TMD) and Baie des Assassins (BdA. Given the limitations of national-level maps for certain localized applications, and the limitations of existing and/or lack of ecosystem-specific maps for AOIs, we build on two previous studies for AAB and MHJ [25,61] and employ Landsat data to produce detailed, contemporary, maps of TMD and BdA. The applications of existing (AAB; MHJ) and new (TMD; BdA) AOI-specific maps to mangrove conservation initiatives through Blue Ventures' Blue Forests programme are discussed.

EXPERIMENTAL SECTION

Study Area

The study area includes the entire distribution of Madagascar's mangroves (Figure 1). Madagascar's ~5000 km coastline is one of the most extensive shallow marine areas in the Western Indian Ocean (WIO), wherein mangrove ecosystems are located almost entirely along the west coast. While not as species-rich as other regions, at least eight true mangrove species occur, supporting unique flora and fauna, much of which is currently endangered or at-risk [12,68].

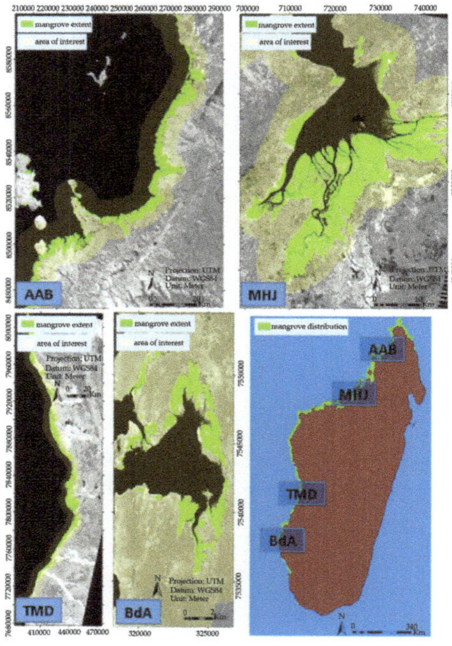

Figure 1: The study area includes the entire distribution of Madagascar's mangroves (shown in green, *circa* 2010, bottom right) (data obtained with permission from Giri [69]). The combined marine and terrestrial extent within 7 km from the coast in four primary mangrove ecosystems (*i.e.*, Ambaro-Ambanja Bays (AAB), Mahajamba Bay (MHJ), Tsiribihina-Manambolo Delta (TMD) and the Baie des Assassins (BdA)) represent four specific areas of interest (AOIs). The location of each AOI is shown in the country-wide inset (bottom right). Mangrove extent (*circa* 2010) within each AOI is shown in green (data obtained with permission from Giri [69]).

Madagascar's mangrove ecosystems are governed by a complex legal framework involving forestry, land planning, fisheries and environmental laws [78,79,80,81,82,83,84,85]. The most relevant texts regarding mangrove management and use are found in the forestry and environmental sectors. Under forestry laws, mangroves are within the public domain of the state, but local communities can be granted user rights for their domestic use of related resources, including timber [78,80,82]. Under environmental laws, mangroves fall within the sensitive area category, in which an environmental impact assessment is compulsory prior to any type of investment, and where commercial timber extraction has been forbidden since 2000 [79,81]. More recently, in 2014 a Decree was issued by the Government of Madagascar to ban the extraction, transportation, stocking and sale of timber specifically in mangrove ecosystems [85]. In practice, the Forestry Administration continues to grant communities user rights on mangrove timber, despite the two-above mentioned bans. Local management rights can be established through either Protected Areas or Natural Resource Management Transfer regulations; though this process is complex, expensive and time consuming. Regardless of management rights, comprehensive and effective management, which impedes mangrove degradation and loss, is lacking for almost all mangrove ecosystems in Madagascar. Due to the lack of governance, an increasing demand in fuel-wood and timber, as well as the expansion of agricultural land, the pressure on mangrove ecosystems from anthropogenic activities continues to rise [25]. If degradation and conversion continue unimpeded, biodiversity will be imperiled, greenhouse gas emissions will increase, and many important services provided to coastal communities will be seriously compromised [86].

Owing to the lack of effective mangrove management in Madagascar and local people's high dependence on the numerous goods and services mangroves provide, the marine conservation NGO Blue Ventures (www.blueventures.org) initiated its Blue Forests programme in 2011, to work with coastal communities, partner NGOs, Malagasy and foreign Universities, and government bodies at all levels to establish incentivized models for community-based sustainable mangrove forest and fisheries management. The national-level mangrove dynamics data detailed herein and complementary socioeconomic information highlighted three priority AOIs along the west coast of Madagascar for Blue Ventures' Blue Forest programme: Ambaro-Ambanja Bays (AAB), Mahajamba Bay (MHJ) and the Baie des Assassins (BdA) (Figure 1). The Tsiribihina and Manambolo Deltas (TMD) form a fourth AOI, where the localized mapping detailed herein is supporting WWF Madagascar West Indian Ocean Programme Office's work in the region. Each AOI encompasses the marine and terrestrial extent below 30 meters (m) elevation within 7 km of the coast.

AAB (centered at latitude 13°26′S, longitude 48°30′E) is comprised of two bays lined with extensive mangroves, which in aggregate form a contiguous ecosystem. AAB is most influenced by the Sambirano, Mananjeba, Mahavavy and Ifasy Rivers, with their headwaters in the Tsarantanana Massif mountain range to the southeast, including Maromokotro, Madagascar's highest peak at 2876 m. Situated at the convergence of the Sofia and Mahajamba rivers (centered at latitude 15°24′S, longitude 47°05′E), the tidal plains of MHJ encompass an extensive mangrove ecosystem. MHJ experiences a semidiurnal tidal range of 1.5–3 m (4–4.5 m during spring tide), an average surface water salinity range of 25–45 parts per thousand, and a mean annual precipitation of 1500 mm [73,74]. TMD (centered at latitude 19°36′S, longitude 44°27′E) is comprised of two deltas lined with extensive mangroves, forming a contiguous ecosystem. TMD is fed by the Mahajilo, Sakay, Kitsamby, Mania and Sakeny Rivers [87], with their headwaters in the Central Highlands (Ankaratra Massif [88]). The BdA (centered at latitude 22°12′S, longitude 43°17′E) is comprised of a single bay with a comparatively modest amount of mangroves, representing a contiguous ecosystem. A north-south rainfall gradient along the west coast results in a comparative abundance of precipitation in the north (e.g., AAB, MHJ and TMD), which contributes to higher stature mangrove trees than further south (e.g., BdA) [68,89].

Inventory and Comparison of Existing Data-Sets

Several studies have resulted in single- or multi-date national-level mangrove distribution maps for Madagascar (*i.e.*, [68,70,71,72]). Mayaux *et al.* [70] delineated mangrove and five other vegetation cover types, *circa* 1998/1999, using 1 km Satellite Pour l'Observation de la Terre (SPOT) VEGETATION data. The Critical Ecosystem Partnership Fund (CEPF) [71] Madagascar Mapping Project mapped mangroves and 14 other vegetation types, *circa* 2001, with Moderate-resolution Imaging Spectroradiometer (MODIS) and Landsat data. Harper *et al.* [72] employed Landsat data to map forest cover, including mangroves, for 1973, 1990 and 2000. Giri and Muhlhausen [68] and Giri [69] utilized Landsat data to map mangrove and non-mangrove for 1973, 1990, 2000, 2005 and 2010. National-level dynamics calculated from these maps through 2005 are presented in Giri and Muhlhausen [68]). To date, no studies have reported on the nation-wide dynamics of Madagascar's mangroves beyond 2005.

Of the four specific AOIs, prior to 2014, AOI-specific maps existed for only MHJ and the BdA. For MHJ, Rasolofoharinoro *et al.* [73] employed Satellite Pour l'Observation de la Terre (SPOT) data to map seven intertidal vegetation zones for 1986 and 1993, which included four mangrove classes (*i.e.*, pioneering; dense mature; decaying; back (e.g., mangrove/terrestrial interface)). Pasqualini *et al.* [74] combined SPOT and radar data to delineate 10 coastal ecosystem types, *circa* 1993, including four morphologically (*i.e.*, frontal; interfluvial) and dynamics (*i.e.*, mature; recessive) related mangrove classes. Guillet *et al.* [75] utilized Landsat and SPOT data to map 12 coastal ecosystem classes, *circa* 1973, 1989, 2000 and 2006, including four mangrove types (*i.e.*, sparse *Avicennia marina*; average density *Avicennia marina*; dense *Avicennia marina*; and dense *Rhizophora mucronata*). For BdA, Roy *et al.* [90] used Quickbird (DigitalGlobe, Longmont, CO, USA) data to produce a map of mangrove and surrounding marine classes, *circa* 2005. These maps (*i.e.*, [73,74,75,90] are no longer contemporary and/or unable to provide the thematic detail required for specific applications (e.g., carbon stock estimation).

As described in later sections, to improve upon the shortcomings associated with available maps for MHJ, Jones *et al.* [61] used Landsat data acquired in 2011 to produce a detailed, contemporary AOI-specific map of MHJ. Using similar methodologies, Jones *et al.* [25] produced an AOI-specific map for AAB from Landsat data acquired in 2010. At the time of writing, detailed contemporary maps existed for no other specific mangrove ecosystems in Madagascar; highlighting the need to produce new maps for TMD and BdA to complement those produced for AAB and MHJ.

National-Level Mangrove Distribution and Dynamics; Ecosystem-Level Dynamics

National-level data-sets were compared to determine which offered the most comprehensive (*i.e.*, in terms of their coverage of actual known mangrove area) and contemporary (*i.e.*, most recent) mangrove distributional information. Contemporary Landsat data, finer spatial resolution imagery viewable in Google Earth (Google, Mountain View, CA, USA) and extensive field observations were used to guide qualitative comparisons (Figure 2). Building on Giri and Muhlhausen [68], who reported mangrove extent *circa* 2005 and dynamics from 1990 to 2005, the maps deemed most comprehensive and contemporary were employed to establish contemporary (*i.e.*, beyond 2005) mangrove extent and quantify national-level mangrove dynamics (*i.e.*, loss, persistence and gain) from 1990 to 2010 using the Idrisi Land Change Modeler; a land planning and decision support tool (Clark Labs, Worcester, MA, USA). To investigate mangrove extent and dynamics at a finer scale, national-level mangrove data-sets were partitioned in to primary mangrove ecosystems, defined herein as individual ecosystems occupying at least 1000 ha and containing true mangroves: salt-tolerant halophytic trees and/or shrubs occurring entirely in tidal and inter-tidal areas [91].

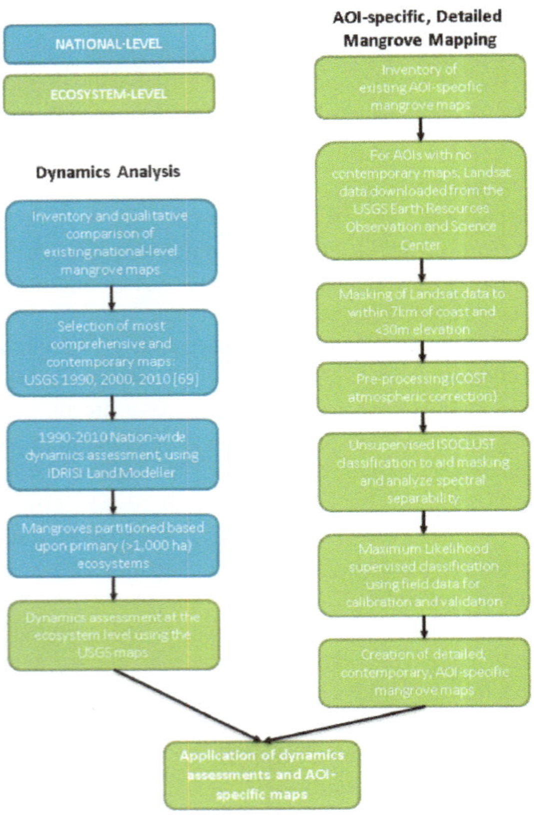

Figure 2: Workflow. Blue boxes represent national-level processing including the entire distribution of Madagascar's mangroves. Green boxes represent ecosystem-level processing, including 30 primary mangrove ecosystems, with focus on four AOIs (*i.e.*, Ambaro-Ambanja Bays (AAB), Mahajamba May (MHJ), Tsiribihina-Manambolo Delta (TMD) and Baie des Assassins (BdA)). The results of nation-wide and ecosystem-specific dynamics from 1990 to 2010, and ecosystem-specific maps for TMD and BdA are presented in this study for the first time. The production of ecosystem-specific maps builds on previous studies which employed similar methods for AAB and MHJ (*i.e.*, Jones *et al.* [25]; Jones *et al.* [61]).

AOI-Specific Ecosystem-Level Mapping

Acquisition and Pre-Processing of Remotely Sensed Data

Following the methodology described in Jones *et al.* [25] for AAB, and Jones *et al.* [61] for MHJ, for mapping efforts in TMD and BdA, 30 m spatial resolution Landsat scenes were downloaded from the United States Geological Survey (USGS) Earth Resources Observation and Science Center (Sioux Falls, SD, USA) (Table 1). All Landsat scenes were orthorectified to a Shuttle Radar Topography Mission (SRTM) digital elevation model (DEM). Atmospheric correction was undertaken using the Cos(t) model [92], which estimates the impact of absorption by atmospheric gases and Rayleigh scattering, minimizes systematic haze, and converts image units to at-surface reflectance.

The bounding extent for all AOIs were established using distance to coastline (*i.e.*, 7 km) as an assumed mangrove habitat requirement [55,58,93]. Reducing the image extent based on habitat requirement through masking can increase classification accuracy by decreasing spectral confusion amongst target classes [57]. Emulating studies that establish the effectiveness of SRTM data to estimate mangrove forest canopy height (e.g., [94,95,96,97,98]), an SRTM height mask further eliminated unnecessary scene components based on a 30 m threshold, above which mangrove habitat were observed in each AOI not to exist [25,61].

Table 1: Summary of single-date Landsat scenes used for detailed, contemporary mapping of the four AOIs. Maps produced for Ambaro-Ambanja Bays (AAB) and Mahajamba Bay (MHJ) are originally described in Jones *et al.* [25] and Jones *et al.* [61]; whereas maps produced for Tsiribihina-Manambolo Delta (TMD) and Baie des Assassins (BdA) result from this study. Tide (m) indicates the average tidal height above mean sea level in or near the AOI during image acquisition

AOI	Sensor	Date	Path/row	Tide (m)
AAB	Landsat 7 (L7) ETM	09/06/2010	159/69	1.9
MHJ	Landsat 5 (L5) TM	29/07/2011	160/71	1.7
TMD	Landsat 8 (L8) OLI	28/07/2014	161/73; 161/74	1.5
BdA	Landsat 8 (L8) OLI	23/04/2014	161/75	2.3

Initial Mapping; Definition and Refinement of Mangrove and Surrounding Land-Cover Categories

Adhering to the methods described in Jones *et al.* [25] for AAB, and Jones *et al.* [61] for MHJ, to produce initial maps for TMD and BdA, an unsupervised iterative self-organizing classification algorithm (*i.e.*, ISOCLUST) was employed to group pixels in to dominant cover types based on shared spectral properties in bands 1–5 and 7 (*i.e.*, TM and ETM) or bands 2–7 (*i.e.*, OLI). Unsupervised classification for mapping mangroves and closely related ecosystem types is proven, and has produced both preliminary and final maps described in numerous studies [25,57,59,61,68,96,97,99,100,101,102]. Unsupervised classification results further facilitated masking based on areas dominated by water, cloud and/or shadow. Making reference to existing maps and finer spatial resolution imagery viewable in Google Earth, aggregation and iterative labelling was used to define and refine mangrove and surrounding land-cover types. Mangrove types varied slightly between AOIs, but were based on spectral differences attributable to ecological properties such as canopy-cover, stature and density (Table 2). All mangrove classes represent true mangroves and are assumed to be dominated by salt-tolerant halophytic trees and/or shrubs occurring entirely in tidal and inter-tidal areas [91]. Mangrove classes dominated by trees are considered forested stands; whereas classes dominated by shrubs/stunted trees do not meet international or national definitions for forest. Herein, the term mangrove ecosystem refers to a contiguous ecological unit containing both forest and non-forest areas of mangrove. Forested classes within a particular mangrove ecosystem are collectively referred to as mangrove forest.

In accordance with the approach described in Jones *et al.* [25] for AAB, and Jones *et al.* [61] for MHJ, to ensure the representativeness of and refine mangrove type and surrounding land-cover categories for TMD and BdA maps, preliminary field surveys were conducted in all four AOIs. Stratified random sampling was employed to target ha-sized reference plot locations within strata defined by the unsupervised classification results. Within mangrove reference plots established within each strata (Table 2), the height, diameter, canopy-cover, and crown dimensions were recorded for representative examples of each mangrove species present. Within non-mangrove reference plots, field notes and photographs recorded variability and confirmed representativeness. Within all reference plots, plot center was established with a Garmin GPSmap 62sc GPS unit left recording during the duration of measurements.

Table 2: Mapped classes, descriptions and 3 × 3 pixel calibration and validation reference areas for each of the four AOIs. Mapped classes for TMD and BdA build on previously published studies for AAB ([25]) and MHJ ([61])

Area	Class	Description	Calibration	Validation	Total
AAB	Savannah	Dry grass, exposed soil, extremely sparse trees/shrubs	10	6	16
	Woodland	Dry grass and scattered trees/shrubs; canopy <30% closed	8	6	14
	Active cultivation	Dominated by pre-harvest agriculture (e.g., rice; sugar cane)	10	6	16
	Closed-canopy terrestrial forest	Stands of trees with well-formed canopies >60% closed	12	6	18
	Open-canopy terrestrial forest	Stands of trees/shrubs with canopies 30%–60% closed	8	6	14
	Closed-canopy mangrove	Tall, mature stands; canopy >60% closed	20	10	30
	Open-canopy mangrove I	Short-medium stands of trees/shrubs; canopy 30%–60% closed; moderately influenced by background soil/mud	15	9	24
	Open-canopy mangrove II	Stunted/short stands, shrub-dominant, very sparse; canopy ≥10% closed	11	6	17
	Deforested mangrove	Mosaic of stumps, scattered trees; canopy <30% closed	8	6	14
	Exposed soil	Inactive agri/aquacultural fields; extremely patchy savannah; extremely dry mud-flats	10	5	15
	Exposed mud	Mangrove/ocean interface; river sediment; wet mud-flats	8	5	13
		Total	***120***	***71***	***191***
MHJ	Active cultivation	Dominated by pre-harvest agriculture (e.g., rice)	14	7	21
	Closed-canopy terrestrial forest	Stands of trees with well-formed canopies >60% closed	14	7	21
	Open-canopy terrestrial forest	Stands of trees/shrubs with canopies 30%–60% closed	16	8	24
	Closed-canopy mangrove I	Tall, mature stands; canopy >80% closed	22	10	32
	Closed-canopy mangrove II	Tall, mature stands; canopy >60% closed	22	10	32
	Open-canopy mangrove I	Short-medium stands of trees/shrubs; canopy 30%–70% closed; moderately influenced by background soil/mud	14	7	21
	Open-canopy mangrove II	Short-medium stands of trees/shrubs; canopy 30%–70% closed; significantly influenced by background soil/mud	12	7	19
	Open-canopy mangrove III	Stunted/short stands, shrub-dominant, very sparse; canopy <30% closed	14	7	21
	Exposed soil	Inactive agri/aquacultural fields; sparsely vegetated, soil-dominated areas; dry mud-flats	21	10	31
	Exposed mud	Mangrove/ocean interface; river sediment; wet mud-flats; inactive aquaculture ponds	14	7	21
		Total	***163***	***80***	***243***
TMD	Dense mangrove	Tall, mature stands; canopy >70% closed	16	6	22
	Sparse mangrove	Short-medium stands of trees/shrubs; canopy 30%–70% closed; moderately influenced by background soil/mud	12	5	17
	Bare type I	Exposed soil; scrub mangrove; mud-flats; fallow agriculture; patchy grass/bushland	14	6	20
	Dense terrestrial forest	High-stature terrestrial trees with well-formed canopies	12	6	18
	Sparse terrestrial forest	Mixed-stature terrestrial trees with relatively open canopies interspersed with wooded grassland-bushland	10	5	15
	Wooded grassland-bushland	Grassland and/or bush-land	10	5	15
	Active agriculture	Dominated by pre-harvest agriculture (e.g., rice)	3	1	4
	Bare type II	Typically dry soil and/or sand; extremely dry patchy grass/bushland	10	5	15
		Total	***87***	***39***	***126***
BdA	Spiny forest	Moderate-high stature, relatively closed-canopy stands	9	3	12
	Barren/exposed	Dominance of rock, sand, dry soil; interspersed with sparse vegetation	9	3	12
	Cultivated/degraded/woodland	Active or fallow cultivation; degraded and/or sparse terrestrial forest; woodland	9	3	12
	Burnt	Areas which have recently experienced fire	9	3	12
	Water dominated	Water dominant; mud-flats	10	4	14

Closed-canopy mangrove	Tall, mature stands; canopy >60% closed	9	3	12
Open-canopy mangrove I	Short-medium stands of trees/shrubs; canopy 30%–70% closed; influenced by background soil/mud	9	3	12
Open-canopy mangrove II	Stunted/short stands; shrub-dominant, very sparse; canopy <30% closed	9	3	12
	Total	*73*	*25*	*98*

Supervised Image Classification

As with previously reported mapping efforts in AAB and MHJ, for TMD and BDA, reference plots facilitated both calibrating the spectral properties of different mangrove and surrounding land-cover types for image classification (*i.e.*, calibration) and assessing resulting map accuracy (*i.e.*, validation). By exploiting the familiarity gained with the appearance and location of target classes, supplemental reference areas were located in finer spatial resolution imagery viewable in Google Earth for all mapped categories. Reference areas were spread throughout and randomly partitioned to facilitate both calibration and validation (Table 2). In adherence with methods described in Jones *et al.* [25] for AAB, and Jones *et al.* [61] for MHJ, for TMD and BdA, supervised classification of Landsat data was employed to produce maps of class distributions using the maximum likelihood (ML) algorithm. Numerous studies have shown the effectiveness of ML for classifying mangrove habitat with Landsat-like data [25,58,60,61,73, 95,103,104,105,106,107,108]. The accuracies of resulting maps were quantified using confusion matrices, which cross-tabulate independent validation data against mapped classes. The Kappa index of agreement further assessed how much better than random each map was [109].

RESULTS AND DISCUSSION

Overview of Existing National-Level Maps and Data-Sets

A qualitative comparison of existing national-level mangrove data-sets (*i.e.*, [68,70,71,72]) concluded that those produced by the USGS (*i.e.*, [69]) offered the most comprehensive historic and contemporary areal estimates of Madagascar's mangroves (Figure 3). Crucially, the USGS produced maps were the only ones to provide relatively contemporary (*i.e.*, 2010) distribution and focus solely on mangroves, with all other national-level maps representing multiple forest types and time periods nearly or greater than 10 years earlier. Thus, the USGS data-sets were utilized for all consequent dynamics analyses. Additional details regarding comparisons between Madagascar's existing national-level mangrove data-sets are available in Giri and Muhlhausen [68], Jones *et al.* [25] and Jones *et al.* [61]. Further details outlining loss calculated for Madagascar's mangroves using USGS maps through to 2005 is available in Giri and Muhlhausen [68].

Figure 3: A Landsat TM Blue-Green-Red 354 false color composite from July, 2011 is shown in the top left panel wherein mangroves appear primarily in shades of red. Mangrove coverage from three national data-sets (*i.e.*, Giri [69]; CEPF [71]; Harper *et al.*[72]) is overlain on the Landsat composite in the three other panels. This comparison illustrates how data produced by Giri [69] provide the most comprehensive (in terms of representing actual known mangrove area) and contemporary (in terms of year) mangrove coverage; as is illustrated by the variably represented mangrove covered islands in the middle-right of each panel. The area shown in all four panes is a portion of MHJ.

Mangrove Distribution and Dynamics: 1990–2010

National-Level Distribution and Dynamics

According to the USGS-produced Landsat-derived maps of national-level mangrove coverage from 1990, 2000 and 2010 [69], as of 2010 Madagascar contained approximately 213,000 ha of mangrove ecosystems distributed primarily along the west coast, with scattered, isolated pockets found on the north east coast (Figure 1). Analysis of the 1990, 2000 and 2010 USGS maps indicates that from 1990 to 2010, there was a country-wide net loss (*i.e.*, loss-gain) of 21% (*i.e.*, 57,359 ha).

Ecosystem-Level Dynamics

Considered independently, there are nearly 100 distinct, non-contiguous mangrove ecosystems in Madagascar; however, over 200,000 ha (*circa* 2010) are represented by 30 primary ecosystems, each greater than 1000 ha in size. Of these primary ecosystems, according to the USGS maps, as of 2010, MHJ had the greatest extent (*i.e.*, 26,667 ha), followed by AAB (*i.e.*, 25,664) and TMD (*i.e.*, 20,242 ha) (Table 3; Figure 4). BdA, ranked as Madagascar's 26th largest mangrove ecosystem, containing 1362 ha as of 2010. The extent in ha for 1990, 2000 and 2010 according to USGS-produced maps as partitioned by primary mangrove ecosystems are provided in Table 3 and Figure 4.

As previously reported in Jones *et al.* [25], and Jones *et al.* [61], in AAB, the analysis of USGS maps suggest a loss of 7659 ha (23.7%) and gain of 995 ha (3.1%) from 1990 to 2010 (Figure 5). Deforestation is mostly occurring on or near the peninsular base, which separates the two Bays. As described in Jones *et al.* [61], for MHJ, analyzing the USGS maps indicate 1251 ha were lost (4.5%) and 150 gained (0.5%) from 1990 to 2010 (Figure 5). As compared with other primary mangrove ecosystems in Madagascar, notably AAB, MHJ's mangroves have remained comparatively stable. Comparisons with terrestrial data (*i.e.*, [72]) also imply that from 2000 to 2005 alone, loss in the terrestrial forests surrounding MHJ exceeded mangrove loss from 2000 to 2010. However, mangrove loss in MHJ does appear to be increasing, particularly in the east. For TMD, the analysis of the USGS maps indicated a net loss of 12,612 ha (38.4%) from 1990 to 2010 (Figure 5). For BdA, analyzing the USGS maps indicated that from 1990 to 2010 there was a net loss of 360 ha (20.9%) (Figure 5).

Table 3: The extent in hectares for Madagascar's primary mangrove ecosystems for 1990, 2000 and 2010, based on partitioning USGS-produced Landsat-derived national-level mangrove cover maps for 1990, 2000 and 2010 [69] in to primary (*i.e.*, >1000 ha) mangrove ecosystems. Extents are ordered from largest to smallest based on 2010 values

	Mangrove Extent (Hectares)		
Mangrove ecosystem	**1990**	**2000**	**2010**
Mahajamba Bay (MHJ)	27778	27577	26677
Ambaro-Ambanja Bays (AAB)	32328	30321	25664
Tsiribihina and Manambolo Deltas (TMD)	32854	24651	20242
Antsohihy	17081	16065	13838
Tambohorano	21140	12781	13418
Sahamalaza	12107	11063	10956
Mahavavy su Sud	10615	10870	10654
Mahajanga	12375	11814	9574
Mangoky	14684	12247	9431
Morondava-Bosy	8743	7500	6123
Kamendriky-Tsilambana	6102	6102	5924
Mahabo-Andramy	4721	5939	5905
Maintirano	8937	4644	5900
Boeny	3870	3888	3867
Baly-Soalala	3687	3683	3507
Besalampy	6247	5097	3287
Rigny-Irody	3224	3232	3231
Morombe	3652	2952	3035
Mariarano	2472	2412	2330
Narinda	2249	2058	2036
Sohany	2470	1984	2025
Belo sur Mer	2603	2387	1917
Vilamatsa	1881	1881	1847
Kabatomena	2458	1882	1529
Reharaka	2229	1528	1406
Baie des Assassins (BdA)	1723	1301	1362
Manampatra	1405	1404	1327
Morovasa	1253	1199	1199
Mangolovo	1415	981	1172
Ambondrombe	1462	1349	1109
Total	**253,765**	**220,792**	**200,492**

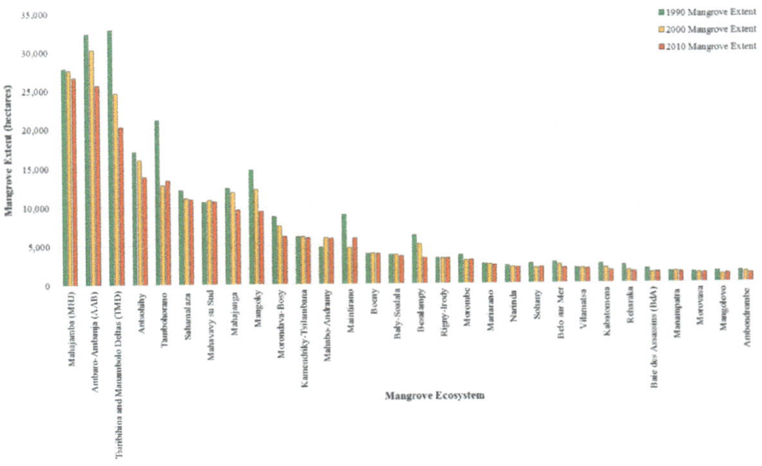

Figure 4: The extent in hectares for Madagascar's primary mangrove ecosystems for 1990, 2000 and 2010, based on partitioning USGS-produced Landsat-derived national-level mangrove cover maps for 1990, 2000 and 2010 [69] in to primary (*i.e.*, >1000 ha) mangrove ecosystems. Extents are ordered from largest to smallest based on 2010 values.

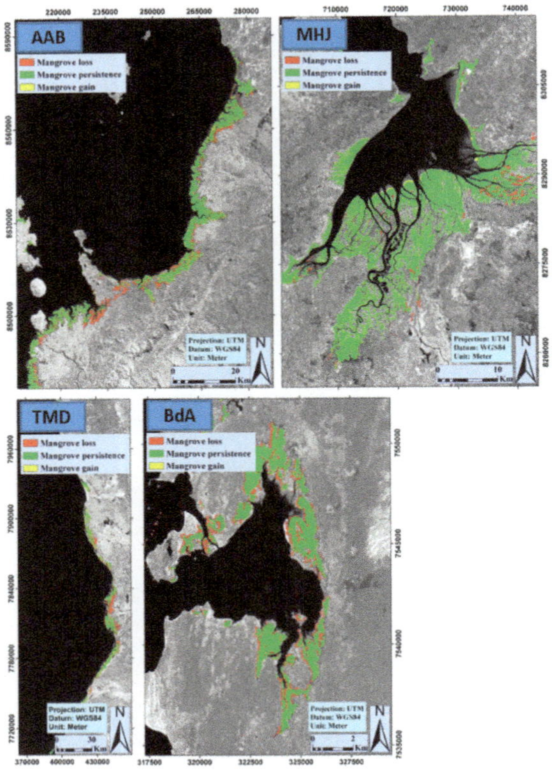

Figure 5: Dynamics from 1990 to 2010 for each of four AOIs calculated using the USGS-produced, Landsat-derived mangrove cover maps for 1990, 2000 and 2010 [69]. Background images are Landsat NIR bands from AOI-specific dates provided in Table 1.

AOI-Specific Ecosystem-Level Mapping Results

Spectral Separability and Classification Results

As demonstrated through results initially presented in Jones *et al.* [25] for AAB, and Jones *et al.* [61] for MHJ (Figure 6), for TMD and BdA, all mapped classes are spectrally separable using specific portions of the electromagnetic spectrum as represented by certain Landsat bands. In particular, the near-infrared (NIR) and short-wave infrared (SWIR) were useful for distinguishing between mangrove types and differentiating mangroves from surrounding mapped categories. In the NIR (0.76–0.90 micrometers (μm)), the spectral distinctiveness of mangrove classes was probably driven by vegetative reflectance relating to the transitional red-edge, internal vegetation structure, and leaf dry-matter content [15,69,99,110]. In the SWIR (1.55–1.75 and 2.08–2.35 μm), differences in reflectance were driven by vegetation and soil moisture content, and canopy-level biogeochemical constituents likely facilitated mangrove differentiation [100]. These results further support previous studies, which demonstrate that SWIR wavelengths help distinguish mangroves from surrounding terrestrial vegetation [25,111]. Mangroves types were also further differentiated in the visible bands (*i.e.*, blue: 0.45–0.52 μm, green: 0.53–0.61 μm, and red: 0.63–0.69 μm).

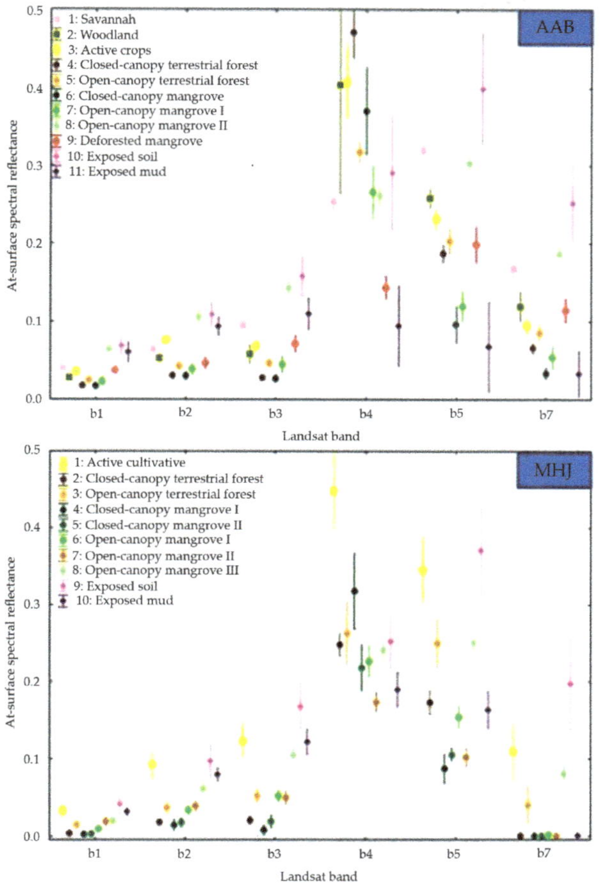

Figure 6: The mean spectral reflectance (±1 standard deviation) of mapped classes for Ambaro-Ambanja Bays (AAB) and Mahajamba Bay (MHJ) (adapted from [25,61]).

In support of results originally described in Jones *et al.* [25] for AAB, and Jones *et al.* [61] for MHJ, ML classification for TMD and BdA resulted in highly accurate AOI-specific maps (Table 4; Figure 7). Our results further indicate that while common as a source of classification error in mangrove studies, confusion with other vegetation classes (e.g., terrestrial forest types) was mostly avoided. As compared with single-class contemporary (*i.e.*, 2010) mangrove coverage provided by the USGS national-level map, it is clear that our classification results provide more detailed and comprehensive representation of mangroves (Figure 8). In total, USGS maps imply 25,664, 26,677, 20,242 and 1362 ha of mangroves for AAB, MHJ, TMD and BdA, respectively. In contrast, AOI-specific maps imply 45,680, 45,107, 28,513, and 1652 ha of mangrove ecosystem for AAB, MHJ, TMD and BdA, respectively. When considering mangrove forest only and excluding classes dominated by shrub/scrub, the amount of mangroves in each AOI is reduced but still markedly higher than USGS estimates (Figure 9). The USGS map under-represents mangrove stands which are naturally lower stature and more open or highly degraded, and completely omits most scrub/shrub-dominated areas and certain narrow linear strips. The under-representation results in conservative distribution estimates, which in turn impacts ranking primary mangrove ecosystems by extent, as undertaken in this study. This is exemplified by MHJ being estimated as Madagascar's largest mangrove ecosystem based on the USGS map; whereas our AOI-specific map indicates that AAB was slightly larger.

Under-representation of lower stature or sparser mangrove areas associated with national-level USGS maps can also exaggerate dynamics by indicating loss in areas that may actually represent degradation or sparse areas of mangrove on the edge of the limits of detection by the automated algorithms used to create the data. Field observations confirm that certain areas appearing as loss in the USGS maps were in actuality occupied by sparse or degraded mangroves. In addition, given that numerous Landsat images were mosaicked together for each temporal increment to create the USGS maps, the influence of different dates and thus tidal conditions may have also exaggerated dynamics. The potential for exaggerated dynamics in the USGS maps needs to be taken in to account when considering the national-level dynamics presented in this study; though concerns of exaggerated loss must be tempered with an acknowledgement of an under-representation of sparse and degraded mangroves, which leads to initial under-estimates of mangrove distributions from which dynamics are calculated. Bearing in mind the limitations of the single mangrove class represented by the USGS maps, they also provide no context regarding surrounding land-cover categories, including extensive mud flats that once were or could again become mangrove ecosystems. In aggregate, these shortcomings highlight the importance of detailed, contemporary ecosystem-specific localized mapping; which is bolstered by USGS maps being >four years old, providing outdated representation for ecosystems which continue to experience increased and widespread degradation and deforestation.

Table 4: (**a–d**) Classification accuracies, including overall accuracies, Kappa statistics and per-class user's and producer's accuracies. Results for AAB and MHJ were previously reported in Jones *et al.* [25] and Jones *et al.* [61]

(a)

Area:	Class:	1	2	3	4	5	6	7	8	9	10	11	Total	User's (%)	Commission (%)
AAB:	Savannah (1)	54	3	0	0	0	0	0	0	0	1	0	58	93	7
	Woodland (2)	0	39	0	0	0	0	0	0	0	0	0	39	100	0
	Active cultivation (3)	0	0	51	0	0	0	0	0	0	0	0	51	100	0
	Closed-canopy terrestrial forest (4)	0	0	0	54	0	0	0	0	0	0	0	54	100	0
	Open-canopy terrestrial forest (5)	0	4	0	0	54	0	0	0	0	0	0	58	93	7
	Closed-canopy mangrove (6)	0	0	0	0	0	79	9	0	0	0	0	88	90	10
	Open-canopy mangrove I (7)	0	0	0	0	0	11	72	0	2	0	0	85	85	15
	Open-canopy mangrove II (8)	0	0	0	0	0	0	0	52	0	0	0	52	100	0
	Deforested mangrove (9)	0	0	0	0	0	0	0	0	60	0	0	60	100	0
	Exposed soil (10)	0	8	3	0	0	0	0	2	1	53	0	67	79	21
	Exposed mud (11)	0	0	0	0	0	0	0	0	0	0	54	54	100	0
	Total	54	54	54	54	54	90	81	54	63	54	54	666		
	Producer's (%)	100	72	94	100	100	88	89	96	95	98	100		Overall Accuracy = 93.4%	
	Omission (%)	0	28	6	0	0	12	11	4	5	2	0		Kappa = 0.9	

(b)

Area:	Class:	1	2	3	4	5	6	7	8	9	10	Total	User's (%)	Commission (%)
MHJ:	Active cultivation (1)	63	0	0	0	0	0	0	0	0	0	63	100	0
	Closed-canopy terrestrial forest (2)	0	62	0	0	0	0	0	0	0	0	62	100	0
	Open-canopy terrestrial forest (3)	0	1	72	0	0	3	0	1	0	0	77	94	6
	Closed-canopy mangrove I (4)	0	0	0	90	2	0	0	0	0	0	92	98	2
	Closed-canopy mangrove II (5)	0	0	0	0	88	0	0	0	0	0	88	100	0
	Open-canopy mangrove I (6)	0	0	0	0	0	60	0	0	0	0	60	100	0
	Open-canopy mangrove II (7)	0	0	0	0	0	0	60	0	0	0	60	100	0
	Open-canopy mangrove III (8)	0	0	0	0	0	0	0	62	0	0	62	100	0
	Exposed soil (9)	0	0	0	0	0	0	0	0	90	0	90	100	0
	Exposed mud (10)	0	0	0	0	0	0	3	0	0	63	66	95	5
	Total	63	63	72	90	90	63	63	63	90	63	720		
	Producer's (%)	100	98	100	100	98	95	95	98	100	100		Overall Accuracy = 98.6%	
	Omission (%)	0	2	0	0	2	5	5	2	0	0		Kappa = 0.9	

(c)

Area:	Class:	1	2	3	4	5	6	7	8	Total	User's (%)	Commission (%)
TMD:	Dense mangrove (1)	54	0	0	0	0	0	0	0	54	100	0
	Sparse mangrove (2)	0	45	0	0	0	0	0	0	45	100	0
	Bare type I (3)	0	0	52	0	0	0	0	0	52	100	0
	Dense forest (4)	0	0	0	54	0	0	0	0	54	100	0
	Sparse forest (5)	0	0	0	0	45	0	0	0	45	100	0
	Wooded grassland-bushland (6)	0	0	0	0	0	45	0	0	45	100	0
	Active agriculture (7)	0	0	0	0	0	0	9	0	9	100	0
	Bare type II (8)	0	0	2	0	0	0	0	48	50	96	4
	Total	54	45	54	54	45	45	9	48	354		
	Producer's (%)	100	100	96	100	100	100	100	100		Overall Accuracy = 99.4%	
	Omission (%)	0	0	4	0	0	0	0	0		Kappa = 0.99	

(d)

Area:	Class:	1	2	3	4	5	6	7	8	Total	User's (%)	Commission (%)
BdA:	Spiny forest (1)	27	0	0	0	0	0	0	0	27	100	0
	Barren/exposed (2)	0	30	0	0	0	0	0	0	30	100	0
	Cultivated/degraded/woodland (3)	0	0	27	0	0	0	0	0	27	100	0
	Burnt (4)	0	0	0	27	0	0	0	0	27	100	0
	Water dominated (5)	0	0	0	0	42	0	0	0	42	100	0
	Closed-canopy mangrove (6)	0	0	0	0	0	27	0	0	27	100	0
	Open-canopy mangrove I (7)	0	0	0	0	0	0	27	0	27	100	0
	Open-canopy mangrove II (8)	0	0	0	0	0	0	0	27	27	100	0
	Total	27	30	27	27	42	27	27	27	234		
	Producer's (%)	100	100	100	100	100	100	100	100		Overall Accuracy = 100	
	Omission (%)	0	0	0	0	0	0	0	0		Kappa = 1	

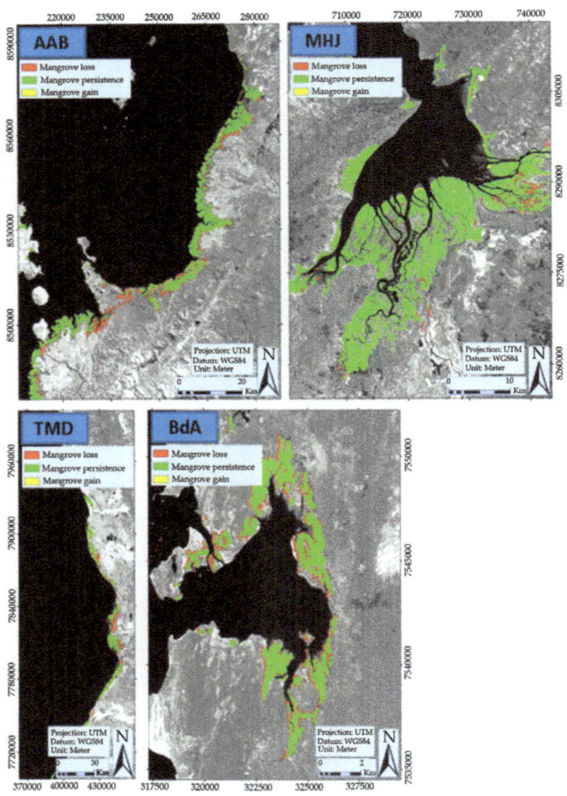

Figure 7: The results of detailed, contemporary, ecosystem-level mapping for four AOIs (*i.e.*, Ambaro-Ambanja Bays (AAB), Mahajamba bay (MHJ), Tsiribihina Manambolo Delta (TMD), and Baie des Assassins (BdA)). The water dominated class is excluded for BdA and is excluded from the figure. For each AOI, the background image is a Landsat NIR band (*i.e.*, band 4: Landsat 5 TM and Landsat 7 ETM+ ; band 5: Landsat 8 OLI) as described in Table 2. Mapping results for AAB and MHJ were originally presented in Jones *et al.* [25] and Jones *et al.* [61].

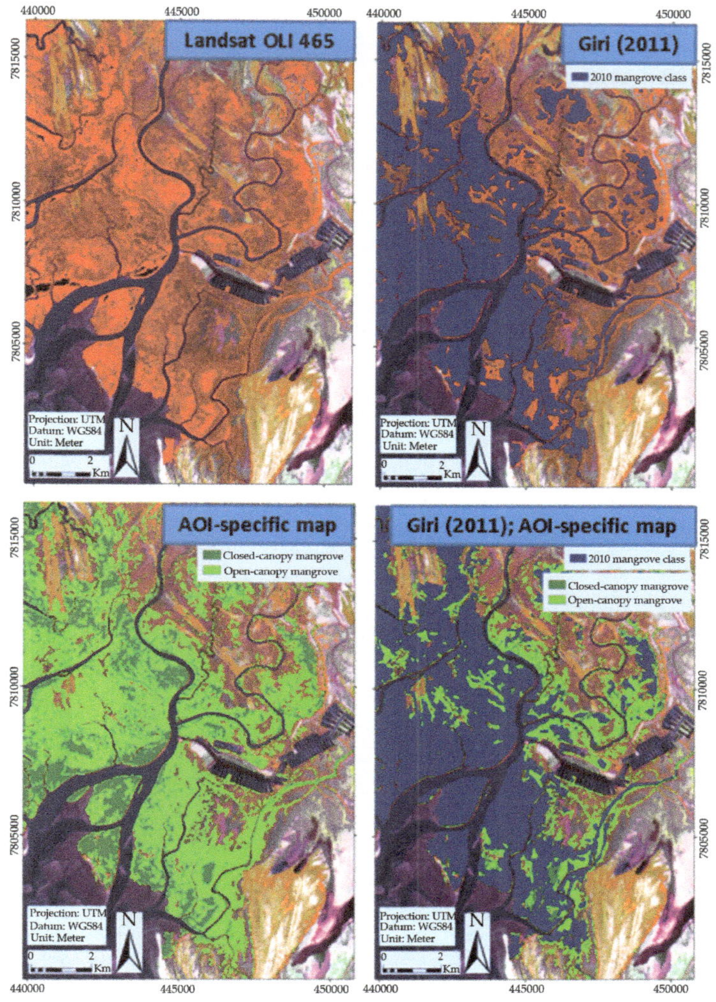

Figure 8: A Landsat OLI 465 false color composite from July, 2014, over the Tsiribihina-Manambolo Delta (TMD) AOI, is shown in the top left panel, wherein mangroves appear primarily in shades of red. The other panels compare USGS-produced national-level mangrove coverage *circa* 2010 (Giri [69]) to one of the detailed, contemporary, AOI-specific maps produced through this study. The comparison demonstrates how the USGS map (Giri [69]) underrepresents naturally lower stature, more open or highly degraded mangroves and at times omits scrub/shrub and narrow linear strips of mangroves.

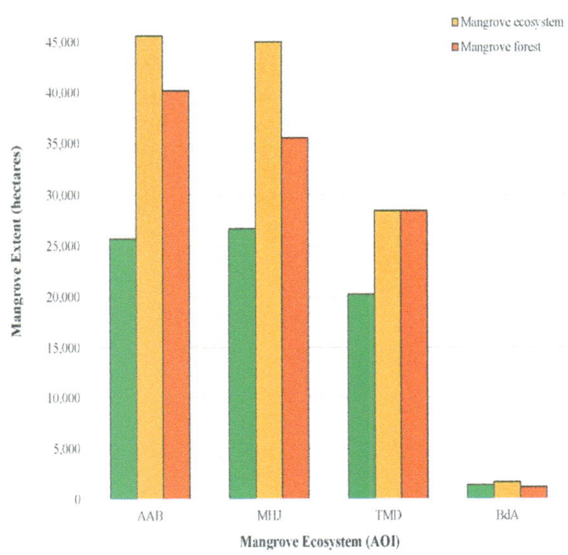

Figure 9: The extent of mangroves based on the 2010 national-level USGS-produced Landsat derived mangrove cover map [69] *versus* the results of AOI-specific contemporary mapping is shown for each AOI. For AOI-specific mapping results, extent is shown based on all mangrove classes including shrub/scrub dominated (*i.e.*, mangrove ecosystem) and based on only classes meeting the definition of forest (*i.e.*, mangrove forest).

Application of Mapping Results

The results of the national- and ecosystem-level dynamics analysis described in Section 3.2 have supported Blue Venture's Blue Forests programme in selecting areas in critical need of mangrove conservation and restoration. The contemporary AOI-specific maps described in Section 3.3 are being applied for ecological characterization and carbon stock estimation, to provide baselines for historical mapping/future monitoring, and to support the establishment of carbon projects.

As described in Jones *et al.* [25] for AAB, and Jones *et al.* [61] for MHJ, localized mangrove maps were used to stratify mangroves and systematically establish carbon plots based on adaptations of methods proposed by the Centre for International Forestry Research (CIFOR) as outlined in Kauffman and Donato [24]. In total, 55 and 51 carbon plots were established for AAB and MHJ, respectively. Tree measurements summed at the plot-level allowed for summarizing the primary ecological characteristics of each mapped mangrove type, and allowed breaking mangrove classes in to sub-types based on dominant ecological traits (Table 5). Open-canopy areas were typically comprised of sparse and mostly stunted/shrub, low stature mangroves with very open canopies or moderately-dense stands of medium stature trees with relatively open canopies. In contrast, closed-canopy areas were typified by higher stature trees of variable density with well-formed canopies. Exceptions to these typical ecosystem characteristics included, (1) areas dominated by extremely dense, closed-canopy medium/near-tall trees and (2) mature stands which were either highly degraded or naturally open, both of which can spectrally appear as open-canopy mangrove. This overlap highlights a distinct limitation of our maps, which is the inability to reliably detect and distinguish mangrove degradation. While field observations confirm that degradation exists in all AOIs, of both natural and anthropogenic in origin, the full extent of degradation remains uncertain. While mangrove conversion (*i.e.*, deforestation) can be reliably mapped and monitored with moderate resolution data (e.g., Landsat) using established methods, accurately detecting and tracking the subtle sub-pixel modification (*i.e.*, degradation) of mangrove ecosystems remains a vexing challenge. Finer spatial resolution imagery can be employed to accurately detect tree-level degradation, but these data remain prohibitively expensive for small or not-for-profit organizations. Directly incorporating other remotely sensed data–sets, such as those representing forest structure (e.g., LiDAR; radar), into the classification process could greatly increase explanatory power and help further sub-divide existing classes based on degradation; though access to such complimentary data-sets remains limited. Even if able to accurately detect degradation, it would remain difficult to confidently partition natural *versus* anthropogenic. Despite the limitations of our mangrove classes, they are spectrally distinct and field observations confirm that they are ecologically different and meaningful.

For AAB and MHJ, the tree diameter and height measurements collected in plots were used as input in allometric equations to calculate above-ground biomass and estimate carbon stocks. Tree below-ground biomass was calculated with a generalized equation presented in Komiyama *et al.* [112]. Equations found in Kauffman and Donato [24] were used to estimate the biomass of standing dead wood. Soil samples taken at the center of each plot allow for the determination of soil organic carbon (SOC), which is still ongoing. For AAB, preliminary SOC was estimated using a modified Walkley-Black method [113,114,115] from samples sent to the Laboratoire des Radio Isotopies (LRI) in Antananarivo, Madagascar. Estimates of carbon calculated based on plot measurements in all AOIs were scaled to the ha-level.

As presented in Jones *et al.* [61,62], and representing the first total carbon stock estimates for a mangrove ecosystem in Madagascar, total carbon stock estimates in AAB varied from 126.42 to 570.72 $Mg{\cdot}C{\cdot}ha^{-1}$, with an overall mean of 356.36 (±16.96) $Mg{\cdot}C{\cdot}ha^{-1}$ (Figure 10). Average vegetation carbon was highest within the tall-stature closed-canopy mangroves (114.8 (±9.3) $Mg{\cdot}C{\cdot}ha^{-1}$). In comparison, open-canopy I and open-canopy II mangroves had average vegetation C values of 43.6 (±7.3) $Mg{\cdot}C{\cdot}ha^{-1}$ and 19.3 (±4.6) $Mg{\cdot}ha^{-1}$ respectively. Mean SOC values, based on soil depth up to 100 cm, ranged from 165.2 (±29.1) $Mg{\cdot}C{\cdot}ha^{-1}$ for open-canopy II mangroves, to 278.80 (±20.99) $Mg{\cdot}C{\cdot}ha^{-1}$ for open-canopy I mangroves, and 309.87 (±19.36) $Mg{\cdot}C{\cdot}ha^{-1}$ for closed-canopy mangroves. Total carbon stock estimates of trees (*i.e.*, above- and below-ground) in MHJ varied from 2.97 to 279.49 $Mg{\cdot}C{\cdot}ha^{-1}$, with an overall mean of 100.96 (±10.49) $Mg{\cdot}C{\cdot}ha^{-1}$ (Jones *et al.* [61,62]) (Figure 11). The closed-canopy I class, wherein tree stature was largest and tree density highest, had the highest carbon values (166.82 (±15.38) $Mg{\cdot}C{\cdot}ha^{-1}$).

The differences in carbon stock estimates between closed- and open-canopy classes in AAB and MHJ (Figure 11) reflect variable forest stature and density and supports that larger, taller, denser trees contain significantly greater amounts of carbon. The difference in carbon stocks between the two AOIs is influenced by disproportionate amounts of higher stature trees in MHJ as compared with AAB. Collectively, these estimates support a growing body of evidence that mangroves are amongst the most carbon-dense forests in the tropics, with similar above- and larger below-ground stocks than terrestrial tropical upland systems [14,17,21,22,23,116,117,118,119,120,121].

Table 5: Mangrove class, species dominance, average tree height (m) (± standard error (SE)), average dbh (cm) (±SE) and average trees per hectare (ha^{-1}) (± SE) for mapped mangrove categories. Adapted from Jones *et al.* [25] and Jones *et al.* [61]

Class	Code	Description	Species Dominance	N	Average Tree Height (m)	Average dbh (cm)	Average Number of Trees (ha^{-1})
(a) Ambaro-Ambanja Bays (AAB)							
Closed-canopy mangrove	CC	Intact, tall, mature stands	*A. marina*	1	8.6	14.9	1250
			C. tagal	3	7.3 (± 1.2)	10.1 (± 0.5)	2625 (± 318)
			R. mucronata	14	7.0 (± 1.3)	10.1 (± 3.0)	4719 (± 1133)
			S. alba	1	5.6	10.6	5300
			Mixed species	2	6.7 (± 1.6)	11.3 (± 2.5)	1825 (± 248)
		Very dense medium-tall stands	R. mucronata	2	4.8 (+ 0.1)	7.8 (± 1.1)	5600 (± 1838)
Open-canopy mangrove I	OC I	Medium stands	*C. tagal*	7	4.6 (± 0.8)	7.5 (± 1.6)	3300 (± 8.49)
			R. mucronata	6	4.2 (± 0.6)	7.3 (± 1.3)	2160 (± 498)
			Mixed species	2	4.8 (± 0.1)	9.5 (± 2.0)	1800 (± 141)
		Naturally open/very degraded tall	*Mixed species*	4	5.7 (± 0.3)	10.1 (± 1.2)	1525 (± 35)
		very dense short stands	C. tagal	5	2.5 (± 0.3)	5.1 (± 0.9)	2780 (± 750)
Open-canopy mangrove II	OC II	Stunted, scrub ecosystems	*A. marina*	4	1.7 (± 0.5)	4.6 (± 0.2)	1306 (± 554)
(b) Mahajamba Bay (MHJ)							
Closed-canopy mangrove I	CC I	Tall, mature stands; canopy >80% closed	*A. marina*	7	10.24 (± 0.52)	13.68 (± 1.01)	1571 (± 255)
			R. mucronata	2	5.62 (± 0.55)	7.27 (± 1.69)	4900 (± 1500)
			S. alba	1	9.39	8.31	5100
			Mixed species	3	12.48 (± 1.40)	18.18 (± 1.56)	1108 (± 208)
Closed-canopy mangrove II	CC II	Tall mature stands; canopy >60% closed	*A. marina*	10	7.68 (± 0.56)	12.95 (± 1.26)	895 (± 102)
			Mixed species	2	7.74 (± 0.04)	12.45 (± 0.08)	1412 (± 12)
Open-canopy mangrove I	OC I	Short-medium stands; canopy 30%–70% closed; moderately influenced by background soil/mud	*A. marina*	6	3.32 (± 0.16)	4.85 (± 0.43)	1417 (± 226)
			R. mucronata	1	3.21	7.39	2200
			X. granatum	1	5.41	10.84	1300
			Mixed species	5	4.33 (± 0.44)	7.62 (± 0.86)	1185 (± 237)
Open-canopy mangrove II	OC II	Short-medium stands; canopy 30%–70% closed; significantly influenced by background soil/mud	*C. tagal*	2	3.39 (± 0.18)	6.18 (± 0.12)	963 (± 238)
			R. mucronata	4	4.63 (± 0.30)	7.85 (± 1.66)	1388 (± 449)
Open-canopy mangrove III	OC III	Stunted, short stands, very sparse; canopy < 30% closed; dominated by exposed soil/mud	*A. marina*	7	2.31 (± 0.17)	3.96 (± 0.18)	1089 (±134)

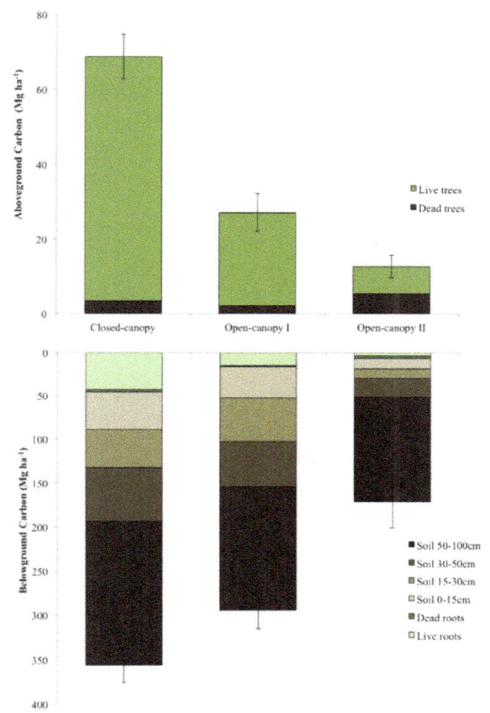

Figure 10: Total vegetation carbon and soil organic carbon stocks for the Ambaro-Ambanja Bays, Northwest Madagascar. Error bars represent ± SE of total carbon stocks. Adapted from Jones *et al.* [62] with permission of Springer).

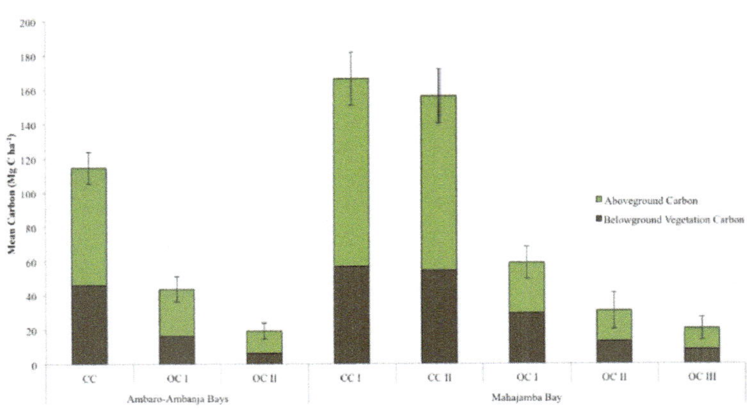

Figure 11: Above- and below-ground mangrove vegetation carbon estimates for the Ambaro-Ambanja Bays (AAB) [62] and Mahajamba Bay (MHJ) [61], Madagascar. Error bars represent ± SE of total vegetation carbon stocks. Adapted from Jones *et al.* [61].

Mapping and modelling of historical ecosystem loss is a critical element in both the assessment of feasibility and the development of carbon projects. The generation of carbon credits through the conservation, restoration and reduced-impact use of mangroves has the potential to catalyze and fund sustainable mangrove management in coastal communities, augmenting existing livelihoods, preparing for climate change and safeguarding biodiversity. While national-level datasets are very valuable for the initial selection of potential project sites, the limitations outlined in Section 3.3.1 prevent them from being suitable for ecosystem-specific/project-scale historical deforestation analysis in Madagascar. The existing (*i.e.*, AAB and MHJ) and new (*i.e.*, TMD and BdA) AOI-specific maps presented in here can be used as a starting point for more detailed project-scale deforestation analysis, and also as a baseline for future project monitoring.

Working with local communities, the University of Antananarivo and other partners in AAB, the Blue Forests programme is using the data described above to assess the feasibility of a community-centered mangrove carbon project validated under the VCS standard (www.v-c-s.org), which focusses on the locally-led conservation and sustainable use of AAB's mangrove resources.

In TMD, in addition to supporting WWF Madagascar West Indian Ocean Programme Office's broader work in the region, the mapping detailed herein was also central to assessing the feasibility of a mangrove carbon project in the region].

In BdA, Blue Ventures is supporting communities to develop a Plan Vivo mangrove carbon project. Plan Vivo initiatives differ from other types of forest carbon projects. Whilst they are still measured and valued according to their impact on greenhouse gas emissions, Plan Vivo project design must be community-led. Fifty-two Plan Vivo projects have been initiated throughout the world to date, yet in only three countries (*i.e.*, Kenya, Columbia and Madagascar) are there projects focusing specifically on mangroves.

By piloting and developing mangrove carbon projects in close contact with in-country government institutions, the Blue Forests programme is supplying robust greenhouse gas emissions reductions estimates and helping to integrate mangroves into Madagascar's national REDD+ strategy.

CONCLUSIONS

This study presents, for the first time, national-level and mangrove ecosystem specific dynamics for Madagascar from 1990 to 2010, providing an unprecedented overview of mangrove loss. While not without their limitations, our AOI-specific contemporary maps offer numerous improvements over national-level USGS data-sets, providing detailed and accurate coverage of spectrally and ecologically distinct mangrove types and surrounding land-cover categories for two new specific Madagascan mangrove ecosystems (*i.e.*, TMD and BdA); building on and complementing similar maps for AAB and MHJ. Taken as a whole, these maps are the first of their kind for AAB and TMD, and provide updated information and improved thematic detail for MHJ and BdA. The methods used are easily replicable and employ freely available Landsat data. Factoring in their strengths, the primary weakness of our AOI-specific maps is their inability to represent degradation. The ability to accurately detect degradation would be greatly augmented through the incorporation of complementary remotely sensed data-sets (e.g., LiDAR; radar; finer spatial resolution optical imagery) in further mapping and monitoring.

A continuation and acceleration of modification and deforestation of Madagascar's mangrove ecosystems will jeopardize if not halt key ecosystem services. The extent and consequences of the ripple effects from continued loss and degradation in to surrounding ecosystems is unknown. To help safeguard the long-term status of Madagascar's mangrove forests, improved resource management is required. The AOI-specific maps described in this study are being applied through Blue Ventures' Blue Forests programme towards standardizing replicable methods for improved, community-led mangrove resource management throughout Madagascar's mangroves and beyond. Resulting carbon stock estimates for AAB and MHJ include the first total (*i.e.*, above- and below-ground including soil) estimates published for Madagascar (*i.e.*, [25]), and the first total-tree estimates published for MHJ (*i.e.*, [61]. While carbon stock estimates for MHJ are limited to trees, they are still high, consistent with regional estimates, and highlight the importance of this ecosystem towards climate change mitigation. Ongoing soil analysis will also finalize SOC values for AAB and MHJ, and total carbon stock estimates for MHJ, TMD and BdA, resulting in comprehensive carbon stock estimates for all four AOIs.

AOI-specific maps are directly helping to explore the feasibility of mangrove carbon projects in AAB and TMD and support the establishment of a Plan Vivo project in BdA. In addition to carbon projects, the results presented in this study provide an invaluable updated scientific baseline for policy makers and civil society to identify efficient mangrove conservation strategies at the ecosystem or national scale.

ACKNOWLEDGMENTS

This research was funded by the Global Environmental Facility's Blue Forest project and grants from the Western Indian Ocean Marine Science Association, the John D. and Catherine T. MacArthur Foundation and the Darwin Initiative. All TMD analysis was done in collaboration with the World Wildlife Fund's Madagascar West Indian Ocean Programme Office. The authors greatly appreciate the logistical support for field work in MHJ provided by AQUALMA. Many thanks to Raymond Raherindray, Zo Andriamahenina, Ismael Ratefinjanahary, Jaona Ravelonjatovo, Ferdinand Botsy, Tina Haingonirina and Holy Andriamitantsoa for assisting with field missions. Thanks also to Kate England, Bienvenue Zafindrasilivonona, Sylvia Paulot, Pierre-Francois Roy and Rado Rakotomanana for contributions to socio-economic research and analysis. Special thanks to community members from within all AOIs for their guidance, hospitality and assistance with field work. Additional thanks to extremely helpful anonymous reviewers.

AUTHOR CONTRIBUTIONS

Trevor Jones took the lead on designing and undertaking the dynamics assessment, AOI- specific mapping, and writing the manuscript. Leah Glass assisted with the dynamics assessment and AOI-specific mapping and helped provide oversight for the manuscript, contextualizing loss. Samir Gandhi assisted with geospatial data pre-processing, the dynamics assessment, AOI-specific mapping and formatting the manuscript.

Lalao Ravaoarinorotsihoarana helped design, and lead field missions, summarize resulting data, and further helped with carbon stock estimates. Aude Carro provided specific input regarding contextualizing mangrove loss. Lisa Benson assisted with calculating, compiling and presenting carbon stock estimates. Harifidy Rakoto Ratsimba helped design and provided oversight of field missions, and assisted with the dynamics assessment and AOI-specific mapping. Chandra Giri assisted with the dynamics assessment and AOI-specific mapping. Dannick Randriamanatena assisted with all logistics, field work and mapping related to TMD. Garth Cripps helped provide overall oversight on design and implementation.

REFERENCES

1. Lugo, A.E.; Snedaker, S.C. The ecology of mangroves. *Annu. Rev. Ecol. Syst.* 1974, *5*, 39–64.
2. Blasco, F.; Bellan, M.F.; Chaudhury, M.U. Estimating the extent of floods in Bangladesh—Using SPOT data. *Remote Sens. Environ.* 1992, *39*, 167–178.
3. Marshall, N. Mangrove conservation in relation to overall environmental considerations. *Hydrobiologia* 1994, *285*, 303–309.
4. Primavera, J.H. Socio-economic impacts of shrimp culture. *Aquac. Res.* 1997, *28*, 815–827.
5. Kathiresan, K.; Bingham, B. Biology of mangroves and mangrove ecosystems. *Adv. Mar. Biol.* 2001, *40*, 81–251.
6. Alongi, D.M. Present state and future of world's mangrove forest. *Environ. Conserv.* 2002, *29*, 331–349.
7. Mumby, P.J.; Edwards, A.J.; Arias-Gonzáles, E.; Lindeman, K.C.; Blackwell, P.G.; Gall, A.; Gorczynska, M.I.; Harborne, A.R.; Pescod, C.L.; Renken, H.; *et al.* Mangrove enhance the biomass of coral reef fish communities in the Caribbean. *Nature* 2004, *427*, 533–536.
8. Dahdouh-Guebas, F.; Jayatissa, L.P.; di Nitto, D.; Bosire, J.O.; Lo Seen, D.; Koedam, N. How effective were mangroves as a defence against the recent tsunami? *Curr. Biol.* 2005, *15*, R443–R447.
9. Barbier, E.B. Natural barriers to natural disasters: Replanting mangroves after tsunami. *Front. Ecol. Environ.* 2006, *4*, 124–131.
10. Food and Agricultural Organization (FAO). *The World's Mangroves 1980–2005*; FAO Forestry Paper 153; FAO: Rome, Italy, 2007.
11. Alongi, D.M. Mangrove forests: Resilience, protection from tsunamis, and responses to global climate change. *Estuar. Coast. Shelf Sci.* 2008, *76*, 1–13.
12. Nagelkerken, I.; Blaber, S.J.; Bouillon, S.; Green, P.; Haywood, M.; Kirton, L.G.; Meynecke, J.-O.; Pawlik, J.; Penrose, H.M.; Sasekumar, A.; *et al.* The habit function of mangroves for terrestrial and marine fauna: A review. *Aquat. Bot.*2008, *89*, 155–185.
13. Alongi, D.M. Carbon payments for mangrove conservation: Ecosystem constraints and uncertainties of sequestration potential. *Environ. Sci. Policy* 2011, *14*, 462–470.
14. Donato, D.C.; Kauffman, J.B.; Murdiyarso, D.; Kumianto, S.; Stidham, M.; Kanninen, M. Mangroves among the most carbon-rich forests in the tropics. *Nat. Geosci.* 2011, *4*, 293–297.
15. Kuezner, C.; Bluemel, A.; Gebhardt, S.; Quoc, T.V.; Dech, S. Remote sensing of mangrove ecosystems: A review. *Remote Sens.* 2011, *3*, 878–928.
16. Pendleton, L.; Donato, D.C.; Murray, B.C.; Crooks, S.; Jenkins, W.A.; Sifleet, S.; Craft, C.; Fourqurean, J.W.; Kauffman, J.B.; Marba, N.; *et al.* Estimating global "blue carbon" emissions from conversion and degradation of vegetated coastal ecosystems. *PLoS ONE* 2012, *7*, e43542.
17. Kauffman, J.B.; Heider, C.; Norfolk, J.; Payton, F. Carbon stocks of intact mangroves and carbon emissions arising from their conversion in the Dominican Republic. *Ecol. Appl.* 2014, *24*, 518–527.
18. Thompson, B.S.; Clubbe, C.P.; Primavera, J.H.; Curnick, D.; Koldeway, H.J. Locally assessing the economic viability of blue carbon: A case study from Panay Island, the Philippines. *Ecosyst. Serv.* 2014, *8*, 128–140.
19. Giri, C.; Long, J.; Abbas, S.; Mani Murali, R.; Qamer, F.M.; Pengra, B.; Thau, D. Distribution and dynamics of mangrove forests of South Asia. *J. Environ. Manag.* 2015, *148*, 101–111.
20. Huxham, M.; Emerton, L.; Kairo, J.; Munyi, F.; Abdirizak, H.; Muriuki, T.; Nunan, F.; Briers, R.A. Applying Climate Compatible Development and economic valuation to coastal management: A case study of Kenya's mangrove forests. *J. Environ. Manag.* 2015, *157*, 168–181.
21. Kauffman, J.B.; Heider, C.; Cole, T.G.; Dwire, K.A.; Donato, D.C. Ecosystem carbon stocks of Micronesian mangrove forests. *Wetlands* 2011, *31*, 343–352.
22. Adame, M.F.; Kauffman, J.B.; Medina, I.; Gamboa, J.N.; Torres, O.; Caamal, J.P.; Reza, M.; Herrera-Silveira, J.A. Carbon stocks of tropical coastal wetlands within the Karstic landscape of the Mexican Caribbean. *PLoS ONE* 2013, *8*, e56569.
23. Wang, G.; Dongsheng, G.; Peart, M.R.; Chen, Y.; Peng, Y. Ecosystem carbon stocks of mangrove forest in Yingluo Bay, Guangdon Province of South China. *For. Ecol. Manag.* 2013, *310*, 539–546.
24. Kauffman, J.B.; Donato, D.C. *Protocols for the Measurement, Monitoring and Reporting of Structure, Biomass and Carbon Stocks in Mangrove Forests*; Working Paper 86; CIFOR: Bogor, Indonesia, 2012.
25. Jones, T.G.; Ratsimba, H.R.; Ravaoarinorotsihoarana, L.; Cripps, G.; Bey, A. Ecological Variability and Carbon Stock Estimates of Mangrove Ecosystems in Northwestern Madagascar. *Forests* 2014, *5*, 177–205.
26. Valiela, I.; Bowen, J.L.; York, J.K. Mangrove forests: One of the world's threatened major tropical environments. *Bioscience* 2001, *51*, 807–815.
27. Duke, N.C.; Meynecke, J.O.; Dittmann, S.; Ellison, A.M.; Anger, K.; Berger, U.; Cannicci, S.; Diele, K.; Ewel, K.C.; Field, C.D.; *et al.* A world without mangroves? *Science* 2007, *317*, 41–42.
28. Spalding, M.; Kainuma, M.; Collins, L. *World Atlas of Mangroves*; Earthscan: London, UK, 2010.

29. Friess, D.A.; Webb, E.L. Variability in mangrove change estimates and implications for the assessment of ecosystem provision. *Glob. Ecol. Biogeogr.* 2013, *23*, 715–725.

30. Alongi, D.M. The Impact of Climate Change on Mangrove Forests. *Curr. Clim. Chang. Rep.* 2015, *1*, 30–39.

31. Farnsworth, E.J.; Ellison, A.M. The global conservation status of mangroves. *Ambio* 1997, *26*, 328–334.

32. Primavera, J.H. Development and conservation of Philippine mangroves: Institutional issues. *Ecol. Econ.* 2000, *35*, 91–106.

33. Dahdouh-Guebas, F. The use of remote sensing and GIS in the sustainable management of tropical coastal ecosystems. *Environ. Dev. Sustain.* 2002, *4*, 93–112.

34. Primavera, J.H. Mangroves, fishponds, and the quest for sustainability. *Science* 2005, *310*, 57–59.

35. Gopal, B.; Chauhan, M. Biodiversity and its conservation in the Sundarban Mangrove Ecosystem. *Aquat. Sci.* 2006, *68*, 338–354.

36. Primavera, J.H. Overcoming the impacts of aquaculture on the coastal zone. *Ocean Coast. Manag.* 2006, *49*, 531–545.

37. Gilman, E.L.; Ellison, J.; Duke, N.C.; Field, C. Threats to mangroves from climate change and adaptation options: A review. *Aquat. Bot.* 2008, *89*, 237–250.

38. Walters, B.B.; Rönnbäck, P.; Kovacs, J.M.; Crona, B.; Hussain, S.A.; Badola, R.; Primavera, J.H.; Barbier, E.; Dahdouh-Guebas, F. Ethnobiology, socio-economics and management of mangrove forests: A review. *Aquat. Bot.* 2008, *89*, 220–236.

39. Webb, E.L.; Jachowski, N.R.A.; Phelps, J.; Friess, D.A.; Than, M.M.; Ziegler, A.D. Deforestation in the Ayeyarwady Delta and the conservation implications of an internationally-engaged Myanmar. *Glob. Environ. Chang.* 2014, *24*, 321–333.

40. Sitoe, A.A.; Mandlate, L.J.C.; Guedes, B.S. Biomass and Carbon Stocks of Sofala Bay Mangrove Forests. *Forests* 2014, *5*, 1967–1981.

41. Field, C.D. Impact of expected climate change on mangroves. *Hydrobiologia* 1995, *295*, 75–81.

42. Krauss, K.W.; Lovelock, C.E.; McKee, K.L.; Lopez-Hoffman, L.; Ewe, S.M.L.; Sousa, W.P. Environmental drivers in mangrove establishment and early development: A review. *Aquat. Bot.* 2008, *89*, 105–127.

43. Chan, H.T.; Baba, S. *Manual on Guidelines for Rehabilitation of Coastal Forests Damaged by Natural Hazards in the Asia-Pacific Region*; International Society for Mangrove Ecosystems (ISME) and International Tropical Timber Organization (ITTO): Okinawa, Japan, 2009; p. 66.

44. Suzuki, T.; Zijlema, M.; Burger, B.; Meijer, M.C.; Narayan, S. Wave dissipation by vegetation with layer schematization in SWAN. *Coast. Eng.* 2012, *59*, 64–71.

45. Di Nitto, D.; Neukermans, G.; Koedman, N.; Defever, H.; Pattyn, F.; Kairo, J.G.; Dahdouh-Guebas, F. Mangroves facing climate change: Landward migration potential in response to projected scenarios of sea level rise. *Biogeosciences*2014, *11*, 857–871.

46. Polidoro, B.A.; Carpenter, K.E.; Collins, L.; Duke, N.C.; Ellison, A.M.; Ellison, J.C.; Farnsworth, E.J.; Fernando, E.S.; Kathiresan, K.; Koedam, N.E.; *et al.* Mangrove extinction risk and geographic areas of global concern. *PLoS ONE* 2010, *5*, e10095.

47. Daru, B.H.; Yessoufou, K.; Mankga, L.T.; Davies, J. A Global Trend towards the Loss of Evolutionarily Unique Species in Mangrove Ecosystems. *PLoS ONE* 2013, *8*, 1–9.

48. Heumann, B.W. Satellite remote sensing of mangrove forests: Recent advances and future opportunities. *Prog. Phys. Geogr.* 2011, *35*, 87–108.

49. Manson, F.J.; Loneragan, N.R.; McLeod, I.M.; Kenyon, R.A. Assessing techniques for estimating the extent of mangroves: Topographic maps, aerial photographs, and Landsat TM images. *Mar. Freshw. Res.* 2001, *52*, 787–792.

50. Ruiz-Luna, A.; Berlanga-Robles, C.A. Land use, land cover changes and coastal lagoon surface reduction associated with urban growth in northwest Mexico. *Landsc. Ecol.* 2003, *18*, 159–171.

51. Cornejo, R.H.; Koedam, N.; Luna, A.R.; Troell, M.; Dahdouh-Guebas, F. Remote sensing and ethno-botanical assessment of the mangrove forest changes in the Navachiste-San Ignacio-Macapule lagoon complex, Sinaloa, Mexico. *Ecol. Soc.* 2005, *10*, 16.

52. Beland, M.; Goita, K.; Bonn, F.; Pham, T.T.H. Assessment of land-cover changes related to shrimp aquaculture using remote sensing data: A case study in the Giao Thuy District, Vietnam. *Int. J. Remote Sens.* 2006, *27*, 1491–1510.

53. Giri, C.; Pengra, B.; Zhu, Z.L.; Singh, A.; Tieszen, L.L. Monitoring mangrove forest dynamics of the Sundarbans in Bangladesh and India using multi-temporal satellite data from 1973 to 2000. *Estuar. Coast. Shelf Sci.* 2007, *73*, 91–100.

54. Giri, C.; Zhu, Z.; Tieszen, L.L.; Singh, A.; Gillette, S.; Kelmelis, J.A. Mangrove forest distributions and dynamics (1975–2005) of the tsunami-affected region of Asia. *J. Biogeogr.* 2008, *35*, 519–528.

55. Liu, K.; Li, X.; Shi, X.; Wang, S.G. Monitoring mangrove forest changes using remote sensing and GIS data with decision-tree learning. *Wetlands* 2008, *28*, 336–346.

56. Paling, E.I.; Kobryn, H.T.; Humphreys, G. Assessing the extent of mangrove change caused by Cyclone Vance in the eastern Exmouth Gulf, northwestern Australia. *Estuar. Coast. Shelf Sci.* 2008, *77*, 603–613.

57. Giri, C.; Ochieng, E.; Tieszen, L.L.; Zhu, Z.; Singh, A.; Loveland, T.; Masek, J.; Duke, N. Status and distribution of mangrove forests of the world using earth observation satellite data. *Glob. Ecol. Biogeogr.* 2011, *20*, 154–159.

58. Alsaaideh, B.; Al-Hanbali, A.; Tateishi, R.; KoBayashi, T.; Hoan, N.T. Mangrove forests mapping in the southern part of Japan using Landsat ETM+ with DEM. *J. Geogr. Inf. Syst.* 2013, *5*, 369–377.

59. Li, M.S.; Mao, L.J.; Shen, W.J.; Liu, S.Q.; Wei, A.I. Change and fragmentation trends of Zhanjiang mangrove forests in southern China using multi-temporal Landsat imagery (1977–2010). *Estuar. Coast. Shelf Sci.* 2013, *130*, 111–120.

60. Nguyen, H.; McAlpine, C.; Pullar, D.; Johansen, K.; Duke, N. The relationship of spatial-temporal changes in fringe mangrove extent and adjacent land-use: Case study of Ken Giang coast, Vietnam. *Ocean Coast. Manag.* 2013, *76*, 12–22.

61. Jones, T.G.; Ratsimba, H.R.; Ravaoarinorotsihoarana, L.; Glass, L.; Benson, L.; Teoh, M.; Carro, A.; Cripps, G.; Giri, C.; Gandhi, S.; *et al.* The Dynamics, Ecological Variability and Estimated Carbon Stocks of Mangroves in Mahajamba Bay, Madagascar. *J. Mar. Sci. Eng.* 2015, *3*, 793–820.

62. Jones, T.G.; Ratsimba, H.R.; Carro, A.; Ravaoarinorotsihoarana, L.; Glass, L.; Teoh, M.; Benson, L.; Cripps, G.; Giri, C.; Zafindrasilivonona, B.; *et al.* The mangroves of Ambanja and Ambaro Bays, northwest Madagascar: Historical dynamics, current status and deforestation mitigation strategy. In *Estuaries: A Lifeline of Ecosystem Services in Western Indian Ocean*; Diop, S., Scheren, P., Eds.; Springer International Publishing: Cham, Switzerland, 2016; in press.

63. Giri, S.; Mukhopadhyay, A.; Hazra, S.; Mukherjee, S.; Roy, D.; Ghosh, S.; Ghosh, T.; Mitra, D. A study on abundance and distribution of mangrove species in Indian Sundarban using remote sensing technique. *J. Coast. Conserv.* 2014, *18*, 359–367.

64. Jhonnerie, R.; Siregar, V.P.; Nababan, B.; Prasetyo, L.B.; Wouthuyzen, S. Random Forest Classification for Mangrove Land Cover Mapping Using Landsat 5 TM and Alos Palsar Imageries. *Procedia Environ. Sci.* 2015, *24*, 215–221.

65. Ramdani, F.; Rahman, S.; Setiani, P. Inexpensive Method to Assess mangroves Forest through the Use of Open Source Software and Data Available Freely in Public Domain. *J. Geogr. Inf. Syst.* 2015, *7*, 43–57.

66. Moffett, K.B.; Nardin, W.; Silvestri, S.; Wang, C.; Temmerman, S. multiple stable states and catastrophic shifts in coastal wetlands: Progress, challenges, and opportunities in validating theory using remote sensing and other methods. *Remote Sens.* 2015, *7*, 10184–10226.

67. Nardin, W.; Locatelli, S.; Pasquarella, V.; Rulli, M.C.; Woodcock, C.E.; Fagherazzi, S. Dynamics of a fringe mangrove forest detected by Landsat images in the Mekong delta, Vietnam. *Earth Surf. Process. Landf.* 2015. In press.

68. Giri, C.; Muhlhausen, J. Mangrove forest distributions and dynamics in Madagascar (1975–2005). *Sensors* 2008, *8*, 2104–2117.

69. Giri, C. *National-Level Mangrove Cover Data-Sets for 1990, 2000 and 2010*; United States Geological Survey: Sioux Falls, SD, USA, 2011.

70. Mayaux, P.; Gond, V.; Bartholome, E. A near-real time forest-cover map of Madagascar derived from SPOT-4 VEGETATION data. *Int. J. Remote Sens.* 2000, *21*, 3139–3144.

71. Critical Ecosystem Partnership Fund (CEPF). *Madagascar Vegetation Mapping Project*; CEPF: Arlington, VA, USA, 2007.

72. Harper, G.J.; Steininger, M.K.; Tucker, C.J.; Juhn, D.; Hawkins, F. Fifty years of deforestation and forest fragmentation in Madagascar. *Environ. Conserv.* 2007, *34*, 325–333.

73. Rasolofoharinoro, M.; Blasco, F.; Bellan, M.F.; Aizpuru, M.; Gauquelin, T.; Denis, J. A remote sensing based methodology for mangrove studies in Madagascar. *Int. J. Remote Sens.* 1998, *19*, 1873–1886.

74. Pasqualini, V.; Iltis, J.; Dessay, N.; Lointier, M.; Guelorget, O.; Polidori, L. Mangrove mapping in North-Western Madagascar using SPOT-XS and SIR-C radar data. *Hydrobiologia* 1999, *413*, 127–133.

75. Guillet, M.; Renou, E.; Robin, M.; Debaine, F.; Ratsivalaka, S. Suivi et analyse de l'evolution de la mangrove de Mahajamba (Nord-ouest de Madagascar). In Proceedings of the International Pluridisciplinary Conference, Lille, France, 16–18 January 2008.

76. Raharimahefa, T.; Kusky, T.M. Environmental monitoring of Bombetoka Bay and the Betsiboka Estuary, Madagascar, using multi-temporal satellite data. *J. Earth Sci.* 2010, *21*, 210–226.

77. Rakotomavo, A.; Fromard, F. Dynamics of mangrove forests in the Mangoky River delta, Madagascar, under the influence of natural and human factors. *For. Ecol. Manag.* 2010, *259*, 1161–1169.

78. *Law No. 97–017 on the Revision of Forestry Legislation*; Articles 2 & 41. Government of Madagascar (GoM): Antananarivo, Madagascar, 1997.

79. *Inter-Ministerial Order No. 4355/97 on the Definition of the Sensitize Areas*; Articles 2 & 3. Government of Madagascar (GoM): Antananarivo, Madagascar, 1997.

80. *Decree No. 98–781 Defining The Conditions of Application of the National Forest Policy*; Articles 34 & 35. Government of Madagascar (GoM): Antananarivo, Madagascar, 1998.

81. *Prohibiting Any Extractive Activity of Wood Resources in Sensitive Areas*; Order No. 12.704/2000 of 20 November, 2000. Government of Madagascar (GoM): Antananarivo, Madagascar, 2000.

82. *Decree No. 2005–849 of 13 December 2005 Revising the General Conditions of Application of Law No. 97–017 of 8 August 1997, Revising Forestry Legislation*; Government of Madagascar (GoM): Antananarivo, Madagascar, 2005; Chapter 4.

83. *Law No. 2008–013 on the Public Domain, Government Gazette of 2008*; Article 3(a). Government of Madagascar (GoM): Antananarivo, Madagascar, 2008.

84. Government of Madagascar (GoM). *Order No. 2055–2009 Creating Biologically Sensitive Shrimp Zones in Zone A in Ambaro Bay*; Government of Madagascar (GoM): Antananarivo, Madagascar, 2009.

85. *Prohibiting Any Exploitation of Mangrove Wood on the National Territory*; Inter-Ministerial Order No. 32.100/2014 of 24 October, 2014; Government of Madagascar (GoM): Antananarivo, Madagascar, 2014.

86. Jones, T.G. Shining a light on Madagascar's mangroves. *Madag. Conserv. Dev.* 2013, *8*, 4–6.
87. Hughes, R.H.; Hughes, J.S. Region 6: Madagascar. In *A Directory of African Wetlands*, 1st ed.; IUCN: Gland, Switzerland; Cambridge, UK; UNEP: Nairboi, Kenya; WCMC: Cambridge, UK, 1992; pp. 793–806.
88. Vences, M.; Andreone, F.; Glaw, F.; Raminosoa, N.; Randrianirina, J.E.; Vieites, D.R. Amphibians and reptiles of the Ankaratra Massif: Reproductive diversity, biogeography and conservation of a montane fauna in Madagascar. *Ital. J. Zool.* 2002, *69*, 263–284.
89. Rasofolo, M.V. Use of mangroves by traditional fishermen in Madagascar. *Mangroves Salt Marshes* 1997, *1*, 243–253.
90. Roy, R.; Purkis, S.; Dunn, S. *Mapping Velondriake: The Application of Bathymetric and Marine Habitat Mapping to Support Conservation Planning, Southwest Madagascar*; Blue Ventures Internal Report; Blue Ventures: London, UK, 2009.
91. Tomlinson, P.B. *The Botany of Mangroves*; Cambridge University Press: Melbourne, Australia, 1986.
92. Chavez, P.S. Image-based atmospheric corrections: Revisited and improved. *Photogramm. Eng. Remote Sens.* 1996, *62*, 1025–1036.
93. Kirui, K.B.; Kairo, J.G.; Bosire, J.; Viergever, K.M.; Rudra, S.; Huxham, M.; Briers, R.A. Mapping of mangrove forest land cover change along the Kenya coastline using Landsat imagery. *Ocean Coast. Manag.* 2013, *83*, 19–24.
94. Simard, M.; Zhang, K.Q.; Rivera-Monroy, V.H.; Ross, M.S.; Ruiz, P.L.; Castaneda-Moya, E.; Twilley, R.R.; Rodriguez, E. Mapping height and biomass of mangrove forests in Everglades National Park with SRTM elevation data. *Photogramm. Eng. Remote Sens.* 2006, *72*, 299–311.
95. Fatoyinbo, T.E.; Simard, M.; Washington-Allen, R.A.; Shugart, H.H. Landscape-scale extent, height, biomass, and carbon estimation of Mozambique's mangrove forests with Landsat ETM+ and Shuttle Radar Topography Mission elevation data. *J. Geophys. Res. Biogeosci.* 2008, *113*.
96. Simard, M.; Rivera-Monroy, V.H.; Mancera-Pineda, J.E.; Castaneda-Moya, E.; Twilley, R.R. A systematic method for 3D mapping of mangrove forests based on Shuttle Radar Topography Mission elevation data, ICEsat/GLAS waveforms and field data: Application to Cienaga Grande de Santa Marta, Colombia. *Remote Sens. Environ.* 2008, *112*, 2131–2144.
97. Fatoyinbo, T.E.; Simard, M. Height and biomass of mangroves in Africa from ICESat/GLAS and SRTM. *Int. J. Remote Sens.* 2013, *34*, 668–681.
98. Aslan, A.; Rahman, A.F.; Warren, M.; Robeson, S.M.; Darusman, T. Combined use of active and passive remote sensing for mapping distribution and biomass of coastal mangroves. In Proceeding of American Geophysical Union Fall Meeting, San Francisco, CA, USA, 15–19 December 2014.
99. Sinclair, T.T.; Hoffer, R.M.; Schreiber, M.M. Reflectance and internal structure of leaves from several crops during a growing season. *Agron. J.* 1971, *63*, 864–868.
100. Elvidge, C.D. Visible and near-infrared reflectance characteristics of dry plant materials. *Int. J. Remote Sens.* 1990, *11*, 1775–1795.
101. Bhattarai, B.; Giri, C. Assessment of mangrove forests in the Pacific region using Landsat imagery. *J. Appl. Remote Sens.*2011, *5*, 053509.
102. Long, J.B.; Giri, C. Mapping the Philippines' mangrove forests using Landsat imagery. *Sensors* 2011, *11*, 2972–2981.
103. Aschbacher, J.; Ofren, R.; Delsol, J.P.; Suselo, T.B.; Vibulsresth, S.; Charrupat, T. An integrated comparative approach to mangrove vegetation mapping using advanced remote sensing and GIS technologies: Preliminary results. *Hydrologica* 1995, *295*, 285–295.
104. Gao, J.A. Hybrid method toward accurate mapping of mangroves in a marginal habitat from SPOT Multispectral data. *Int. J. Remote Sens.* 1998, *19*, 1887–1899.
105. Green, E.P.; Clark, C.D.; Mumby, P.J.; Edwards, A.J.; Ellis, A.C. Remote sensing techniques for mangrove mapping. *Int. J. Remote Sens.* 1998, *19*, 935–956.
106. Gao, J.A. Comparative study on spatial and spectral resolutions of satellite data in mapping mangrove forests. *Int. J. Remote Sens.* 1999, *20*, 2823–2833.
107. Saito, H.; Bellan, M.F.; Al-Habshi, A.; Aizpuru, M.; Blasco, F. Mangrove research and coastal ecosystem studies with SPOT-4 HRVIR and TERRA ASTER in Arabian Gulf. *Int. J. Remote Sens.* 2003, *24*, 4073–4092.
108. Tong, P.H.; Auda, Y.; Populus, J.; Aizpura, M.; Habshi, A.A.; Blasco, F. Assessment from space of mangroves evolution in the Mekong Delta, in relation to extensive shrimp farming. *Int. J. Remote Sens.* 2004, *25*, 4795–4812.
109. Jensen, L.S.; Mueller, T.; Tate, K.R.; Ross, D.J.; Magid, J.; Nielsen, N.E. Soil surface CO2 flux as an index of soil respiration *in situ*: A comparison of two chamber methods. *Soil Biol. Biochem.* 1996, *28*, 1297–1306.
110. Comley, B.W.T.; McGuinness, K.A. Above- and below-ground biomass, and allometry, of four common northern Australian mangroves. *Aust. J. Bot.* 2005, *53*, 431–436.
111. Curran, P.J. Remote sensing of foliar chemistry. *Remote Sens. Environ.* 1989, *30*, 271–278.
112. Komiyama, A.; Poungparn, S.; Kato, S. Common allometric equations for estimate the tree weight of mangroves. *J. Trop. Ecol.* 2005, *21*, 471–477.
113. Schumacher, B. *Methods for the Determination of Total Organic Carbon (TOC) in Soils and Sediments*; Ecological Risk Assessment Support Center, Office of Research and Development, US Environmental Protection Agency: Washington, DC, USA, 2002.
114. De Vos, B.; Letterns, S.; Muys, B.; Deckers, J.A. Walkley-Black analysis of forest soil organic carbon: Recovery, limitations and uncertainty. *Soil Use Manag.* 2007, *23*, 221–229.
115. Meersmans, J.; van Wesemael, B.; van Molle, M. Determining soil organic carbon for agricultural soils:

A comparison between the Walkley & Black and the dry combustion methods (North Belgium). *Soil Use Manag.* 2009, *25*, 346–353.

116. Ray, R.; Ganguly, D.; Chowdhury, C.; Dey, M.; Das, S.; Dutta, M.K.; Mandal, S.K.; Majumder, N.; De, T.K.; Mukhopadhyay, S.K.; *et al.* Carbon sequestration and annual increase of carbon stock in a mangrove forest. *Atmos. Environ.* 2011, *45*, 5016–5024.
117. Chen, L.; Zeng, X.; Tam, N.F.Y.; Lu, W.; Luo, Z.; Du, X.; Wang, J. Comparing carbon sequestration and stand structure of monoculture and mixed mangrove plantations of *Sonneratia caseolaris* and *S. apetala* in Southern China. *For. Ecol. Manag.* 2012, *284*, 222–229.
118. Donato, D.C.; Kauffman, J.B.; Mackenzie, R.A.; Ainsworth, A.; Pfleeger, A.Z. Whole-island carbon stocks in the tropical Pacific: Implications for mangrove conservation and upland restoration. *J. Environ. Manag.* 2012, *97*, 89–96.
119. Fujimoto, K.; Imaya, A.; Tabuchi, R.; Kuramoto, S.; Utsugi, H.; Murofushi, T. Belowground C storage of Micronesian mangrove forests. *Ecol. Res.* 1999, *14*, 409–413.
120. Jardine, S.L.; Siikamäki, J.V. A global predictive model of carbon in mangrove soils. *Environ. Res. Lett.* 2014, *9*, 104013.
121. Liu, H.; Ren, H.; Hui, D.; Wang, W.; Liao, B.; Cao, Q. Carbon stocks and potential carbon storage in the mangrove forests of China. *J. Environ. Manag.* 2014, *133*, 86–93.

CHAPTER 8

MAPPING INVASIVE TAMARISK (TAMARIX): A COMPARISON OF SINGLE-SCENE AND TIME-SERIES ANALYSES OF REMOTELY SENSED DATA

Paul H. Evangelista, Thomas J. Stohlgren, Jeffrey T. Morisette and Sunil Kumar

Keywords: Landsat 7 ETM+, Maxent, phenology, remote sensing, spatial modeling, Tamarix, time-series

INTRODUCTION

Mapping invasive plant species has become a high priority for resource managers and researchers across the United States. Ground surveys are still commonly used for most mapping projects despite intensive labor requirements, associated high economic costs, and incomplete coverage of the landscape [1,2]. Improved methods to accurately determine the current distribution of invaders are required to better assess their environmental impacts, formulate effective control strategies, and forecast potential spread. Remote sensing has played an important, but limited, role in detecting and mapping invasive plants [3,4,5]. It is more commonly applied to mapping weeds in agricultural environments where species richness and diversity are minimal [6,7]. Detecting a specific plant species in forests, rangelands, riparian areas and natural landscapes using remote sensing techniques has proved to be a greater challenge. Large-scale infestations, where invaders are clearly the dominant species and environmental heterogeneity is reduced, tend to be easier to detect remotely [3,8,9].

One invasive plant that is especially problematic in the western U.S. is tamarisk (*Tamarix* spp.). Tamarisk is a small shrub-like tree that tends to grow in large monotypic stands along riparian corridors. A native to Eurasia and North Africa, tamarisk was introduced to North America in the late 1800s [10,11]. There are eleven species of tamarisk and at least three hybrids that have since become well established in the Americas, from Canada to Argentina [12]. Tamarisk's impacts to native ecosystems in the U.S. include the desiccation of water tables, displacement of native plant communities, alteration of soil chemistry and ecosystem processes, and loss of critical wildlife habitat [13,14,15]. Infestations often begin discretely among species-rich riparian ecosystems before abruptly overwhelming and displacing competitor species. Once tamarisk establishes dominance, control and restoration efforts can be extremely labor intensive and costly. Detecting tamarisk in the early stages of infestation and mapping its distribution are essential to resource management and stewardship. Remote sensing of tamarisk distributions has been only marginally effective [16,17,18,19]. New airborne and satellite sensors promise to improve our ability to spectrally distinguish tamarisk from other species, but the techniques remain inadequate and are not yet economically practical.

The detection of invasive plants using remote sensing may be improved if the target species has phenological attributes that are distinctive from native vegetation. For example, leafy spurge (*Euphorbia esula* L.) has yellow-green inflorescences that are spectrally unique when compared to associated flora [4,20,21]. Similarly, yellow starthistle (*Centaurea solstitialis*), tamarisk (*Tamarix* spp.), yellow hawkweed (*Hieracium pratense*), oxeye daisy (*Chrysanthemum leucanthemum*), and Chinese tallow (*Sapium sebiferum*) also have distinctive colorations that can facilitate remote sensing [22,23,24]. Other species have been detected by their extended growing periods. Broom snakeweed (*Gutierrezia sarothrae*), a perennial sub-shrub, has been remotely sensed during its early-season

Citation: Paul H. Evangelista, Thomas J. Stohlgren, Jeffrey T. Morisette and Sunil Kumar, Mapping Invasive Tamarisk (Tamarix): A Comparison of Single-Scene and Time-Series Analyses of Remotely Sensed Data, doi:10.3390/rs1030519.

greening [25]. Cheatgrass (*Bromus tectorum*), an invasive annual grass, has also been successfully detected because it germinates in winter months prior to most native grasses [26]. The distinctiveness of any phenological attribute can vary widely with regional climate, latitudinal gradients, and species richness within an ecosystem. As a result, the timing of acquiring remotely sensed data is critical and can be difficult to predict.

Time-series analyses of remotely sensed data have been widely used for mapping land cover and ecosystem types and are increasingly being used for detecting broad-scale invasions [27,28] and monitoring the impacts of mitigation treatments [29,30]. A few studies using time-series analyses have reported success in identifying cheatgrass. Bradley and Mustard [26] demonstrated how inter-annual data collected from Landsat and Advanced Very High Resolution Radiometer (AVHRR) can detect cheatgrass responses to precipitation. Peterson [31] was able to distinguish cheatgrass from other vegetation by using scenes from Landsat 7 ETM+ on two different dates within a single year. In both cases, the researchers were able to exploit subtle phenological differences (i.e., extended growing season, rapid response) between the invaders and associated native flora within a growing season. To determine the optimal time of the year to conduct remote sensing surveys, Everitt and Deloach [22] used a time-series of conventional color and color infrared aerial photographs for tamarisk and native riparian vegetation. From their study sites in Texas and Arizona, they found that tamarisk could best be identified in late fall and early winter months when foliage turned a yellow-orange color before dropping.

Remote sensing of tamarisk has had limited success, and conventional methods (e.g., supervised and unsupervised classification) have not proved to be reliable. Ge *et al.* [32] analyzed color aerial photographs at 1-m^2 resolution using a texture analysis for tamarisk in northern California. The photographs were acquired in April and the mean grey-level values were calculated for 60 pixels representing eight cover types. They found that color alone could not distinguish tamarisk from associated vegetation; however, the use of textural classifiers greatly improved separability of cover types. In southern California, Hamada *et al.* [18] used discriminant analyses of presence/absence data and hierarchical clustering with hyperspectral imagery collected in October. Overall accuracy of their research varied by scene and minimum patch size, and results tended to over classify tamarisk distribution. Akasheh *et al.* [33] used an iterative classification procedure to map riparian vegetation with high-resolution multi-spectral airborne sensors in July on the Rio Grande River, New Mexico. The vegetation in their study site was largely dominated by four riparian species: tamarisk, cottonwood, willow, and Russian olive. Using 24 validation plots for tamarisk, they were able to achieve 86% classification accuracy.

These studies demonstrate an evolution of remote sensing and image processing for detecting tamarisk and other invasive species. The development of new airborne and satellite sensors and platforms, coupled with advanced statistical software, geographic information systems (GIS), and predictive models, give researchers a variety of tools to detect and predict the distribution of invasive species [34,35]. In this study, we explored the application of maximum entropy modeling with remotely sensed data to map the distribution of tamarisk while incorporating strategies that have been previously proven to be effective in vegetation mapping. Because conventional methods of remote sensing have limited applications for detecting tamarisk, our analyses were conducted using the Maxent software (v3.2.1; www.cs.princeton.edu/~schapire/maxent/), which uses presence points to predict the potential range and habitat distribution of a species [36,37]. In several recent studies, Maxent has been found to be especially useful for mapping invasive species [38,39] and ranked high when compared to other models for predicting tamarisk distributions [35]. We tested six satellite scenes and derived vegetation indices from different months of the growing season to detect tamarisk using single-scene and time-series analyses. Our objectives were to compare each analysis and determine which month of the growing season was the best time for detecting tamarisk. Lastly, we examined the effectiveness of several vegetation indices derived from remote sensing data that have proven useful in other remote sensing studies.

METHODS

Study Area

This study was conducted in the lower Arkansas River in Southwestern Colorado. The Arkansas River is the sixth longest river in the continental U.S. Its headwaters begin in the Rocky Mountains of central Colorado and it flows 2,364 km east through Kansas, Oklahoma and Arkansas before emptying into the Mississippi River. In Colorado, the river drops 1,400 m in elevation from its origin to the edge of the Great Plains near the city of Pueblo [40]. From Pueblo, the river flows approximately 400 km to the Kansas state line and sustains a wide belt of irrigated agriculture through a series of ditches and channels. Elevations and mean annual precipitation range from 1,417 m and 29.7 cm at Pueblo to 1,021 m and 38.3 cm at the state line [41]. Tamarisk was first reported in this region in 1913, and observers noted the species' rapid spread as early as 1921 [41]. Today, tamarisk infestation along the Arkansas River between Pueblo and the state line is estimated to be more than 120 km^2 resulting in the estimated loss of 47,000 acre-feet of water annually [42].

Our study area was defined by the boundaries of the Landsat7 Enhanced Thematic Mapper Plus (ETM+) scene (Path 32, Row 34) used in our analyses. The study area includes approximately 175 km of the Arkansas River in southeastern Colorado between the town of Avondale and the City of Lamar (Figure 1). Also included in the scenes are John Martin Reservoir, the lower sections of the Purgatoire and Apishapa rivers, developed agricultural lands, and a significant area of semi-arid rangeland that extends into New Mexico.

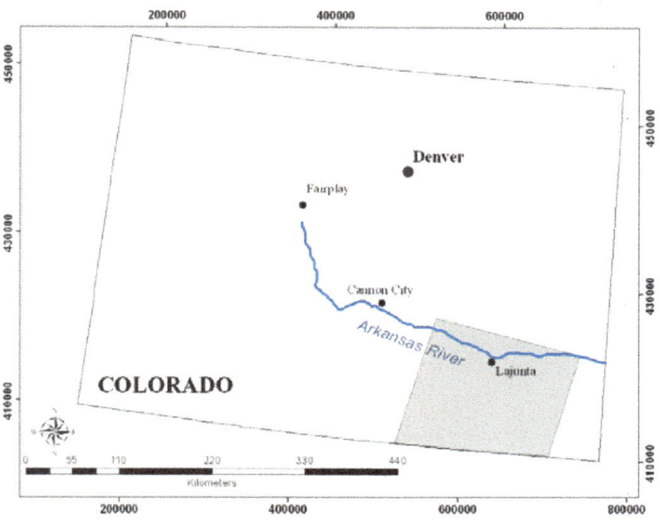

Figure 1: Map of the Arkansas River in Colorado. The study area is highlighted in grey.

Field Data

This study relied on the results of an intensive inventory and mapping effort by the Tamarisk Coalition (www.tamariskcoalition.org) and other groups completed in 2007. Inventory and mapping were coordinated with the U.S. Geological Survey's (USGS) efforts to establish a national on-line database conforming to the weed mapping standards developed by the North American Weed Management Association (http://www.nawma.org/). This study utilized existing aerial photography, satellite imagery, and supplemental information available from multiple resource management agencies, research institutions and non-government organizations [42]. Stands of tamarisk were inventoried by field crews to confirm infestation density, maturity, height, accessibility, presence of native species, and other site characteristics. Stands of tamarisk were inventoried using global positioning systems (GPS) handheld units and transformed into spatially referenced geographical information systems (GIS) polygons with the aid of aerial photography. Over 1,633 miles of the Arkansas River, its tributaries, wetlands, and canals and dry land stands in the Arkansas watershed were surveyed [42], including the area in Colorado occupied by our imagery. These data are now stored and available on the National Institute of Invasive Species Science website (www.niiss.org). From these polygon data, we randomly generated 400 presence points from tamarisk polygons that had a percent basal cover >50%. Of these, 250 points were used for training (62.5%) the model, while 150 points were reserved for testing (37.5%). Some of our model testing required the use of absence points. An additional 150 tamarisk absence points were randomly generated using GIS for areas outside tamarisk polygons. These areas represent dry upland, agriculture, rangeland, and riparian land cover types.

Remotely Sensed Data

We used six Landsat 7 ETM+ scenes for our analyses. Each scene and derived vegetation indices were processed using ERDAS Imagine v9.0 [43] and ArcGIS 9.1 [44] software. The scenes, selected for their seasonal variation, were acquired on April 16, 2000; May 11, 2003; June 23, 2001; August 12, 2002; September 7, 2000; and October 23, 1999. From each scene, bands 1-5 and band 7 were used in our analyses. Additionally, we generated several vegetation indices from each scene that are commonly used for estimating vegetation and land-cover features. Normalized difference vegetation index (NDVI) is a non-linear transformation of the ratio between the visible (red) and near-infrared bands (NIR; [45]). The NDVI is commonly used to measure vegetation canopy characteristics such as biomass, leaf area index, and canopy cover [46,47,48]). We calculated NDVI using the following expression:

$$\text{NDVI}=(\text{band } 4 - \text{band } 3)/(\text{band } 4 + \text{band } 3) \quad (1)$$

The Ratio Vegetation Index (RVI) was calculated by dividing near infrared (band 4) by visible red (band 3) reflectance values [49]. The RVI and NDVI are very similar in that they are a measure of the slope of the line between the origin of red-NIR space and the red-NIR value of each pixel. The major difference between the two is the range of values given by the calculations. Some studies have used RVI and NDVI together [50], while other studies have elected to use one or the other [7].

Tasselled Cap transformations were conducted for each scene using the coefficients reported by Huang *et al.* [51]. Originally developed for understanding changes in crop development, tasselled cap transformations are weighted composites of the six Landsat bands into three orthogonal bands that have been useful in measuring soil brightness (tasselled cap, band 1), vegetation greenness (tasselled cap, band 2), and soil/vegetation wetness (tasselled cap, band 3, [52]). These transformations have been described as a guided and scaled principal components analysis and have been shown to be useful in identifying forest attributes such as species composition, age class, and structure [43,53,54].

We also calculated the Soil-Adjusted Vegetation Index (SAVI; [55]), a ratio-based index developed to minimize the effects of the soil background. The formula used for calculating SAVI is:

$$\text{SAVI} = [(\text{band } 4 - \text{band } 3) \times (1 + L)/(\text{band } 4 + \text{band } 3 + L)] \quad (2)$$

where L is a correction factor ranging from 0 (high vegetation cover) to 1 (low vegetation cover). Each month had twelve potential predictor variables used for single-scene analyses, while the time-series analysis used a total of seventy-two potential predictor variables.

Data Analyses

We conducted our analyses using the Maxent software v.3.2 (www.cs.princeton.edu/~schapire/maxent/), which is a general-purpose method for estimating probability of distributions based on the principle of maximum entropy [36,37]. Maxent uses presence-only data to define known conditions within the parameters of the independent variables to predict a species' distribution and excludes all conditions that are unfounded or undefined. The model is nonlinear, nonparametric, and not sensitive to multicollinearity. Besides having several evaluation features built into the program, Maxent also provides the percent contribution of each predictive variable. Several recent studies have found the Maxent model to perform as well, or better, when compared to other modeling methods [35,39,56].

Each monthly scene, and the associated vegetation indices, were analyzed and tested independently (six models); while the time-series analyses were conducted using two modeling procedures. The first time-series model used all 72 predictor variables generated from the six scenes. From these results, we restricted the models to only those variables that had a predictive contribution >1.0% (n = 17) and used them as predictor variables for our second time-series model. The 17 variables were tested for cross-correlations using SYSTAT (version 12; SYSTAT Software, Port Richmond, California, USA). For variables that were highly correlated (Pearson correlation coefficient >0.80), we removed the ones that had the least predictive contribution in first-time series analysis model. This further reduced our number of variables to seven potential predictors.

We tested the models with threshold-dependent and threshold-independent measures using features available with Schroder's ROC/AUC software (http://brandenburg.geoecology.uni-potsdam.de/users/ schroeder/download.html). The ROC/AUC software was specifically developed for assessing the predictive performance of habitat models and requires presence and absence data. Threshold-dependent evaluations, using the ROC/AUC software, were measured by specificity, sensitivity, and Cohen's maximized Kappa [57]. Specificity is the proportion of true-positive and false-positive absences, and sensitivity is the proportion of true-positives and false-positive presences. The maximized Kappa statistic (K) measures the proportion of correctly classified points (i.e., presence, absence) after accounting for the probability of chance agreement. Kappa statistic values range from –1 to +1, where +1 would be perfect agreement, and any values less than 0 would indicate a performance no better than random [57,58]. Landis and Koch [59] ranked analysis performances as poor when Kappa values are <0.40, good when the Kappa values range from 0.40 to 0.75, and excellent when Kappa values are >0.75.

The threshold-independent evaluation required a Receiver Operating Characteristic (ROC) curve, where 'sensitivity' is plotted against '1-specificity' for all possible thresholds [60]. From the ROC analysis, the Area under the ROC Curve (AUC) is calculated using presence and absence observations to measure the probability that a random, positive point falls within the predicted range of occurrence, and a random negative point falls outside [61]. The AUC value can vary from 0.5 (no better than random) to 1.0 (perfect discrimination, [62]). AUC evaluations for each model are presented in the results.

RESULTS AND DISCUSSION

All the models generally performed well, further highlighting the applicability of remote sensing and vegetation indices for detecting tamarisk. The best model was the first time-series analysis that used all 72 variables (Figure 2). The AUC value for this model was 0.96, while the percent of correct predictions was 0.90 and kappa statistic was 0.79 (Table 1). The next best models were the second time-series analyses with the reduced number of variables, and the October single-scene analysis. The AUC values for these models were 0.93 and 0.89, respectively. The percent of correct predictions and kappa statistics for the second time-series analyses and October single-scene analyses were 0.84 and 0.85, and 0.69 and 0.71, respectively. The June, August, and September single-scene analyses had slightly lower but similar evaluations. Generally, the results from the single-scene analyses improved toward the later part of the growing season and into the fall months when most native plants go into dormancy.

Table 1: Tamarisk model evaluations of single-scene and time-series analyses of Landsat 7 scenes and associated vegetation indices generated using the ROC/AUC Calculator

Scene analysis	Num. variables	AUC	Sensitivity	Specificity	% Correct	Kappa
April	12	0.89	0.75	0.89	0.82	0.64
May	12	0.88	0.83	0.84	0.83	0.67
June	12	0.92	0.93	0.76	0.84	0.69
August	12	0.91	0.91	0.79	0.85	0.70
September	12	0.91	0.83	0.89	0.86	0.71
October	12	0.89	0.77	0.94	0.85	0.71
Time-series[1]	72	0.96	0.93	0.86	0.90	0.79
Time-series[2]	7	0.93	0.85	0.84	0.84	0.69

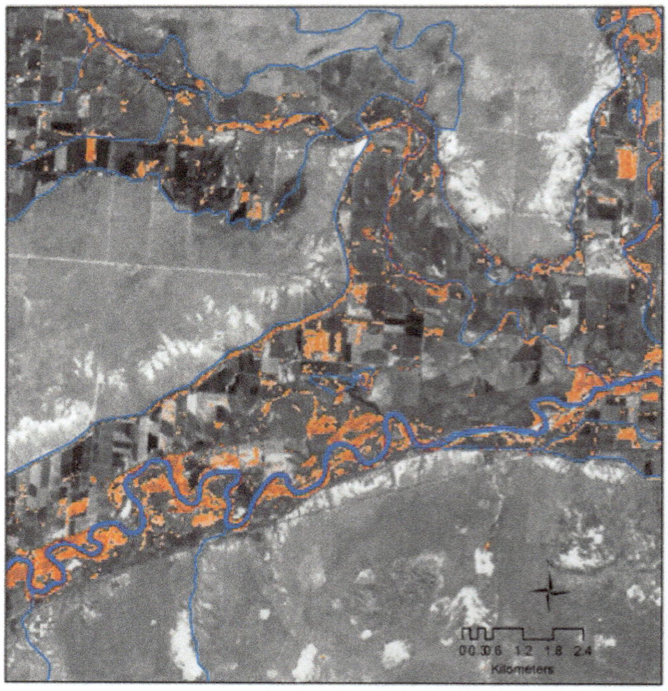

Figure 2: An enlarged view of tamarisk detected along the Arkansas River and irrigation ditches near the town of Riverdale in southeastern Colorado. The results shown here are from a time-series analysis that used 72 remotely sensed data sets from Landsat 7 ETM+. Tamarisk infestations are shown from moderate (orange) to high (red).

The best predictor variables for the first time-series analyses were the June tasselled cap wetness (25.8%), September tasselled cap wetness (16.4%), and October wetness (11.6%, Table 2). Similarly, the best predictors for the second time-series analyses were the June tasselled cap wetness (63.1%), April NDVI (9.7%), and October band 3 (7.8%). It should be noted that the September tassled cap band 3 was found to be highly correlated with the June tasselled cap wetness, thus it was omitted from the second time-series analyses. With the exception of the April NDVI, the best predictors for both time-series analyses were from the months that performed the best of the six single-scene analyses. Of the eight models, seven had at least one tasselled cap transformation as one of the top predictors.

Table 2: The predictive contributions of the top three variables for each model generated from Maxent

Scene Analysis	Variable	Contribution (%)
April	band 7	25.5
	band 4	20.6
	NDVI	17.8
May	band 7	37.7
	tasselled cap wetness	31.9
	band 1	9.9
June	tasselled cap wetness	78.5
	band 1	8.6
	band 4	5.5
August	tasselled cap wetness	59
	band 4	13.6
	band 1	9.6
September	tasselled cap wetness	42.2
	band 5	16.1
	band 7	14.3
October	band 3	30.2
	tasselled cap wetness	21.4
	band 7	17.9
Time-series[1]	(June) tasselled cap wetness	25.8
	(Sept) tasselled cap wetness	16.4
	(Oct) band 3	11.6
Time-series[2]	(June) tasselled cap wetness	63.1
	(April) NDVI	9.7
	(Oct) band 3	7.8

Our results suggest that the time-series analyses can better distinguish phenological differences between tamarisk and native flora than a single-scene analysis (Table 1). Spectral data from the months of June, September, and October consistently had the greatest capability for detecting tamarisk in all of our models. We can speculate that the peak in tamarisk green-up and its purple-pink flowers contribute to the spectral uniqueness during June, while its extended growing season and the yellow foliage late in the year are conspicuous in September and October. Collectively, data from these months produced the strongest results in our time-series analysis; we are encouraged to see that they each performed exceptionally well with the single-scene analyses. Our findings are also in agreement with results from previous remote sensing studies on tamarisk [22,33]. Although spectral data from these months performed best in our study area, they may not necessarily be the best candidates in other ecosystems, geographic regions, or spatial scales.

Vegetation indices used in our analyses made considerable predictive contributions to our final results. Most notable were the tasselled cap transformations for soil/vegetation wetness (tasselled cap band 3). The tasselled cap wetness index has been shown in other studies to be reliable for detecting change in forest structure and biomass using single-scene and time-series analyses [54,63,64]. Tamarisk biomass in our study site is quite extensive and significantly higher than native vegetation [35]. Further study may be required to test the effectiveness of the tasselled cap wetness index on low densities of tamarisk; however, our results indicate the index may perform well with the remote sensing of large tamarisk stands or when used in a time-series analysis.

We were also encouraged by the performance of the Maxent model in analyzing remote sensing data. We are confident that our model results could be improved if we integrated additional geospatial variables that characterize the physical landscape (e.g., distance from water, slope), but we elected to analyze monthly scenes and vegetation indices exclusively to better identify temporal trends with tamarisk's phenology in relation to other vegetation.

CONCLUSIONS

Our study revealed several important factors that may significantly improve the use of remotely sensed data for detecting tamarisk and other invasive species. Our time-series analysis using all 72 variables performed best, while our second time-series analysis using the seven best predictors that were not correlated had only a slight reduction in model performance. This would suggest that variables that have lower predictive contributions can still improve results when used collectively. We have identified at least three different times during the growing season when phenological attributes of tamarisk can help distinguish the species from native vegetation. We have also demonstrated that phenological differences may be better detected using a time-series analysis than a single-scene analysis. Biomass indicators, such as the tasselled cap wetness index, may prove useful for remotely sensing large tamarisk infestations or for detecting landscape change in a time-series analysis.

Our study also demonstrates that the Maxent model may prove to be a powerful new tool for analyzing remotely sensed data. Originally designed as a model for predicting species distribution and potential habitat, Maxent has several features that support remote sensing analysis. Maxent is not sensitive to multicollinearity, which is commonly associated with remotely sensed data sets. It also can be used with presence-only data and requires no additional partitioning of background data.

Lastly, Maxent easily integrates other geospatial data types (elevation, distance from water) that may enhance prediction and detection efforts.

There are several caveats in regard to this study that need to be considered. First, tamarisk presence points used to train and test our models were randomly generated indirectly from GIS polygons derived from field surveys. Similarly, absence points required for some of our evaluations were randomly generated in areas where tamarisk was not reported to occur. These methods may increase the risk of sampling error compared to other sampling designs. Although we are confident in the data quality from the Tamarisk Coalition's survey efforts, we must recognize that this approach may not be the ideal sampling design for training Maxent models and remotely sensed data. For future efforts, we suggest that training and testing data be acquired by direct field observations with geographical coordinates collected on site. Specifically, each presence or absence point used for the analysis should be confirmed and recorded in the field to minimize sampling error and increase data quality.

Second, the training data selected for the analysis required that tamarisk have a minimum basal area of 50%, which we believed was an appropriate threshold for our tests given the resolution of Landsat 7 ETM+ scenes (i.e., 30 m^2), the extent of our study area, and the uncertainty related to the field data. Thus, the presence data used to train our models were derived from areas where tamarisk was the dominant vegetation type. This presents obvious limitations for mapping tamarisk distributions and detecting early infestations. Our methods, however, can be easily modified to fit different management objectives and used with different types and formats of remotely sensed data. For example, smaller study areas may be better suited for high-resolution imagery, which would enhance the detection of individual plants and low-density infestations. This approach may prove to be too costly for resource management agencies that require surveys across large landscapes. Despite the fact that our results are only applicable where tamarisk is dominant, they are still valuable for planning mitigation efforts. For example, the use of bio-control agents and aerial spraying of herbicides is most economically feasible on large areas where tamarisk is dominant. Regardless, we must recognize that remote sensing of individual plants inhabiting diverse ecosystems still presents many challenges and limitations.

Despite these concerns, we believe that the methods tested in this research have produced promising results that will enhance our ability to remotely sense tamarisk and other invasive plant species. We have also found the Maxent model to perform positively when used with remotely sensed data. Further refinement of our methods is needed, including tests that employ different sensors across multiple spatial scales.

ACKNOWLEDGEMENTS

We would like to acknowledge the Tamarisk Coalition for the data contributions that made this study possible. We would also like to thank the Natural Resource Ecology Laboratory at Colorado State University and U.S. Geological Survey Fort Collins Science Center for analyses support; NASA Goddard Space Flight Center for providing remote sensing data; and the Colorado State Forest Service and Colorado Division of Wildlife for field assistance and logistical support. Any use of trade, product, or firm names is for descriptive purposes only and does not imply endorsement by the U.S. Government. Jim Graham and Alycia Crall were helpful for editorial comments on earlier versions of the manuscript. Finally we would like to thank N. Ament, T. Carlson, A. Randell, C. Tate, and S. VanLandingham. To all we are grateful.

REFERENCES AND NOTES

1. Crosier, C.S.; Stohlgren, T.J. Improving biodiversity knowledge with data set synergy: a case study of nonnative plants in Colorado. *Weed Technol.* 2004, *18*, 1441–1444.
2. Dewey, S.A.; Anderson, K.A. Distinct roles of surveys, inventories, and monitoring in adaptive weed management. *Weed Technol.* 2004, *18*, 1449–1452.
3. Anderson, G.L.; Everitt, J.H.; Richardon, A.J.; Escobar, D.E. Using satellite data to map false broomweed (*Ericameria austrotexana*) infestations on south Texas rangelands. *Weed Technol.* 1993, *7*, 865–871.
4. Everitt, J.H.; Anderson, G.L.; Escobar, D.E.; Davis, M.R.; Spencer, N.R.; Andrascik, R.J. Use of remote sensing for detecting and mapping leafy spurge (*Euphorbia esula*). *Weed Technol.* 1995, *9*, 599–609.
5. Rowlinson, L.C.; Summerton, M.; Ahmed, F. Comparison of remote sensing data sources and techniques for identifying and classifying alien invasive vegetation in riparian zones. *Water SA* 1999, *25*, 497–500.
6. Medlin, C.R.; Shaw, D.R.; Gerard, P.D.; LaMastus, F.E. Using remote sensing to detect weed infestations in Glycine max. *Weed Sci.* 2000, *48*, 393–398.
7. Lopez-Granados, F.; Garcia-Torres, L. Using remote sensing for identification of late-season grass weed patches in wheat. *Weed Sci.* 2006, *54*, 346–353.
8. Lass, W.L.; Prather, T.S.; Glenn, N.F.; Weber, K.T.; Mundt, J.T.; Pettingill, J. A review of remote sensing of invasive weeds and example of the early detection of spotted knapweed (*Centaurea maculosa*) and babysbreath (*Gypsophila paniculata*) with a hyperspectral sensor. *Weed Sci.* 2005, *53*, 242–251.
9. Laba, M.; Tsai, F.; Ogurcak, D.; Smith, S.; Richmond, M.E. Field determination of optimal dates for the discrimination of invasive wetland plant species using derivative spectral analysis. *Photogramm. Eng. Remote Sensing* 2005, *71*, 603–611.
10. Carleton, M.A. Adaptation of the tamarisk for dry lands. *Science* 1914, *39*, 692–694.
11. DiTomaso, J.M. Impact, biology, and ecology of saltcedar (*Tamarix* spp.) in the southwestern United States. *Weed Technol.* 1998, *12*, 326–336.
12. U.S. Department of Agriculture (USDA) Plants Database. Natural Resources Conservation Service, 2008. Available online: http://plants.usda.gov/ (accessed on August 28, 2009).
13. Christiansen, E.M. The rate of naturalization of Tamarix in Utah. *Am. Midl. Nat.* 1962, *68*, 51–57.
14. Robinson, T.W. Introduction, spread, and aerial extent of saldcedar (*Tamarix*) in the western states. *Professional*

Paper 491-A; U.S. Department of the Interior, Geological Survey: Washington, DC, USA, 1965; p. 11.

15. Harris, D.R. Recent plant invasions in the arid and semi-arid Southwest of the United States. *Ann. Assn. Amer. Geogr.* 1966, *56*, 408–422.
16. Lass, L.W.; Thill, D.C.; Shafii, B.; Prather, T.S. Detecting spotted knapweed (*Centaurea maculosa*) with hyperspectral remote sensing technology. *Weed Technol.* 2002, *16*, 535–545.
17. Hirano, A.; Madden, M.; Welch, R. Hypersptectral image data for mapping wetland vegetation. *Wetlands* 2003, *23*, 436–448.
18. Hamada, Y.; Stow, D.A.; Coulter, L.L.; Jafolla, J.C.; Hendricks, L.W. Detecting Tamarisk species (*Tamarix* spp.) in riparian habitats of Southern California using high spatial resolution hyperspectral imagery. *Remote Sens. Environ.*2007, *109*, 237–248.
19. Asner, G.P.; Jones, M.O.; Martin, R.E.; Knapp, D.E.; Hughes, R.F. Remote sensing of native and invasive species in Hawaiian forests. *Remote Sens. Environ.* 2008, *112*, 1912–1926.
20. Williams, A.P.; Hunt, E.R., Jr. Estimation of leafy spurge cover from hyperspectral imagery using mixture tuned matched filtering. *Remote Sens. Environ.* 2002, *82*, 446–456.
21. Hunt, E.R., Jr.; Everitt, J.H.; Ritchie, J.C.; Moran, M.S.; Booth, D.T.; Anderson, G.L.; Clard, P.E.; Seyfried, M.S. Applications and research using remote sensing for rangeland management. *Photogramm. Eng. Remote Sensing* 2003, *69*, 675–693.
22. Everitt, J.H.; Deloach, C.J. Remote sensing of Chinese tamarisk (*Tamarix chinensis*) and associated vegetation. *Weed Sci.* 1990, *38*, 273–278.
23. Lass, L.W.; Carson, H.W.; Callihan, R.W. Detection of yellow starthistle (*Centaurea solstitialis*) and common St. Johnswort (*Hypericum perforatum*) with multispectral digital imagery. *Weed Technol.* 1996, *10*, 466–474.
24. Lass, L.W.; Callihan, R.H. 1997. The effect of phenological stage on detectability of yellow hawkweed (*Hieracium pretense*) and oxeye daisy (*Chrysanthemum leucanthemum*) with remote multispectral digital imagery. *Weed Technol.* 1997, *11*, 248–256.
25. Peters, A.J.; Reed, B.C.; Eve, M.D.; McDaniel, K.C. Remote sensing of broom snakeweed (*Gutierrezia sarothrae*) with NOAA-10 spectral image processing. *Weed Technol.* 1992, *6*, 1015–1020.
26. Bradley, B.A.; Mustard, J.F. Identifying land cover variability distinct from land cover change: cheatgrass in the Great Basin. *Remote Sens. Environ.* 2005, *94*, 204–213.
27. Pavri, F.; Aber, J.S. Characterizing wetland landscapes: a spatiotemporal analysis of remotely sensed data at Cheyenne Bottoms, Kansas. *Phys. Geogr.* 2004, *25*, 86–104.
28. Robinson, T.P.; van Klinken, R.D.; Metternicht, G. Spatial and temporal rates and patterns of mesquite (*Prosopis*species) invasion in Western Australia. *J. Arid Environ.* 2008, *72*, 175–188.
29. Anderson, G.L.; Carruthers, R.I.; Ge, S.K.; Gong, P. Monitoring of invasive Tamarix distribution and effects of biological control with airborne hyperspectral remote sensing. *Int. J. Remote Sens.* 2005, *26*, 2487–2489.
30. Everett, J.H.; Yang, C.; Fletcher, R.S.; Deloach, C.J.; Davis, M.R. Using remote sensing to assess biological control of saltcedar. *Southwest. Entomologist* 2007, *32*, 93–103.
31. Peterson, E.B. Estimating cover of an invasive grass (*Bromus tectorum*) using tobit regression and phenology derived from two dates of Landsat ETM+ data. *Int. J. Remote Sens.* 2005, *26*, 2491–2507.
32. Ge, S.; Carruthers, R.; Gong, P.; Herrera, A. Texture analysis for mapping *Tamarix parviflora* using aerial photographs along Cache Creek, California. *Environ. Monit. Assess.* 2006, *114*, 65–83.
33. Akasheh, O.Z.; Neale, C.M.U.; Jayanthi, H. Detailed mapping of riparian vegetation in the middle Rio Grande River using high resolution multi-spectral airborne remote sensing. *J. Arid Environ.* 2008, *72*, 1734–1744.
34. Morissette, J.T.; Jarnevich, C.S.; Ullah, A.; Cai, W.; Pedelty, J.A.; Gentle, J.E.; Stohlgren, T.J.; Schnase, J.L. A tamarisk habitat suitability map for the continental United States. *Front. Ecol. Environ.* 2006, *4*, 11–17.
35. Evangelista, P.H.; Kumar, S.; Stohlgren, T.J.; Jarnevich, C.S.; Crall, A.W.; Norman, B.; Barnett, D.T. Modelling invasion for a habitat generalist and a specialist plant species. *Divers. Distrib.* 2008, *14*, 808–817.
36. Phillips, S.J.; Dudik, M.; Schapire, R.E. A maximum entropy approach to species distribution modeling. In *Proceedings of the 21st International Conference on Machine Learning*; ACM Press: New York, NY, USA, 2004; pp. 655–662.
37. Phillips, S.J.; Anderson, R.P.; Schapire, R.E. Maximum entropy modeling of species geographic distributions. *Ecol. Model.* 2006, *190*, 231–259.
38. Ficetola, G.F.; Thuiller, W.; Miaud, C. Prediction and validation of the potential global distribution of problematic alien invasive species-the American bullfrog. *Divers. Distrib.* 2007, *13*, 476–485.
39. Kumar, S.; Spaulding, S.A.; Stohlgren, T.J.; Hermann, K.A.; Schmidt, T.S.; Bahls, L.L. Potential habitat distribution for the freshwater diatom *Didymosphenia geminata* in the continental US. *Front. Ecol. Environ.* 2009, *7*.
40. Kammerer, J.C. Largest Rivers in the United States. 1990. Available online: http://pubs.usgs.gov/of/1987/ofr87-242/(accessed on August 28, 2009).
41. Lindauer, I.E. A comparison of the plant communities of the South Platte and Arkansas River drainages in eastern Colorado. *Southwest. Naturalist* 1983, *28*, 249–259.
42. Tamarisk Coalition. Riparian Restoration: Assessment of alternative technologies for tamarisk control, biomass reduction and revegetation. 2008. Available online: http://www.tamariskcoalition.org (accessed on August 28, 2009).
43. Leica. *ERDAS Imagine 9.1*; Leica Geosystems Geospatial Imaging; LLC: Atlanta, GA, USA, 1991-2005; Available online: http://www.erdas.com/Leica.

44. ESRI. *ArcGIS 9.1*, ESRI, Redlands, CA, USA. 2004. Available online: http://www.esri.com/index.html (accessed on August 28, 2009).

45. Rouse, J.W.; Haas, R.H.; Schell, J.A.; Deering, D.W. Monitoring vegetation systems in the Great Plains with ERTS. In Proceedings of the Third Earth Resources Technology Satellite-1 Symp., Greenbelt, MD, USA; 1974.

46. Kriegler, F.J.; Malila, W.A.; Nalepka, W.A.; Richardson, W. Preprocessing transformations and their effects on multispectral recognition. In *Proceedings of the Sixth International Symposium on Remote Sensing of Environment*; University of Michigan: Ann Arbor, MI, USA, 1969; pp. 97–131.

47. Myneni, R.B.; Ramakrishna, R.; Nemani, R.; Running, S.W. Estimation of global leaf area index and absorbed par using radiative transfer models. *IEEE Geosci. Remote Sens.* 1997, *35*, 1380–1393.

48. Todd, S.W.; Hoffer, R.M.; Milchunas, D.G. Biomass estimation on grazed and ungrazed rangelands using spectral indices. *Int. J. Remote Sens.* 1998, *19*, 427–438.

49. Jordan, C.F. Derivation of leaf area index from quality of light on the forest floor. *Ecology* 1969, *50*, 663–666.

50. Liu, W.G.; Gao, W.; Gao, Z.Q.; Wang, X.L.; Slusser, J. An analysis between the biomass of ecosystem of oasis and the vegetation index. In *Remote Sensing and Modeling of Ecosystems for Sustainability III*; SPIE: Bellingham, WA, USA, 2006; pp. 62982M-1–62982M-7.

51. Huang, C.; Wylie, B.; Yang, L.; Homer, C.; Zylstra, G. Derivation of a tasselled cap transformation based on Landsat 7 at-satellite reflectance. *Int. J. Remote Sens.* 2002, *23*, 1741–1748.

52. Kauth, R.J.; Thomas, G.S. The tasselled cap-a graphic description of the spectral-temporal development of agricultural crops as seen in Landsat. In *Proceedings of the Symposium on Machine Processing of Remotely Sensed Data*; LARS; Purdue University: West Lafayette, IN, USA, 1976; pp. 41–51.

53. Cohen, W.B.; Spies, T.A.; Fiorella, M. Estimating the age and structure of forests in a multi-ownership landscape of western Oregon, USA. *Int. J. Remote Sens.* 1995, *16*, 721–746.

54. Jin, S.; Sader, S. Comparison of time series tasselled cap wetness and the normalized difference moisture index in detecting forest disturbances. *Remote Sens. Environ.* 2005, *94*, 364–372.

55. Huete, A.R. A Soil-Adjusted Vegetation Index (SAVI). *Remote Sens. Environ.* 1988, *25*, 295–309.

56. Elith, J.; Graham, C.J.; Anderson, R.; Dudik, M.; Ferrier, S.; Guisan, A.; Hijmans, R.; Huettmann, F.; Leathwick, J.; Lehmann, A.; Li, J.; Lohmann, L.; Loiselle, B.; Manion, G.; Moritz, C.; Nakamura, M.; Nakazawa, Y.; Overton, J.; Peterson, A.T.; Phillips, S.; Richardson, K.; Scachetti-Pereia, R.; Schapire, R.; Soberon, J.; Williams, S.; Wisz, M.; Zimmermann, N. Novel methods improve prediction of species distributions from occurrence data. *Ecography* 2006, *29*, 129–151.

57. Cohen, J. A coefficient of agreement of nominal scales. *Educ. Psychol. Meas.* 1960, *20*, 37–46.

58. Allouche, O.; Tsoar, A.; Kadmon, R. Assessing the accuracy of species distribution models: prevalence, Kappa and the true skill statistic (TSS). *J. Appl. Ecol.* 2006, *43*, 1223–1232.

59. Landis, J.R.; Koch, G.C. The measurement of observer agreement for categorical data. *Biometrics* 1977, *33*, 159–74.

60. Pearce, J.; Ferrier, S. Evaluating the predictive performance of habitat models developed using logistic regression. *Ecol. Model.* 2000, *133*, 225–245.

61. Fielding, A.H.; Bell, J.F. A review of methods for the assessment of prediction errors in conservation presence/absence models. *Environ. Conserv.* 1997, *24*, 38–49.

62. Hosmer, D.W.; Lemeshow, S. *Applied Logistic Regression*, 2nd ed.; Wiley: New York, NY, USA, 2000.

63. Franklin, S.; Jagielko, C.; Lavigne, M. Sensitivity of the Landsat enhanced wetness difference index (EWDI) to temporal resolution. *Can. J. Remote Sens.* 2005, *31*, 149–152.

64. Healey, S.; Yang, Z.; Cohen, W.; Pierce, D.J. Application of two regression-based methods to estimate the effects of partial harvest on forest structure using Landsat data. *Remote Sens. Environ.* 2006, *101*, 115–126.

CHAPTER 9

MAPPING OF SOIL EROSION ZONES FOR CONSERVATION —A GEOSPATIAL APPROACH

Usha Chirala, Murali Krishna Gurram, Nooka Ratnam Kinthada

Keywords: Meghadrigedda, GIS and Remote Sensing, Erosion, Siltation of the Reservoir, Soil Conservation

INTRODUCTION

Soil erosion is a major environmental problem which threatens the sustainability and productivity of agricultural and other hydrological systems [1] . Therefore, proper assessment and mapping of soil erosion is very essential for the management of natural resources [2] . Every year, about 75 billion tons of soil is eroded from the land due to human induced land use practices, a rate that is about 13 - 40 times faster than the natural rate of erosion. Over exploitation of land and water resource for various means of developmental activities by rapidly growing population has been inflicting severe damage to soil environment. According to the study conducted by [3] , about 148.9 Mha area representing 45.3% of the Total Geographical Area (TGA) of the country is affected by various forms of soil degradation caused by water, wind, chemical and other physical agencies. Soil erosion is considered as one of the major problems that the Indian agriculture sector is coping with and its negative effects are seen in other spheres as well. It is noticed that, every year India losses millions of tons of soil and its nutrients to the agents of its degradation which adversely affects our national productivity.

A reservoir built across the River Meghadrigedda has a very high value in terms of feeding the major drinking water needs of Visakhapatnam city and agricultural needs in the catchment. Population in the catchment is mostly dependent on agriculture, livestock, and forestry for livelihood. Environmentally, the catchment area has been under tremendous pressure due to increasing population, industrialization and over exploitation of natural vegetation for timber and fuel wood. The catchment has been witnessing rapid urbanization and undergoing large scale topographic transformation which is leading to vegetation loss and soil erosion resulting in silt deposition in the reservoir. The storage capacity of the reservoir is rapidly dwindling due to the excessive siltation and the problem is posing a major risk to the very existence of the reservoir.

STUDY AREA

Meghadrigedda is one of the prominent river catchments originated in the hilly terrain of Visakhapatnam and Vizianagaram districts of Andhra Pradesh. The river joins the Bay of the Bengal in the south near Visakhapatnam Figure 1. A medium scale reservoir was constructed across the river with a gross capacity of 1163 m·ft^3 in the year 1972 near Kamparapalem village, Visakhapatnam. The river catchment is extended in an area of 368 km^2 which includes the reservoir area. Geographically, the catchment extends within the northern latitudes of 17°42" to 17°57" and eastern longitudes of 83°0" to 83°17".

The entire catchment is extended within nine administrative sub-district units (also called mandals) of Visakhapatnam and Vizianagaram districts of the state of Andhra Pradesh. Around 94 km^2 of upper catchment area lies within the Kottavalsa sub-district which comes under the jurisdiction of Vizianagaram district and the rest of the sub-districts comeunder the Visakhapatnam district Table 1. Majorpart of the study area is covered with forests of which considerable extent of the area falls under the reserved forest category, covered by Potukonda, Narava, Nallkonda, Ellappi, Yerrakonda and Dabala reserved forests. The catchment is covered with nearly 521 minor non-

Citation: Chirala, U. , Gurram, M. and Kinthada, N. (2015) Mapping of Soil Erosion Zones for Conservation —A Geospatial Approach
doi: 10.4236/ijg.2015.64026.

perennial/perennial tanks which do not however significantly reduce the flood peaks and their duration. The catchment receives the maximum amount of rainfall during southwest and northeast monsoon seasons which starts from June to December. The river reaches its peak levels mostly during the months of October and November due to the influence of cyclonic storms in the Bay of Bengal. The area enjoys sub-tropical climate and temperatures range from 14°C - 20°C in December and 33°C - 42°C in May. The temperature goes down with the onset of southwest monsoon and tumbles to a mean minimum of 21°C by January after which there will bea reversal trend till the temperature reaches mean maximum of 32.7°C by the end of June. The district receives an annual rainfall of 1202 mm. As per the 2011 census [4] , the provisional population of the district is 42.88 lakhs with a growth rate of 11.89%, which constitutes 5.06% of the population of the state while the geographical area of the district is 11,161 km^2 which is only 4.1% of the area of the state.

MATERIALS AND METHODS USED FOR THE STUDY

Topographical maps of 1:50,000 scale published by Survey of India (SOI) numbered 65 O/1, 65 O/2&3 and 65 O/5 have been georeferenced and the mosaic was used as the base map for the study.

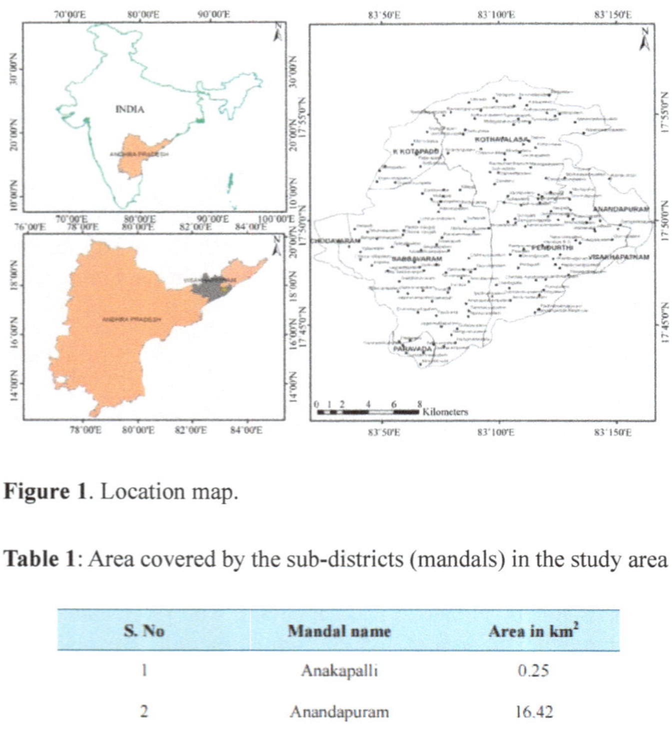

Figure 1. Location map.

Table 1: Area covered by the sub-districts (mandals) in the study area

S. No	Mandal name	Area in km^2
1	Anakapalli	0.25
2	Anandapuram	16.42
3	Chodavaram	9.42
4	K. Kotapadu	21.97
5	Kottavalsa	93.32
6	Paravada	5.80
7	Pendurthi	83.67
8	Sabbavaram	134.33
9	Visakhapatnam	2.82
	Total	**368**

Multi-temporal Remote sensing data obtained using IRS-IB LISS III of 1997, IRS-ID LISS III of 1999, IRS-ID LISS III of 2006 and IRS-ID LISS III of 2009 were used for the study. The LISS III sensor data has the spatial resolution of 23.5 × 23.5 m with 7-bit radiometric resolution. The data has been georeferenced by taking the georeferenced topographic mosaic as a reference.

Various map themes as inputs required for the integration have been prepared by means of interpreting the data through manual digitization. Authenticity of the feature data extracted for all the map themes were rechecked and verified through field checks. The field checks were conducted by collecting samples from nearly 63 villages and 71 hamlets located in the study area and necessary corrections were incorporated.

Drainage and contour layers were extracted from the topographic maps and were used to generate the drainage density and Digital Elevation Model (DEM). The drainage density map is prepared by drawing all the drainage patterns from the topographical map. A grid of 1 km × 1 km is drawn on the entire base map. The cumulative stream lengths are computed in each grid, and the drainage density in each grid is calculated using formula Lu/A.

DEM generated using contours was used to create the slope map for the study area. The slope map gives various groups/ categories of slope as defined in the technical guidelines for integrated study.

The land use/land cover theme reveals the present status of natural and anthropogenic activity which transforms the characteristics of the terrain. The land use/land cover theme for the study area has been delineated from the IRS-ID LISS III data of February, 2009 using the four level classification scheme defined by the National Remote Sensing Center [NRSC] [5] . The analysis was carried out by employing the visual interpretation techniques and the data has been verified with the ground truth data.

The soil map for the study area has been generated using the district soil maps published by National Bureau of Soil Survey and Land Use Planning (NBSS & LUP) by taking the georeferenced remote sensing data as a reference.

WEIGHTED ANALYSIS OF THE SUB-BASINS

In order to perform the spatial analysis to identify and map the zones of high erosion intensity, all the mapped data inputs had to be reclassified according to their soil erosivity characteristics. These reclassified themes are overlaid in the GIS environment to arrive at soil prioritization zones for the entire area which has been divided into nine sub-basins. Various thematic layers such as slope, drainage density, land use/land cover, geomorphology and soil have been used to arrive at soil priority zones. Union of all these layers has been used to generate a theme which will be used to further delineate the areas of soil erosion. The areas which are prone to soil erosion in the study area have been delineated and classified according to the method defined by the National Water Development Program for Rainfed Areas (NWDPRA).

Drainage Density (D_d)

According to the classification proposed by [6] , all the drainage sub-basins of Meghadrigedda have been reclassified into three different zones, such as, poor, medium and excellent based on their $D_d$values Figure 2. Sub- sequently, weightfactors have been assigned to all the sub-basin categories. The weight factors have been assigned in a way that the sub-basins with D_d value more than 1.5 km/km^2 (coded as A) would get the weightage factor of 1. Similarly, sub-basins with D_d between 0.5 - 1.5 km/km^2 (coded as B) have been given the weightage factor of 2 and sub-basins with D_d less than 0.5 km/km^2 (coded as C) are assigned with the weightage factor of 3, respectively Table 2. The reclassification of the sub-basins is done in a way that D_d values indirectly reflects the intensity of soil erosion, it means, higher the D_d, higher the soil erosion and vice-versa.

Slope

Slope is considered as one of the critical parameter for identifying the erosion intensity zones of the study area. The slope theme originally with seven different categories has been reclassified into three categories indicating the intensity of erosion based on the given weight criteria Figure 3. The first category of reclassified slope theme is formed by including the previous three slope categories, such as, 5, 6 and 7, (coded as D) by assigning a weigh- tage factor of 1, and considered as the high erosion category. The 2nd reclassified category of slope, which re- presents zones with medium erosion intensity were formed by reclassifying of the previous 3rd and 4th categories of the slope (coded as E) and assigned a weightage of 2.

Table 2: Codes and weight factors assigned to different zones of drainage density

Drainage density	Polygon code	Weight	Class
>1.5 km	A	1	High erosion
0.5 - 1.5 km	B	2	Medium erosion
<0.5 km	C	3	Low erosion

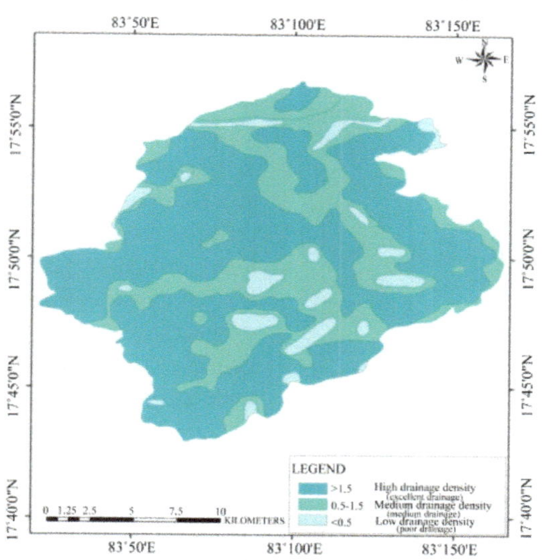

Figure 2: Drainage density map of the study area.

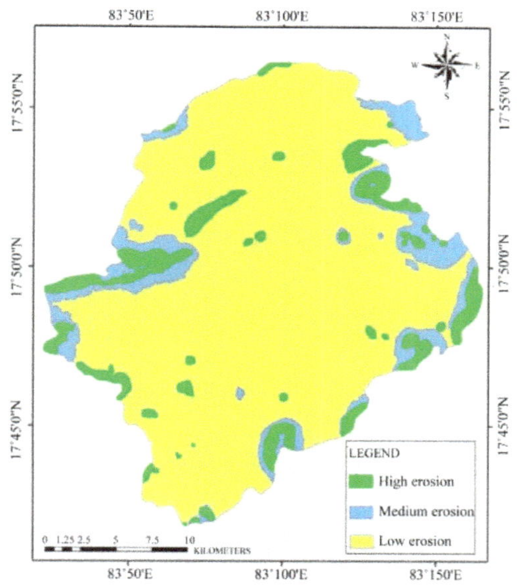

Figure 3: Slope categories of the study area.

Finally, the third category of reclassified slope was formed by grouping of 1st and 2nd category of previous slopes (coded as F) and given a weightage factor of 3, which represents the areas with low erosion intensity. The basis for the weightage criteria is decrease in the value of slope, i.e., areas with low slope gradient represents the zones of no/or less erosion and vice-versa. Table 3 shows regions with slope limits, assigned polygon code, their weightage factors and the erosion class.

Land Use/Land Cover

Land use/land cover is acclaimed as one of the important surface character which plays a vital role in determining the intensity of erosion. Land use/land cover mapping of the Meghadrigedda resulted in various land use/land cover categories, of which cultivation is noticed as the major land use class, followed with plantation, deciduous forest, etc. Of all the land use/land cover categories mapped in the study area, only those categories which have a direct bearing on soil erosion are taken into consideration for delineating erosion prone zones. As such, four different reclassified units of land use/land cover categories with respective codes, weight factors and classes indicating the intensity of soil erosion taken into consideration for the analysis. The first category coded as "G" has been formed by reclassifying of various land use/land cover classes, such as, gullied ravinous, badland, upland with dense scrub, upland with sparse scrub etc., and assigned with a weightage factor of 1 which represents high intensity of erosion. The 2nd category coded as "H" is formed with reclassifying of land use/land cover classes, includes plantation and un-irrigated land and assigned with a weight of 2, which represents areas of medium to high erosion intensity. The 3rd category coded as "I" formed with the class "forest" alone by assigning a weight of 3 representing areas with medium erosion intensity. Finally, the 4th category coded as "J" was formed with classes, such as, tanks, reservoir and irrigated areas with a weight factor of 4 and represents the low erosion intensity areas Figure 4.

Soil

Soil is another vital parameter without which the study of erosion analysis has to be considered incomplete. Soil forms the upper most layer and composed of unconsolidated earth material which is susceptible to erosion. Besides its natural conditions, soil erosion is triggered by anthropogenic activities. The soils of the Meghadrigedda catchment have been reclassified into four categories and each of them has been given a code with weight factors ranging between 1 to 4 according to the intensity of erosion Table 4 and Figure 5. The shallow skeletal soils in the catchment have been coded as "K" with a weight factor of 1 representing the areas of high erosion. Gravelly loams are coded as "L" with a weight factor of 2 representing the areas of medium to high erosion. Similarly, red loams and clays are coded with "M" with a weight factor of 3 representing medium erosion. Finally, clayey soils are coded with "N" with weight factor of 4 representing low intensity of erosion Table 5.

Generation of Map Composites

A composite map is a combination of two or more base maps depicting some or all features from the base maps. A composite map has been generated by overlaying two different map themes on each other. The resultant composite is the union of both the map themes and inherits the characteristics from both. For instance, a polygon denoted as "A" in the drainage map with a weightage factor "1" when combines with a polygon of the slope map denoted as "E" with a weightage factor "2", assumes a new polygon code as "AE" on the composite map with its area being the lesser of the two and with a weightage factor equalizing the sum of the weights of both "A" and "E", i.e., "3".

Table 3: Codes and weight factors assigned to different slope categories

Slope category	Polygon code	Weight	Erosion class
5, 6, 7	D	1	High erosion
3, 4	E	2	Medium erosion
1, 2	F	3	Low erosion

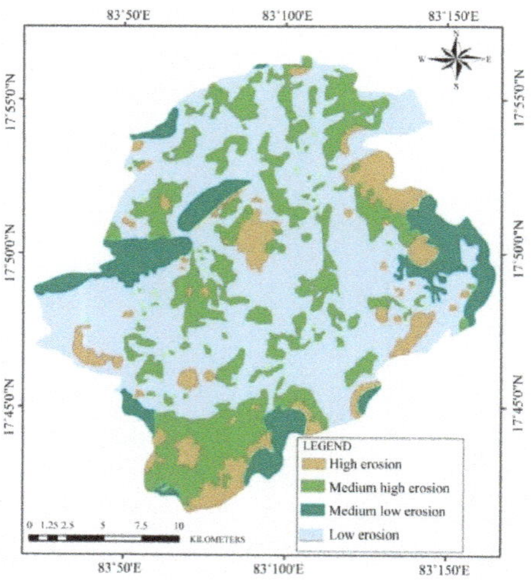

Figure 4: Land use/land cover categories reclassified according to the erosion intensity.

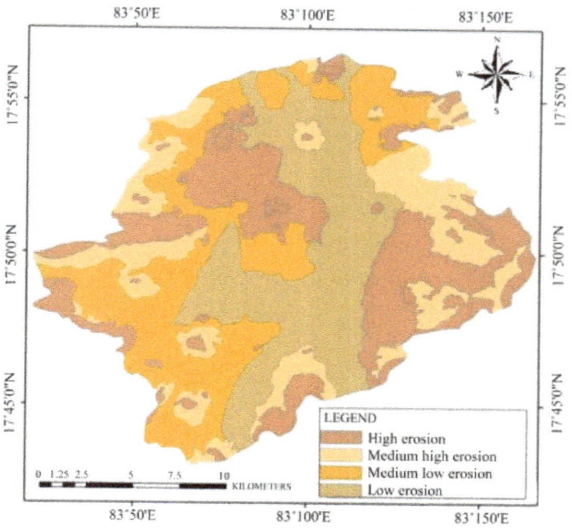

Figure 5: Soil map of the study area reclassified according to the erosion intensity.

Table 4: Codes and weight factors assigned to different soil categories

Soil category	Polygon code	Weight factor	Erosion class
Shallow skeletal soils	K	1	High
Gravelly loams	L	2	Medium high
Red loams and clays	M	3	Medium
Clayey soils	N	4	Low

Table 5: Weights assigned to the zones of composite 1 for classification according to erosion intensity

Combined classes	Weight factors
AD	2
AE	3
AF	4
BD	3
BE	4
BF	5
CD	4
CE	5
CF	6

Map Composite 1

The first composite map was prepared by the integration of slope and drainage density maps. All the resultant (overlapping) feature class combinations such as, AD, AE, AF, BD, BE, BF, CD, CE and CF were delineated. Each combined class was assigned with a weightage factor according to the degree of susceptibility to erosion. The zones which have been identified along with their cumulative weight factors for further analysis as shown in Table 5. Subsequently, the zones which have been categorized into low, medium and high erosion classes are mapped accordingly Figure 6.

Map Composite 2

The second map composite was generated by the integration of the land use/land cover theme with the soil. The resultant map has the combination of different feature classes from land use/land cover and soil themes. The map has resulted in the feature combinations, such as, GK, GL, GM, GN, HK, HL, HM, HN, IK, IL, IM, IN, JK, JL, JM and JN. All the classes have been assigned with certain weights as shown in Table 6 for further analysis. These classes have been categorized, such as low, low to medium, medium to high and high based on their susceptibility to erosion Figure 7.

DISCUSSION: INTEGRATION FOR PRIORITIZATION

Both the reclassified composite themes have been used for the preparation of final erosion map which will be used for the classification of the zones for final prioritization. Weight factors have been assigned to each overlapping feature class after the integration. These weightage factors helps reclassification of eleven resultant classes indicating different degrees of soil erosion intensity into seven prioritized classes and further down to four classes ranging from "very severe" to "low" erosion areas which are prone to different degrees of erosion Figure 6 and Figure 7. The final areas along with the weight factors used for prioritization with respect to their erosion class, level and areal extent in km^2 are presented in Table 7. These seven classes have been prioritized according to the intensity of erosion Table 8 and have been assigned the priorities for conservation Table 9. These zones have been further reclassified into four classes of erosion intensity, namely, very severe, severe, moderate and low. This is done by denoting the very high priority areas as "severe" since they are prone to high intensity of erosion and needs immediate soil conservation measures.

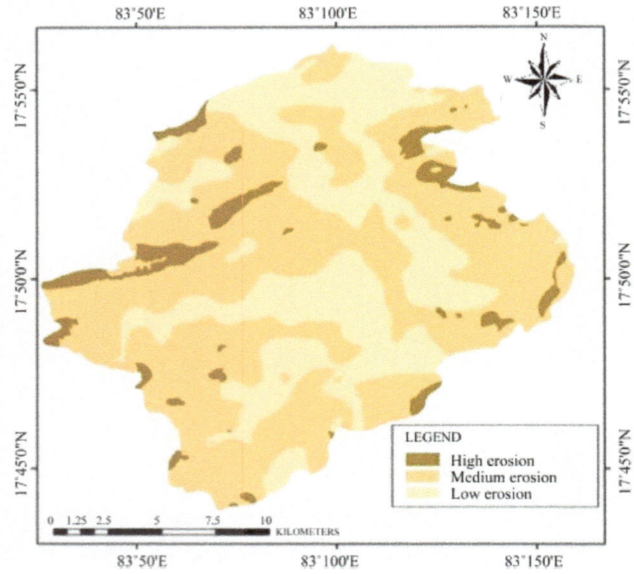

Figure 6: Composite 1 generated by integrating "slope" and "drainage density" classified according to the suceptability to erosion.

Table 6: Weights assigned to the zones of composite 2 for classification according to erosion intensity

Combined classes	Weight factors
GK	2
GL	3
GM	4
GN	5
HK	3
HL	4
HM	5
HN	6
IK	4
IL	5
IM	6
IN	7
JK	5
JL	6
JM	7
JN	8

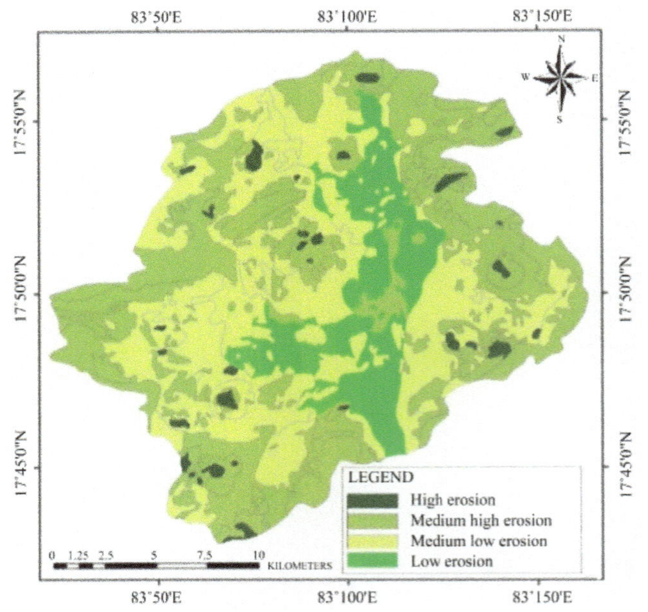

Figure 7: Composite 1 generated by integrating "soil" and "land use/ land cover" classified according to the suceptability to erosion.

Table 7: Weight factors assigned to the integrated zones of the study area

Combined classes	Weight factors
GK	2
GL, HK	3
GM, HL	4
GN, HM, IL, JK	5
HN, IM, JL	6
IN, JM	7
JN	8

Table 8: Different erosional classes indicating the severity of erosion in the study area

S. No	Soil erosion class	Erosion class	Area in km^2	Severity of erosion
1	4	E4	5.46	Very severe erosion
2	5	E4	7.97	Very severe erosion
3	6	E3	27.25	Severe erosion
4	7	E3	32.4	Severe erosion
5	8	E3	22.06	Severe erosion
6	9	E2	40.71	Moderate erosion
7	10	E2	33.53	Moderate erosion

Similarly, the zones which have been given the "least" priority are categorized as "low" erosion intensity zones Table 10.

Various conservations measures are recommended in the Meghadrigedda catchment for arresting the erosion especially in the areas identified with high erosion intensity. The conservation measures have also taken into con- sideration of the prevailing local conditions. A brief descriptive summary of different sub-basins in the Meg- hadrigedda with their physiography characteristics and erosion intensity is given in Table 11. The zones that are identified with very severe erosion covering in an area of 16.58 km^2 i.e., 4.5% i.e. 4.8% of the total area is recommended for social forestry and for growing the fruit bearing trees to conserve the soil.

Table 9: Reclassified soil erosion zones in the study area

S. No	Priority class	Area in km^2	Percentage of area
1	Very high	16.58	4.50
2	Medium high	6.08	1.65
3	High	35.90	9.75
4	High Medium	38.15	10.36
5	Medium	45.93	12.48
6	Low Medium	186.56	50.69
7	Low	38.80	10.54
Total		**368**	**100.00**

Table 10: Prioritized erosional zones along with their areal distribution

Order of priority	Priority class	Area in km^2	Percent area
1	Very severe erosion	16.58	4.5
2	Severe erosion	80.13	21.77
3	Moderate erosion	232.49	63.17
4	Slight erosion	38.80	10.56
Total	**368**	**100**	

Table 11: Sub-basins of Meghadrigedda with description of their physiography, erosion intensity and priority for conservation

Sub-basin	Priority for conservation	Description of the sub-basin, physiography and erosion intensity
1	1	Isolated hillocks are composed of quartzite and migmatite type of rocks as intrusive in khondalite country ricks. Dendritic parallel drainage patterns are prominently seen in khondalite hillocks and plains. It has moderate slopes with more perennial tanks.
2	2	The rock type observed is khondalite. Highly denuded rocks, sub-dendritic drainage pattern rolling topography with moderate slopes. At place sparse and discontinuous drainage is observed down below next to the reservoir area with khondalite rock presence.
3	7	Dendritic drainage developed over drainage country rock with low slopes. Perennial tanks with moderate storage capacity observed.
4	8	Radial, dendritic and sub-dendritic drainage patterns are observed. Khondalite rock with rolling topography is observed.
5	9	Sparse and discontinuous drainage is observed. The discontinuous drainage is due to anthropogenic activity in this watershed. It is located in the reservoir area with natural khondalite rock abutment.
6	4	This is a part of Naravagedda drainage basin which is the main tributary to Meghadrigedda river. The rock types are khondalite and quartzite. Steep slopes are confined in the hilly terrain. Higher rate of denudation and rolling topography are the characteristic features of this sub-basin.
7	5	This sub-basin is the major drainage of the Naravagedda river. Landmass denudation, steep slope with rill erosion and khondalite and migamatite types of rocks are exposed.
8	3	This basin is located in the upper catchment of the river, where the tock types are khondalite and migamatite. The migamatite occurs as intrusive in khondalite massif. Moderate slope with varied type of drainage pattern are common in this sub-basin.
9	6	This sub-basin is the origin of the river Meghadrigedda. Different types of drainage patterns with moderate slope, feather like drainage basin features are observed. Khondalite rock type prevails.

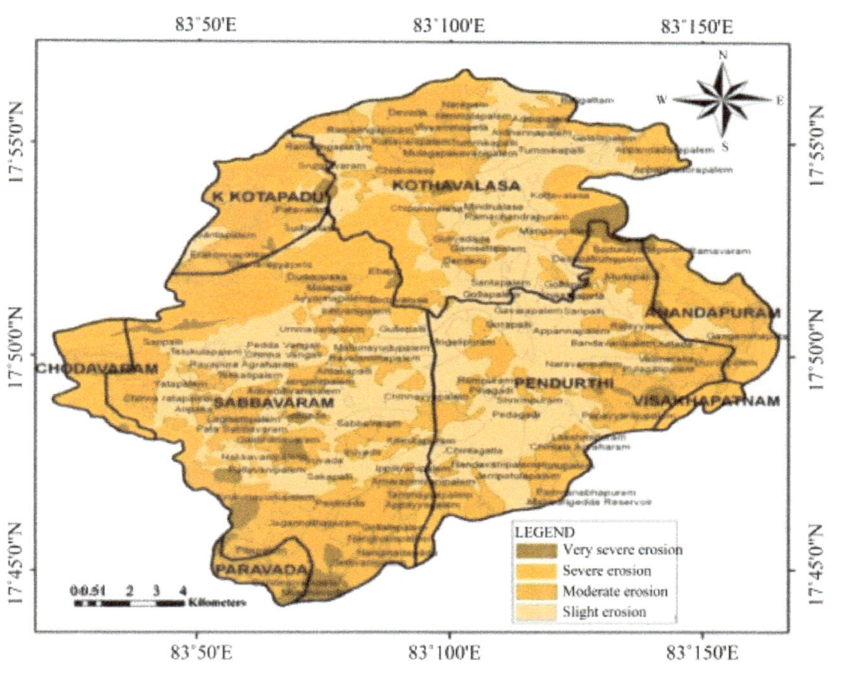

Figure 8: The sub-basins of Meghadrigedda mapped according to the intensity of erosion.

The sub-basins of Meghadrigedda with different erosion intensity zones are shown in Figure 8. The other zones of erosion were recommended with soil conservation measures such as, silt traps, contour bunding and check dams taking into account of the erosional and topographical conditions of the area.

CONCLUSION

Siltation is identified as the major problem in Meghadrigedda reservoir which is gradually reducing its holding capacity year after year and also affecting the water supply system connected to it. The possibility of desiltation of the reservoir by means of dredging is totally ruled out due to various practicality reasons, such as, magnitude of the reservoir, difficulty in handling the process and the cost factor. Various soil conservation measures in the catchment are thought out as the best alternative solution. The measures include social forestry for the high erosion intensity areas. The remaining areas in the catchment are recommended with measures such as, checking dam construction, silting traps, contouring bunding etc. with respect to the prevailing topographical conditions. It is believed that the above recommended measures can be very effective in arresting the burgeoning siltation problem in the Meghadrigedda reservoir.

ACKNOWLEDGEMENTS

Usha Chirala, Murali Krishna Gurram, Nooka Ratnam Kinthada The authors would like to thank the Dept. of GIS, INRIMT, Hyderabad, India for its support in mapping.

REFERENCES

1. Quinton, J.N., Govers, G., Van Oost, K. and Bardgett, R.D. (2010) The Impact of Agricultural Erosion on Biogeochemical Cycling. Nature Geo Science, 3, 311-314. http://dx.doi.org/10.1038/ngeo838
2. Baskan, O., Cebel, H., Akgul, S. and Erpul, G. (2010) Conditional Simulation of USLE/RUSLE Soil Erodiblity Factor by Geostatistics in a Mediterranean Catchment, Turkey. Environmental Earth Science, 60, 1179-1187. http://dx.doi.org/10.1007/s12665-009-0259-2
3. Sehgal, J. and Abrol, I.P. (1994) Soil Degradation in India: Status and Impact. Oxford and IBH Publishing Co., New Delhi, 80.
4. Handbook of Statistics (2011) Chief Planning Officer, Visakhapatnam District, 226. [Paper reference:1]
5. National Remote Sensing Agency (2004) Manual on National Land Use/Land Cover Mapping on 1:250,000 Scale Using Multi-Temporal IRS P6-AWiFS Data. NRSA, Dept. of Space, Govt. of India, Balanagar. http://bhuvan-noeda.nrsc.gov.in/theme/thematic/tools/document/LULC250/1011.pdf
6. Deju, A.R. (1971) Regional Hydrology Fundamentals. Gordon & Breach Publishing Group, 204.

CHAPTER 10

MONITORING THE DISTRIBUTION AND DYNAMICS OF AN INVASIVE GRASS IN TROPICAL SAVANNA USING AIRBORNE LIDAR

Shaun R. Levick, Samantha A. Setterfield, Natalie A. Rossiter-Rachor, Lindsay B. Hutley, Damien McMaster and Jorg M. Hacker

Keywords: Alien plant, gamba grass, invasion, LiDAR, weed mapping

INTRODUCTION

Savannas cover 20% of the global terrestrial land surface and account for 30% of terrestrial net primary production (NPP) [1]. Fire is an integral component of savanna ecology and dynamics, particularly in tropical savannas which typically burn every 1–3 years [2]. Fire can markedly alter the structure and biomass of savanna vegetation, so understanding fire dynamics is critical for managing carbon storage and conserving biodiversity in these habitats. Global changes in land-use and climate threaten many of the ecosystem services that savannas provide, and the spread of alien invasive plants that alter fire regimes presents major additional ecological challenges [3].

Over the past two decades, there has been increased research focus on invasive alien grasses—particularly mapping their distribution, determining patterns of spread, quantifying their ecological impacts and developing more effective management strategies. The focus on this group of invaders is the consequence of the substantial threat that they pose to the structure and function of many of the world's ecosystems [4]. Ecological impacts include the displacement of native species and consequent reduction in native flora and fauna, significant changes to nutrient and water cycling, and substantial changes to fire behaviour and altered fire regimes [5,6].

Introduced crop and pasture plants form the basis of most of Australia's agricultural production, and in north Australia's savannas a number of exotic grass species have spread from the pastoral estate to invade native savanna vegetation [7,8,9]. The species considered the greatest threat to the region's savanna woodlands and forests is *Andropogon gayanus* Kunth. (gamba grass) [10,11]. *A. gayanus* is a perennial C4 grass that forms large tussocks in excess of 4 meters high and displaces the much shorter native vegetation [10]. *A. gayanus* alters nitrogen cycling [5,12] and reduces plant species richness and abundance [10]. Invasion results in fuel loads at least three times higher than native grasses, resulting in significantly more intense fires [8,11,13]. The rate of spread from initial source paddocks of *A. gayanus* has been rapid [14] with invasion now covering ~15,000 km^2. Modelling exercises predict that most of Australia's mesic savanna is suitable for invasion, including ~380,000 km^2 of Australia's Northern Territory, as well as large savanna areas in Queensland and Western Australia [15]. There is an urgent need for effective management of *A. gayanus* in northern Australia.

Accurately mapping the distribution of invasive plants underpins effective management. Distribution mapping is critical for determining the extent and pattern of invasion, and the prioritisation of invaded areas for management actions [16]. In addition, distribution maps are used in predictive modelling of the potential distribution and patterns of spread [17], and for assessing regional impacts [18]. To better understand and better manage the spread and consequences of *A. gayanus*invasion, we need spatially explicit knowledge of where *A. gayanus* occurs and how fast it is spreading. However the savanna region of northern Australia is the largest and most intact savanna system in the world, covering 2 million km^2, so mapping the extent of habitat invaded by *A. gayanus* is challenging.

Satellite remote sensing has played an important role in ecological research since the 1970s and is the most viable option for monitoring large land areas through repeated time

Citation: Shaun R. Levick, Samantha A. Setterfield, Natalie A. Rossiter-Rachor, Lindsay B. Hutley, Damien McMaster and Jorg M. Hacker, Monitoring the Distribution and Dynamics of an Invasive Grass in Tropical Savanna Using Airborne LiDAR, doi:10.3390/rs70505117.

intervals [19]. The use of satellite remote sensing for alien invasive plant studies only started gaining traction in the mid-1990s as spatial and temporal resolutions increased. The application of remote sensing to the study of alien plant invasions is most efficient when the invader presents a novel structure, phenology or biochemistry relative to neighbouring native vegetation [20,21]. Airborne LiDAR has emerged as a powerful tool for the fine-scale structural characterisation of ecosystems [22], including studying the effects of management actions on savanna vegetation dynamics and biodiversity conservation [23,24]. Most ecological studies involving LiDAR focus on woody vegetation, with return pulse densities of 1–4 per m^2 due to necessary compromises between sensor performance, flight altitude and flight speed. Laser returns from the grass layer are generally indistinguishable from the shrub layer at this resolution, or are treated as noise when refining the terrain models.

In this study we explore the potential for very high-resolution airborne LiDAR to map the presence and distribution of A. *gayanus* in the tropical savannas of northern Australia. Our goal was to produce a reliable representation of current *A. gayanus* coverage that can serve as both a benchmark for future rates of spread calculations, and as robust end-members for training satellite based investigations over broader regions.

MATERIALS AND METHODS

Study Species and Site Location

Andropogon gayanus is a perennial C4 grass that is structurally distinct from native grasses of northern Australia. *A. gayanus* forms taller tussocks (typically 2–5 m *cf.* 0.5–3 m for native grasses) with a larger basal area (up to 70 cm diameter *c.f.* up to 20 cm for native grasses) [12]. In the late wet season (March/April), *A. gayanus* undergoes significant growth and soon overtops the native herbaceous layer (Figure 1). In the early dry season (April/May/June), when the dominant annual grasses have begun to senesce, *A. gayanus* remains green and is clearly visible in the landscape.

A. gayanus was introduced into Australia in the 1930s, and subsequently cultivated into an improved pasture with the cultivar 'Kent' released in 1978 [11]. It was planted in paddocks within the study area in the mid-1980s, and spread from these paddocks to adjacent areas was noticed in the 1990s [25]. Significant areas of invasion now occur across northern Australia [15].

The study area for our investigation occurs within the Coomalie Shire region, located approximately 70–100 km south of Darwin, Northern Territory (NT), Australia (Figure 2). The study area has a distinct wet-dry tropical climate. Air temperature is high throughout the year (mean maximum 33 °C), while rainfall is highly seasonal (1662 mm, Batchelor Airport, Bureau of Meteorology, http://www.bom.gov.au) and concentrated in the wet season (November-April). The major vegetation type is savanna woodland dominated by *Eucalyptus miniata* (Cunn. Ex Schauer) and *E. tetrodonta* (F. Muell), with a grass understory dominated by native perennial species such as *Heteropogon contortus* (L.) Roem. & Schult and *Alloteropsis semialata* (R. Br.) Hitchc. or annual grasses, such as *Sorghum intrans* (F.Muell. ex Benth.). The study area is largely under private ownership for pastoral lease or semi-rural development, with other significant areas owned by local Aboriginal communities (the Finniss River Aboriginal Land Trust) or under Government ownership.

Figure 1: Typical vegetation structure in three stages of invasion—(**a**) a native vegetation site; (**b**) a site at the early stage of invasion with *A. gayanus* in understory and an intact Eucalyptus dominated overstory; and (**c**) a site heavily invaded with *A. gayanus* with a decline in overstory trees due to high intensity fires.

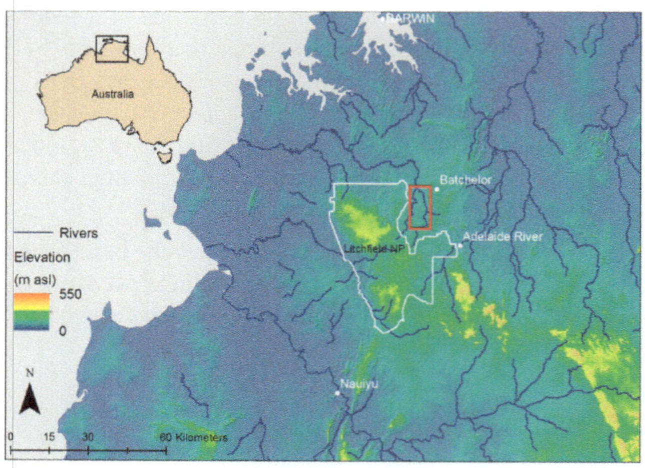

Figure 2: Study site location (red box) in the Batchelor region of the Northern Territory, Australia.

Airborne LiDAR Acquisition and Processing

We mapped 3000 ha of the study landscape with airborne LiDAR in June 2013. The survey was conducted by Airborne Research Australia (ARA) with a full-waveform LiDAR sensor (RIEGL LMS-Q560) operated from a light fixed-wing aircraft (Diamond Aircraft ECO-Dimona). We aimed to achieve a very high-resolution representation of the ecosystem, so overlapping flightlines were used to achieve double coverage of the sites (average flying height 300 m AGL, swath width ~300 m, line spacing 125 m). The RIEGL LMS-Q560 was operated at 240 kHz and 135 lines per second. Our very slow flying speed of less than 40 m/s ensured high point densities along track, with an average pulse density of 13.71 m^2 and an average pulse spacing of 0.28 m.

Raw LiDAR data were processed with RiANALYZE (RIEGL®) for decomposing the full waveforms into discrete returns, and with the ARA RASP open source software (RASP Version 0.98: manual, code and executables available from ARA on request) to orientate the point cloud to Cartesian coordinates and output the geolocated point cloud in the American Society for Photogrammetry and Remote Sensing (ASPRS) standard LAS format. All further point-cloud processing tasks were conducted with the LAStools suite of processing scripts (http://www.lastools.org, Isenburg 2014). The last returns were classified into ground and non-ground points for bare-earth extraction. A digital terrain model (DTM) was constructed from ground returns using a triangulated irregular network approach (TIN) at 0.5 m resolution. The DTM was used to normalize the *z* coordinate of vegetation returns to height above ground level.

A. gayanus Classification and Cover Mapping

We developed a rule-set for *A. gayanus* classification based on prior field-based knowledge of its growth form. *A. gayanus*does not exceed 5 m in height in the field, so we filtered all points less than or equal to 5 m above ground level as a first step in our classification scheme. We excluded points in the first 0.25 m to limit the influence of terrain and woody debris on the classification, so only points in the 0.25–5 m range were considered as potential returns from *A. gayanus*. Commonly occurring native grasses (such as *Sorghum* spp. and *Heteropogon contortus*) and woody shrubs are also present in this height zone, however their distribution is not as uniform as the continuous dense canopy of *A. gayanus* that we observe in the field. As such, we adopted a voxel-based approach (volumetric pixel, see [26] for more details) to quantify the density of LiDAR returns for each cubic meter of space across the study site. This step provided a quantification of the distribution of vegetation both horizontally and vertically in the ecosystem. We then searched for voxels that were vertically contiguous from 0.25 m upwards and flagged them as potential *A. gayanus* returns. We generated a spatial representation of potential *A. gayanus* occurrence across the study site based on voxel continuity with a 1 m horizontal resolution. This presence/absence map was aggregated to 10 m horizontal resolution to provide a cover-based distribution estimate on a 0%–100% scale.

Field-Validation of LiDAR Predictions

The *A. gayanus* classification routine was derived from expert knowledge of the vegetation growth form, and not through sample selection and training. In order to test the robustness of the classification, we conducted two separate validation tests against field data. Firstly, we compared the classification results with field survey data collected from 90 plots in April 2014. These plots measured 10 × 50 m (500 m^2) each with woody stand structure and percentage of *A. gayanus* cover visually estimated. Secondly, we conducted additional fieldwork in November 2014 to explicitly test the accuracy of LiDAR derived *A. gayanus* spatial predictions. We generated 150 validation points in a stratified random manner, assigning 15 points per 10% increment of predicted *A. gayanus* cover. Our 'blind' field team (field team had no prior knowledge of the LiDAR derived predictions) attempted to locate each of the prescribed points by differential GPS in the field, however due to land-ownership constraints and recent fires, only 85 of these points could be visited. An additional 65 points were sampled by the field team to compensate for the unreachable points, ensuring that the full range of *A. gayanus* canopy cover was sampled (total sample size = 150). At each point the presence and percentage cover of *A. gayanus* was estimated within a 10 × 10 m (100 m^2) quadrat centred on the GPS point. We used ordinary least square regression to assess the relationship between field and LiDAR derived *A. gayanus* cover estimates (LiDAR estimates were aggregated to match the 500 m^2 and 100 m^2resolutions of the field transects).

We aggregated the *A. gayanus* cover predictions into cover classes of 20% increments to explore metrics of patch size, shape and spatial configuration. We used a voxel based approach to quantify the mean vegetation canopy height profiles within each cover class to gain insight into the impact of different *A. gayanus* densities on woody vegetation canopy structure.

RESULTS

The airborne LiDAR survey provided excellent three-dimensional representations of the open tropical savanna woodland habitat. Individual trees were well defined in the 3D point cloud, with numerous returns from trunks and branches (Figure 3). Multiple returns were obtained from within the *A. gayanus* layer—rendering it clearly visible in the LiDAR point cloud. The structural differences between native grasses and shrubs (Figure 3a) were visually distinguishable from *A. gayanus*(Figure 3b), primarily through the presence of small gaps in the vegetation layer just above ground level.

Figure 3: Cross-section through the collected LiDAR point cloud, normalised to height above ground level, in (**a**) a native vegetation site, and (**b**) a *A. gayanus* invaded site. The width of the point cloud cross-section is 10 m.

The uniform nature of *A. gayanus* was a key component of the rule-based classification scheme for identifying its occurrence based on height and vertical contiguity (Figure 4a). Spatial representations of *A. gayanus* canopy cover highlighted the clumped nature of its distribution, with complete coverage on some farms (Figure 4b).

Ground-based validation of the *A. gayanus* cover mapping revealed strong correlation between cover estimates obtained in the field and from the air ($R^2 = 0.87$) with moderate error (RMSE = 12.53) at 500 m^2 resolution (Figure 5a). These results were very encouraging, however since the field-data were primarily collected for the characterisation of woody vegetation under high and low *A. gayanus* densities, a large proportion of the reference points were located within the very low and very high cover categories. As such, this validation gives us confidence in our ability to distinguish native savanna from heavily invaded savanna at the 500 m^2 scale, but provides less insight into our ability to map more subtle variations in *A. gayanus*cover. However, the results from our targeted fieldwork at 100 m^2 scale also showed good agreement ($R^2 = 0.79$, RMSE = 14.13) with LiDAR derived estimates of *A. gayanus* cover (Figure 5b). The finer 100 m^2 comparisons did contain more scatter however, and the linear model deviated from the 1:1 line with LiDAR predictions overestimating *A. gayanus* cover relative to field-based estimations.

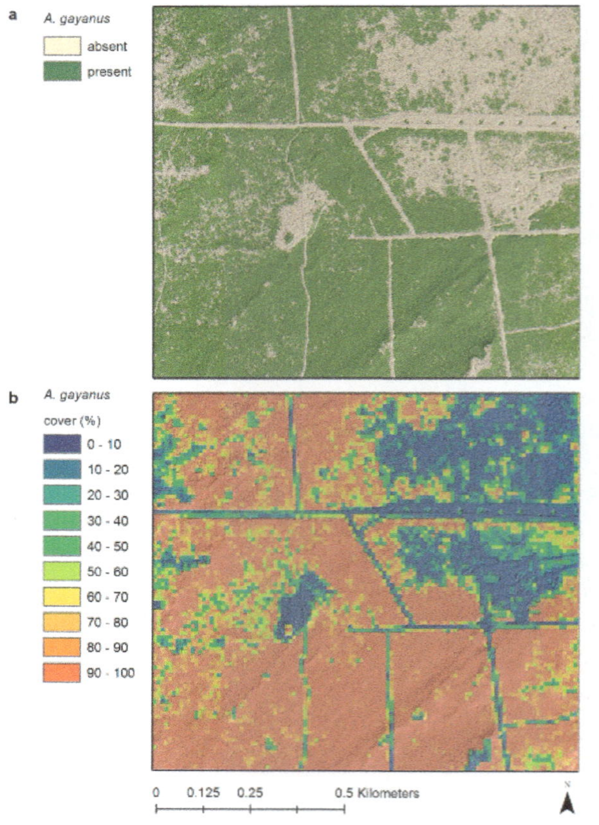

Figure 4: Airborne LiDAR based classification of (**a**) *A. gayanus* occurrence at 1 m scale, and (**b**) *A. gayanus* percentage canopy cover at 10 m scale.

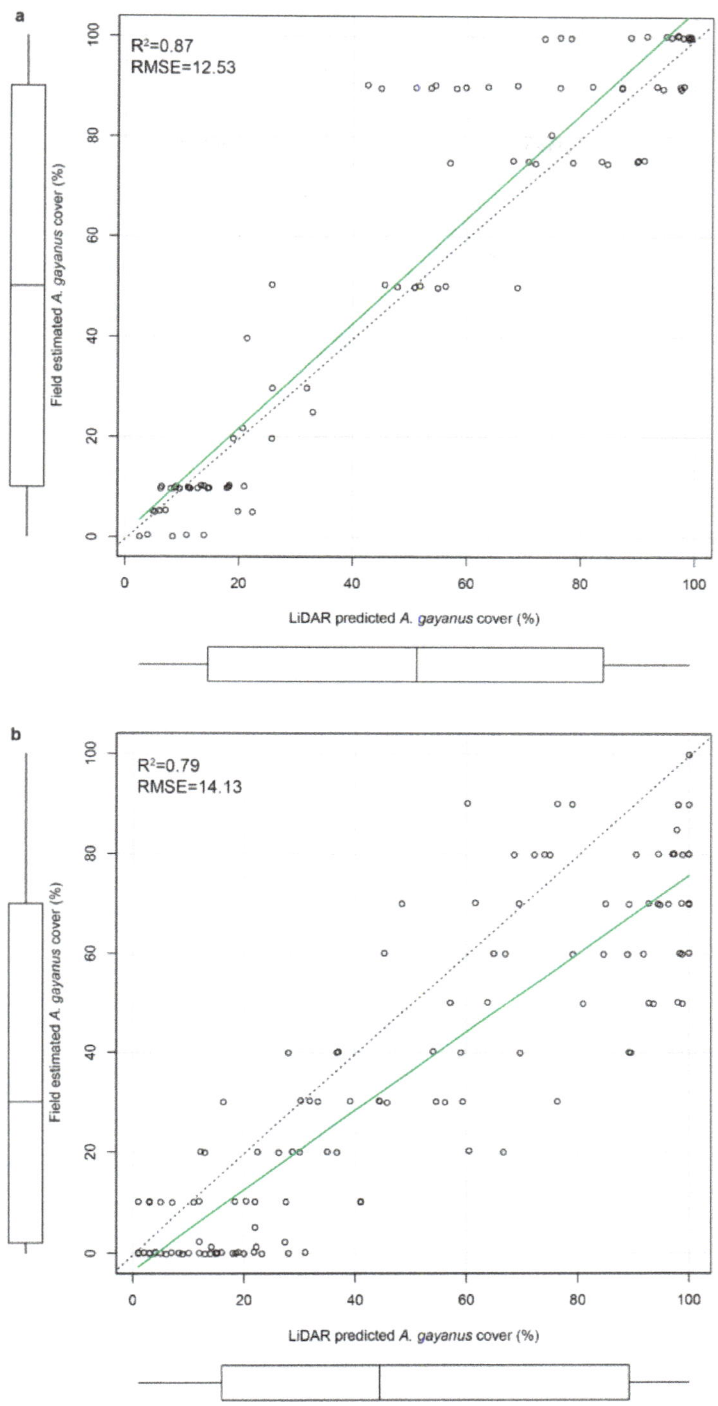

Figure 5: Validation of LiDAR estimated *A. gayanus* coverage against field-based estimates at (**a**) 500 m^2 scale and (**b**) 100 m^2 scale. Solid green line represents the ordinary least squares regression, and the 1:1 line is indicated by the dotted black line.

Mean patch size was largest for patches of low (0%–20%) and high (80%–100%) *A. gayanus* canopy cover, with minimal variation in the intermediate cover categories (Figure 6a). Areas of low *A. gayanus* cover (0%–20%) accounted for almost 45% of the study area, and heavily invaded patches (80%–100%) were present in 13% of the study site (Figure 6b).

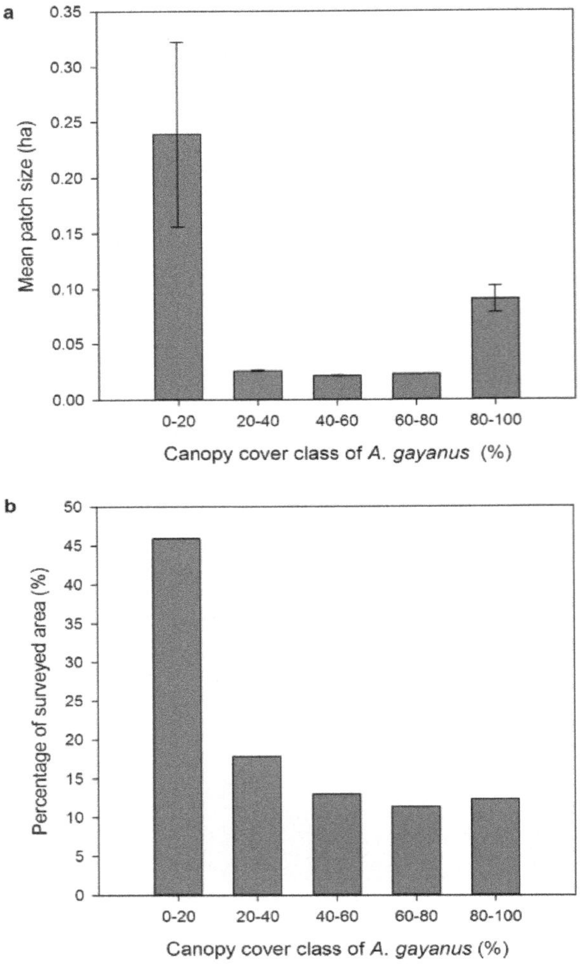

Figure 6: Patch metrics for five classes of *A. gayanus* canopy cover mapped with airborne LiDAR. (**a**) Mean size of spatially contiguous patches of five canopy cover categories. (**b**) Percentage of total area surveyed for each of the five canopy cover categories.

Areas of low (0%–20%) and high (80%–100%) *A. gayanus* cover supported dramatically different woody vegetation structures (Figure 7). The woody layer displayed a bi-modal pattern in areas of low *A. gayanus* cover, with prominent separation between a recruiting layer (<4 m) and an overstory layer with a high proportion of trees in the 8–12 m height classes (Figure 7a). Areas with high invasion contained almost no overstory canopy, with the vast majority of LiDAR returns coming from the grass itself in the <4 m height ranges (Figure 7b).

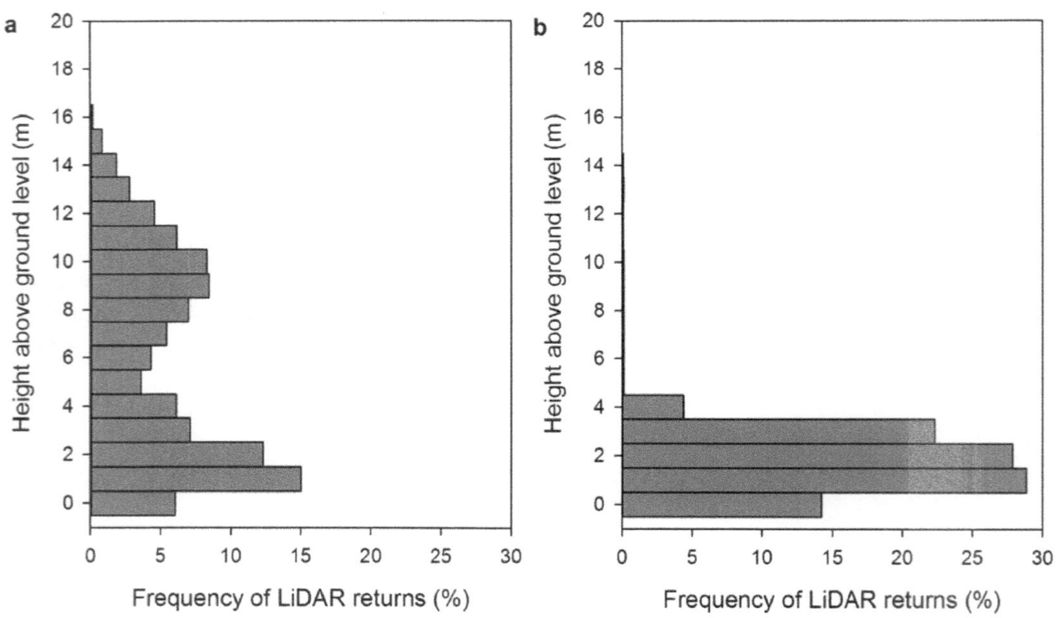

Figure 7: Canopy height profiles of vegetation growing within zones identified as containing (**a**) low *A. gayanus* canopy cover (0%–20%), and (**b**) high *A. gayanus* canopy cover (80%–100%).

DISCUSSION

Airborne LiDAR Based Estimations of A. gayanus *Cover*

Airborne LiDAR has been used extensively for the 3D characterisation of woody vegetation in southern Africa [23,27,28] and eastern Australia [29]. To date, we have not seen it applied to the characterisation of the herbaceous layer in savannas but our findings in this study indicate good potential for retrieving structural information from this zone—provided a sufficiently high pulse density with a small footprint size is achieved. The presence of *A. gayanus* was clearly discernable in the LiDAR point cloud collected during our study (Figure 3) and remotely derived cover estimates were well aligned with those obtained in the field (Figure 4). There were, however, some discrepancies between the LiDAR-field validations conducted at 500 m^2 ($R^2 = 0.87$, RMSE = 12.53) and 100 m^2 ($R^2 = 0.79$, RMSE = 14.13). In general, we would expect better correlation from the larger quadrats as the edge effect is minimized and noise is averaged out. However the differences between our 500 m^2 results (Figure 5a) and 100 m^2 (Figure 5b) do not appear to be artefacts of edge effects, nor GPS positioning (given the high confidence in differential GPS locations with decimeter accuracy). Rather, the deviation of the 100 m^2 fit from the 1:1 line was more systematic, with LiDAR predications overestimating those from the field. Unfortunately, comparisons between these two validations are confounded by the time of year that the fieldwork campaigns were conducted, and the time interval between airborne campaign and field data collection.

The LiDAR campaign was conducted in June 2013, but due to logistical constraints the first field validation (500 m^2) was conducted in April 2014 (10 months post LiDAR survey) and the second field validation (100 m^2) was conducted in November 2014 (16 months post LiDAR survey). *A. gayanus* phenological conditions in June (early dry season) are much closer to those in April (end of wet season) than those in November (end of dry season). Therefore, when *A. gayanus* cover was estimated in the field in November 2014 it had already senesced and therefore had markedly less projected areal cover that when it was assessed in the field in April 2014 or from the air in June 2013. Furthermore, it was noted that a number of sites in November 2014 had been recently burnt. As such, we do not consider the LiDAR predictions to be overestimates, and it is likely that LiDAR provides a more accurate estimate for a particular snapshot in time than what can be obtained in the field—especially when considering the difficulty of measuring *A. gayanus* cover on the ground in the first instance.

Implications for A. gayanus Invasion Ecology and Management

High-resolution and accurate mapping of *A. gayanus* occurrence and canopy cover with airborne LiDAR holds a number of important implications for advancing the ecological understanding and management of its invasion. To date, managers have not had an accurate regional map of the distribution of *A. gayanus* because most of the invasion has occurred in areas that are remote and very difficult to access, and the limited visibility and impenetrability in invaded areas (Figure 1) has meant that systematic ground-based surveys have been limited. The LiDAR mapping provides an accurate map of the area and continuity of increased fuel load resulting from invasion. The high-resolution and accurate mapping results from this study will also underpin both a weed management strategy and a fire fuel management strategy for the region.

In addition to learning more about is spatial distribution and structure (Figure 6), spatially explicit knowledge of *A. gayanus* cover can linked back to LiDAR-based analysis of the woody vegetation structure to quantify the effects of different levels of invasion on structural diversity (Figure 7). The dramatic reduction and loss of diversity in woody canopy structure under high *A. gayanus* cover is alarming from both carbon management and biodiversity conservation standpoints. There is an urgent need for effective management of *A. gayanus* in northern Australia and results from this study indicate that LiDAR can play an important role both in quantifying the effects of invasion and in informing its management.

In most cases, observing alien plants remotely requires working at the limits of at least one type of sensor resolution, be it spatial, temporal or spectral resolution, given the similarity in spectral and/or structural profiles of these native and invasive species [30]. However in this case, the structure of *A. gayanus* differs markedly from that of native vegetation and this signal is readily obtainable from very high-resolution airborne LiDAR. The drawback of this approach however, is the cost associated with collecting these types of data over large areas, although this needs to be assessed relative to costs associated with extensive ground based field surveys in often remote and/or inaccessible areas. However, flying low and slow with multiple flight lines is not feasible over broad scales and satellite based options are required for regional scale monitoring. Most satellite sensors with broad spatial coverage and high temporal resolution lack the spatial resolution required to distinguish individual plants or canopy strata. A multi-scaled approach is therefore needed to integrate information from different airborne and spaceborne sensors. Airborne LiDAR can provide detailed spatial information of *A. gayanus* cover at fine scales, which can then be used: (i) as end-member information for training satellite-based analyses at broader-scales (e.g., from Landsat, WorldView and the forthcoming Sentinel-2 mission); (ii) for informing sub-pixel unmixing to derive fractional cover estimates; (iii) for improved calibration of historical satellite based imagery; and (iv) for developing monitoring tools that provide future mapping and tracking of invasion fronts. Furthermore, by using LiDAR to obtain a robust snapshot of current *A. gayanus* distribution, it will be possible to both measure and test predictions of how fast the grass is expanding its range, by utilizing a series of repeat airborne campaigns in the future. Time-series investigations in ecology utilizing airborne LiDAR are rare due to the costs of airborne campaigns, but the few investigations that do exist in savannas have revealed important spatial insights into how ecosystems change over time [24,28,31]. Of particular interest in this region will be the long-term monitoring of invasion fronts that are well depicted from the LiDAR analysis (Figure 8).

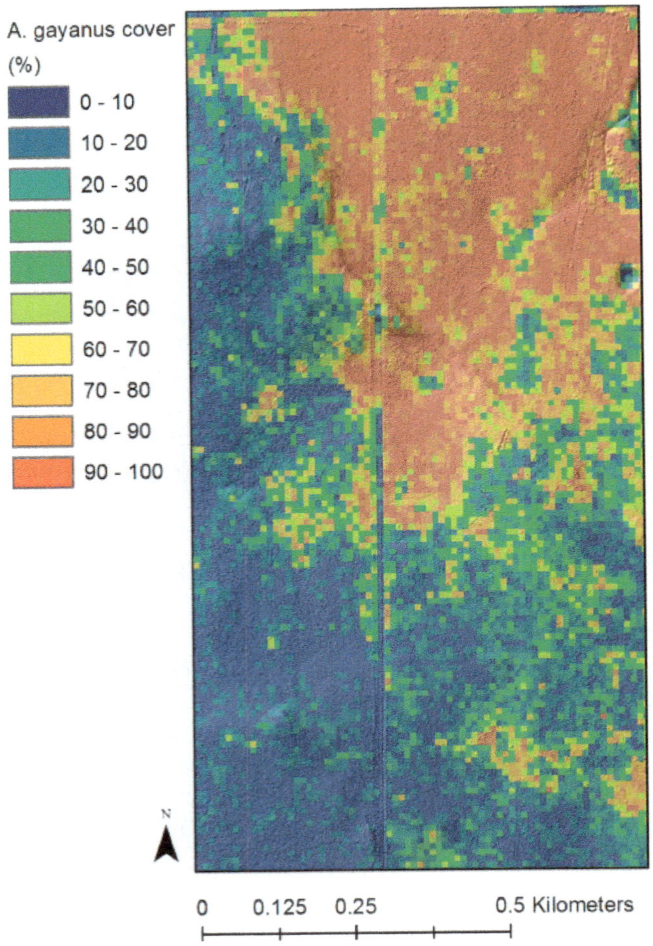

Figure 8: Spatial variability of *A. gayanus* cover mapped using airborne LiDAR showing the advancing invasion front in the region adjacent to Litchfield National Park, NT.

CONCLUSIONS

Our study highlights the potential of high-resolution airborne LiDAR to provide robust estimates of *A. gayanus* canopy cover (R^2 = 0.87, RMSE = 12.53, at 500 m^2 scale and R^2 = 0.79, RMSE = 14.13, at 100 m^2 scale). These airborne mapping results provide a solid benchmark for evaluating the rate and pattern of *A. gayanus* spread from future LiDAR campaigns. In addition, this high-resolution mapping can be used to inform satellite image analyses for the evaluation of *A. gayanus*invasion over broader regional scales. LiDAR sensor technology continues to advance, and in coming years it will be possible to obtain similar structural results shown here from higher altitudes and faster air speeds, with fewer overpasses, or via the use of unmanned airborne vehicles. Current and continued developments in LiDAR remote sensing science hold great potential for invasive plant research, monitoring and management.

ACKNOWLEDGMENTS

The German Federal Ministry of Education and Research (BMBF) and Australian Research Council (ARC) provided funding for travel related expenses as part of a Germany-Australia mobility grant (FIREBIODIV–01DR14010). Funding for data collection and field validation was provided by the Australian Department of Environment's National Environmental Research Program, Northern Australia Hub (NERP-NAH). We are grateful to Mitchel Rudge for assistance in the field.

AUTHOR CONTRIBUTIONS

Samantha A. Setterfield, Natalie A. Rossiter-Rachor and Lindsay B. Hutley conceived the study. Jorg M. Hacker conducted the airborne surveying and pre-processing. Damien MacMaster carried out the field-validation. Shaun R Levick performed the analyses. SRL, SAS, NARR, LBH, and JMH wrote the paper.

REFERENCES

1. Grace, J.; José, J.S.; Meir, P.; Miranda, H.S.; Montes, R.A. Productivity and carbon fluxes of tropical savannas. *J. Biogeogr.* 2006, *33*, 387–400.
2. Beringer, J.; Hutley, L.B.; Tapper, N.J.; Cernusak, L.A. Savanna fires and their impact on net ecosystem productivity in North Australia. *Glob. Chang. Biol.* 2007, *13*, 990–1004.
3. Walther, G.R.; Roques, A.; Hulme, P.E.; Sykes, M.T.; Pyšek, P.; Kühn, I.; Zobel, M.; Bacher, S.; Botta-Dukát, Z.; Bugmann, H.; *et al.* Alien species in a warmer world: Risks and opportunities. *Trends Ecol. Evol.* 2009, *24*, 686–693.
4. D'Antonio, C.M.; Vitousek, P.M. Biological invasions by exotic grasses, the grass/fire cycle, and global change. *Annu. Rev. Ecol. Syst.* 1992, *23*, 63–87.
5. Rossiter-Rachor, N.A.; Setterfield, S.A.; Douglas, M.M.; Hutley, L.B.; Cook, G.D. *Andropogon gayanus* (gamba grass) invasion increases fire-mediated nitrogen losses in the tropical savannas of northern Australia. *Ecosystems* 2008, *11*, 77–88.
6. Cook, G.D.; Grice, A.C. Historical perspectives on invasive grasses and their impact on wildlife in Australia. *Wildl. Soc. Bull.* 2013, *37*, 469–477.
7. Douglas, M.M.; O'Connor, R.A. Weed invasion changes fuel characteristics: Para Grass (Urochloa mutica (Forssk.) T.Q. Nguyen) on a tropical floodplain. *Ecol. Manag. Restor.* 2004, *5*, 143–145.
8. Setterfield, S.A.; Rossiter-Rachor, N.A.; Douglas, M.M.; Wainger, L.; Petty, A.M.; Barrow, P.; Shepherd, I.J.; Ferdinands, K.B. Adding fuel to the fire: The impacts of non-native grass invasion on fire management at a regional scale. *PLoS ONE* 2013, *8*, e59144.
9. Flores, T.A.; Setterfield, S.A.; Douglas, M.M. Seedling recruitment of the exotic grass *Andropogon gayanus* (Poaceae) in northern Australia. *Aust. J. Bot.* 2005, *53*, 243–249.
10. Brooks, K.J.; Setterfield, S.A.; Douglas, M.M. Exotic grass invasions: Applying a conceptual framework to the dynamics of degradation and restoration in Australia's tropical savannas. *Restor. Ecol.* 2010, *18*, 188–197.
11. Setterfield, S.A.; Rossiter-Rachor, N.A.; Hutley, L.B.; Douglas, M.M.; Williams, R.J. Turning up the heat: The impacts of *Andropogon gayanus* (gamba grass) invasion on fire behaviour in northern Australian savannas. *Divers. Distrib.* 2010, *16*, 854–861.

12. Rossiter-Rachor, N.A.; Setterfield, S.A.; Douglas, M.M.; Hutley, L.B.; Cook, G.D.; Schmidt, S. Invasive *Andropogon gayanus* (gamba grass) is an ecosystem transformer of nitrogen relations in Australian savanna. *Ecol. Appl.* 2009, *19*, 1546–1560.
13. Rossiter, N.A.; Setterfield, S.A.; Douglas, M.M.; Hutley, L.B. Testing the grass-fire cycle: Alien grass invasion in the tropical savannas of northern Australia. *Divers. Distrib.* 2003, *9*, 169–176.
14. Adams, V.M.; Setterfield, S.A. Estimating the financial risks of *Andropogon gayanus* to greenhouse gas abatement projects in northern Australia. *Environ. Res. Lett.* 2013, *8*, 025018.
15. Hutley, L.B.; Setterfield, S.A. Savanna. In *Encyclopedia of Ecology*; Jorgensen, S.E., Fath, B.D., Eds.; Academic Press: Oxford, UK, 2008; pp. 3143–3154.
16. Shaw, D.R. Translation of remote sensing data into weed management decisions. *Weed Sci.* 2005, *53*, 264–273.
17. Guisan, A.; Tingley, R.; Baumgartner, J.B.; Naujokaitis-Lewis, I.; Sutcliffe, P.R.; Tulloch, A.I.T.; Regan, T.J.; Brotons, L.; McDonald-Madden, E.; Mantyka-Pringle, C.; *et al.* Predicting species distributions for conservation decisions. *Ecol. Lett.* 2013, *16*, 1424–1435.
18. Balch, J.K.; Bradley, B.A.; D'Antonio, C.M.; Gómez-Dans, J. Introduced annual grass increases regional fire activity across the arid western USA (1980–2009). *Glob. Chang. Biol.* 2013, *19*, 173–183.
19. Pettorelli, N.; Safi, K.; Turner, W. Satellite remote sensing, biodiversity research and conservation of the future. *Philos. Trans. R. Society B: Biol. Sci.* 2014, *369*, 20130190.
20. Huang, C.-Y.; Asner, G.P. Applications of remote sensing to alien invasive plant studies. *Sensors* 2009, *9*, 4869–4889.
21. Singh, K.K.; Davis, A.J.; Meentemeyer, R.K. Detecting understory plant invasion in urban forests using LiDAR. *Int. J. Appl. Earth Obs. Geoinf.* 2015, *38*, 267–279.
22. Lefsky, M.; Cohen, W.; Parker, G.; Harding, D. Lidar remote sensing for ecosystem studies. *BioScience* 2002, *52*, 19–30.
23. Levick, S.R.; Rogers, K.H. Structural biodiversity monitoring in savanna ecosystems: Integrating LiDAR and high resolution imagery through object-based image analysis. In *Object-Based Image Analysis: Spatial Concepts for Knowledge-Driven Remote Sensing Applications*; Blaschke, T., Lang, S., Hay, G., Eds.; Springer-Verlag: Berlin, Germany, 2008; pp. 477–492.
24. Levick, S.R.; Baldeck, C.A.; Asner, G.P. Demographic legacies of fire history in an African savanna. *Funct. Ecol,* 2014, *29*, 131–139.
25. Adams, V.M.; Setterfield, S.A. Optimal dynamic control of invasions: Applying a systematic conservation approach. *Ecol. Appl.* (in press).
26. Popescu, S.C.; Zhao, K. A voxel-based lidar method for estimating crown base height for deciduous and pine trees. *Remote Sens. Environ.* 2008, *112*, 767–781.
27. Asner, G.P.; Levick, S.R.; Kennedy-Bowdoin, T.; Knapp, D. E.; Emerson, R.; Jacobson, J.; Colgan, M.S.; Martin, R.E. Large-scale impacts of herbivores on the structural diversity of African savannas. *Proc. Natl. Acad. Sci. USA* 2009, *106*, 4947–4952.
28. Levick, S.R.; Asner, G.P. The rate and spatial pattern of treefall in a savanna landscape. *Biol. Conserv.* 2013, *157*, 121–127.
29. Lee, A.C.; Lucas, R.M. A LiDAR-derived canopy density model for tree stem and crown mapping in Australian forests. *Remote Sens. Environ.* 2007, *111*, 493–518.
30. Asner, G. Hyperspectral remote sensing of canopy chemistry, physiology and diversity in tropical rainforests. In *Hyperspectral Remote Sensing of Tropical and Sub-Tropical Forests*; Kalacska, M., Sanchez-Azofeifa, G.A., Eds.; Taylor and Francis Group: Oxford, UK, 2008; pp. 261–288.
31. Asner, G.P.; Levick, S.R. Landscape-scale effects of herbivores on treefall in African savannas. *Ecol. Lett.* 2012, *15*, 1211–1217.

CHAPTER 11

SUITABILITY OF MODIS TIME SERIES METRICS TO MAP VEGETATION TYPES IN DRY SAVANNA ECOSYSTEMS

Christian Hüttich, Ursula Gessner, Martin Herold, Ben J. Strohbach, Michael Schmidt, Manfred Keil and Stefan Dech

Keywords: Land cover, plant communities, remote sensing, Kalahari, Random Forest classification, CART, MODIS, Landsat, TRMM, EVI, time series, Batthacharrya distance

INTRODUCTION

Plant communities are basic natural resource management units and provide baseline information for ecological processes and functioning in semi-arid rangelands to evaluate dynamic tendencies and grazing capacity [1]. However, there is a lack of consistent environmental geodata on a national scale in Namibia [2]. In the past, a number of vegetation survey projects have been carried out in parts of Namibia. Volk [3] and Giess [4] conducted general descriptive vegetation surveys on a national scale. Phyto-sociological and vegetation-environmental studies have been realized in selected regions, e.g., in the Khomas Hochland south of Windhoek [5], Central Namib [6,7], Waterberg [8] and in the Etosha Pan by Le Roux *et al.* [9] and Du Plessis *et al.* [10]. Nation-wide estimations of biomass and vegetation cover have been conducted using remote sensing techniques [11,12,13]. Nevertheless, approximately 60%–70% of Namibia's surface has not been analyzed in terms of vegetation composition and community structure [2], which emphasizes the need for implementing standardized bottom-up-approaches in the field of biodiversity assessment and vegetation mapping in Namibia.

New techniques have emerged during the past years to estimate and analyse land-cover and land-cover change (LUCC), land value and land functioning, strongly supported by interdisciplinary approaches of different scientific communities on land change science, remote sensing, geoinformatics, and local scale studies [14,15]. Satellite applications have proven to be an effective tool for land-cover mapping and monitoring, providing consistent spectral, spatially explicit and temporal up-to-date indicators of surface processes and status of biodiversity [16,17].

Phenological characterizations of satellite time series of the Advanced Very High Resolution Radiometer (AVHRR) onboard the National Oceanic and Atmospheric Administration satellites (NOAA) and Moderate Resolution Spectroradiometer (MODIS) onboard Terra/Aqua have been widely used for global land-cover mapping purposes [18] and mapping of southern African Biomes and Bioregions [19]. Statistical metrics computed on MODIS time series have been successfully used as input features to classify land-cover classes characterized through differing phenological patterns, such as Hansen *et al.* [20] for mapping a global continuous fields tree cover and Gessner *et al.* [21] for mapping fractional vegetation cover in Namibia on a regional scale.

Non-parametric machine learning algorithms proved to be an effective tool for land-cover mapping using a high number of input variables. Classification and regression trees (CART), as introduced by Breiman *et al.* [22], have been applied on satellite time series reflectances and vegetation indices, such as the normalized difference vegetation index (NDVI, [23,24,25]). The application of ensembles of classification and regression trees resulted in increased mapping accuracies and more stable classification results [26]. Random Forests, a tree-based ensemble classifier [27], proved to be effective for land-cover classification for mapping agricultural land-use [28] and to map ecotope classes based on hyperspectral imagery [29]. The applications of tree-based classification algorithms have focussed on a per-pixel-basis in the global and regional remote sensing community. Integrating land-cover related object attributes, such as shape and neighbourhood,

Citation: Christian Hüttich, Ursula Gessner, Martin Herold, Ben J. Strohbach, Michael Schmidt, Manfred Keil and Stefan Dech, Suitability of MODIS Time Series Metrics to Map Vegetation Types in Dry Savanna Ecosystems, doi:10.3390/rs1040620.

for classification issues have been recently applied and can enhance thematic depth and mapping accuracies. Object-oriented classification techniques have been combined with Random Forest classification for mapping agricultural lands by Watts and Lawrence [30].

Classification and regression trees (CART, [22]) are well known and applied for global classifications of plant functional types [31,32], mapping fractional vegetation cover mapping on global [20] and regional scales in Namibia [33], and classifying global and regional land-cover maps [23,29].

Results from legend harmonization analyses of the main global land-cover maps showed limited class agreements in highly heterogeneous landscapes, characterized by mixed classes of trees, shrubs and herbaceous vegetation. Thus, semi-arid savanna ecosystems, characterized as lands with herbaceous or woody understories and a forest canopy cover between 10%–30% have the lowest mapping accuracies compared with more homogeneous vegetation types [34,35]. The most recent global land-cover map on a 300 m spatial resolution developed from bi-monthly composites of the Medium Resolution Imaging Spectrometer (MERIS) onboard the ENVISAT platform is provided by the GLOBCOVER project. Validation results over classes related to savanna ecosystems such as closed to open shrubland and grassland (>15% vegetation cover, <5 m height) show insufficient users accuracies [36], which emphasizes the need for adopted regional studies to develop standardized training methods for image classification, and estimate the most important satellite-based features for classification.

Since 2000, the interdisciplinary project BIOTA-Africa (Biodiversity Transect Analysis in Africa) has collected and analyzed data on different levels of detail in Namibia, involving different observations such as plant species composition from *in-situ* data and coarse scale earth observation land-cover data on regional scale. Bottom-up approaches of ecosystem assessment and understanding of landscape functioning can provide basic information to develop reliable regional validated vegetation maps. High resolution satellite time series data provide consistent information on landscape dynamics, disturbances and land change processes. Towards the development of reliable land-cover geo-information, a thorough understanding on what detail can be mapped on each scale and how land-cover information can be integrated has to be provided to make synergistic use of the multiple observation data sources. Recognizing the existing uncertainties, especially in savanna ecosystems, this paper aims to conduct a bottom-up remote sensing-based vegetation mapping to outline the capabilities for developing reliable land-cover products in dry semi-arid ecosystems. Thus, the purpose of this paper is to:

- present an integrated concept for vegetation mapping in a dry savanna ecosystem based on local scale in-situ botanical survey data with high resolution (Landsat) and coarse scale (MODIS) satellite time series data.
- analyse the suitability of intensity-related and phenology-related metrics derived from MODIS time series for single annual and long-term inter-annual classifications from 2001 to 2007.

MATERIAL AND METHODS

Study Region

The study region, as shown in Figure 1, comprises the communal areas in the Eastern Kalahari in Namibia with a geographic extent from 17°30'E to 21°E and 19°45'S to 21°45'S. The area is characterized by a sub-continental climate with summer rain sums of 350–450 mm in the long-term annual average, usually with a high variability. The geology of the study area is dominated by aeolian Kalahari sands with sporadic rock outcrops of sandstone, limestone, schist and dolomite of the Karoo Sequence and the Damara Sequence with a mean altitude of 1,200 m [37]. The landscape can be grouped in three main vegetation types after Giess [4], the Central Kalahari, the Northern Kalahari and the Thornbush shrubland in the western part of the study region. Topographically, the transition between the Central Namibian Highlands and the Kalahari Basin is apparent from the SW-NE oriented incised omirimba (shallow water courses with no visible gradients or visible water course, typical of the arid Kalahari sand plateau [38]). The area can be classified into seven agro-ecological zones (AEZ) of common land management practices, e.g., the Southern Omatako and Fringe plains of the Central Plateau, Kalkveld, pans and stabilized dunes of the Kalahari Sand Plateau. All AEZ are characterized by extensive grazing and limited cropping [39].

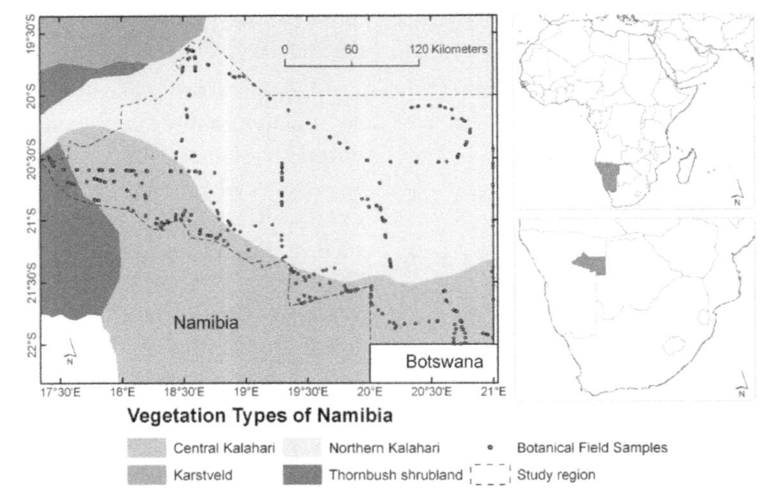

Figure 1: Overview of the study region in communal areas in the Namibian Eastern Kalahari showing the main savanna vegetation types after Giess [4] overlain with the distribution of botanical field samples [40].

Field Survey and Vegetetion Data Processing

Table 1: Synoptic vegetation type legend showing the ten vegetation types synthetised from the phyto-sociological analysis. The structural vegetation type classes after Edwards [41] as well as mean and standard deviation (Sd) of the cover of tree, shrub and herbaceous layer as sampled in the field indicate slight differences in the life form composition. Note the variances of relevé numbers due to limited accessibility and the extended sample size after the merge process with image objects from the segmentation of Landsat imagery

Synoptic Vegetation Type Legend	Vegetation structure types after Edward	Cover tree layer [%]		Cover shrub layer [%]		Cover herb layer [%]		No. of relevés	Smpl. size [pixel]
		Mean	Sd.	Mean	Sd.	Mean	Sd.		
Pterocarpus angolensis – Burkea africana woodlands (Pa-Ba)	Tall moderately closed bushland	11.1	8.5	40	7.7	40	11.6	17	758
Combretum imberbe – Acacia tortilis woodlands (Ci-At)	Tall semi-open woodlands	12.3	7.5	6.7	2.9	26.7	15.3	3	582
Terminalia sericea – Combretum collinum shrub- and bushlands (Ts-Cc)	Short moderately closed bushland	3.3	4.6	40.0	9.1	34.6	12.6	212	3.500
Acacia erioloba – Terminalia sericea bushlands (Ae-Ts)	Tall moderately closed bushland	4.6	7.0	36.7	10.5	30.8	14.3	59	3.480
Hyphaene petersiana plains (Hp_pl)	Short moderately closed bushlands	4.8	3.8	41.9	8.0	27.5	7.6	8	316
Acacia mellifera – Stipagrostis uniplumis shrublands, Typicum (Am-Su)	High moderately closed shrublands	2.9	4.8	40.0	11.6	33.7	14.4	87	3,834
Enneapogon desvauxii – Eriocephalus luederitzianus short shrublands on calcareous omirimba & pans (Ed-El)	Tall semi-open shrublands	2.2	4.4	23.6	15.1	44.7	15.6	9	568
Acacia luederitzii – Ptichtolobium biflorum floodplains of the Omatako Omuramba (Al-Pb)	Short moderately closed bushland	10.3	5.5	31.7	14.4	21.7	9.8	6	770
Terminalia prunioides thickets (Tp_th)	Tall moderately closed thickets	18.3	7.6	38.3	10.4	28.3	12.6	3	220
Eragrostis rigidior – Urochloa brachyura grasslands (Er-Ub)	Tall semi-open shrubland	0.4	0.7	16.1	8.3	72.9	17.5	12	304
Graminoid Crops (Gr_cr)	–	–	–	–	–	–	–	–	130
Bare Areas, Hardpans (pans)	–	–	–	–	–	–	–	–	120

In this study, data of different scales were used and each data type contains scale-specific land-cover information. *In-situ*land-cover descriptions resulting from the reconnaissance survey of the landscapes, soils and vegetation of the Eastern Communal Areas provide a detailed description of land types, vegetation composition and physiognomy and habitat settings. However, extensive effort is needed for updating and the spatial alienability is limited. A number of 422 randomly selected vegetation samples, as described in Strohbach *et al.* [40], were taken during the late growing season from April to May 2004 by applying a stratified sampling with a standardized plot size of 20 m × 50 m following the Braun-Blanquet approach [1]. The survey included GPS reading and the record of floristic composition and habitat information after Edwards [41] using the SOTER methodology [42]. Species and habitat data were archived in the TurboVeg database [43] of the National Botanical Research Institute (NBRI), then classified and refined using the TWINSPAN [44] and PHYTOTAB [45] packages. The resulting vegetation associations were clustered after characteristic, differentiating and typical species [40], to synopsize 12 main vegetation types according to dominant species occurrence, as shown in the synoptic legend in Table 1.

The main vegetation structure types after Edwards [41] are moderately closed to open shrub- and bushlands followed by a thicket and a woodland class. The sample size varies among the vegetation types due to the general inaccessibility of the study area, especially the sparsely populated central and eastern areas. Two additional classes (graminoid crops and bare areas/hardpans) were added to the classification legend to properly represent the major land-cover characteristics in NE Namibia.

Satellite Data

Five Landsat-7 ETM+ scenes with six reflectance bands (1–5, 7) from 0.45–2.35 μm spectral range (path 176 to 178, row 74 to 75) acquired between March and June 2000 with 30 m resolution were used as input for the object-oriented segmentation explained in Section 1.4. The MODIS collection 5 product (MOD13Q1, 16-day composites in sinusoidal projection) at the original 232 m resolution was used for time series analysis. The pre-processing included subsetting to the areal extent of the vegetation survey plots and quality analysis. Low quality data were identified based on the MODIS quality assessment science data sets [46] and gaps were filled by linear interpolation. The quality analysis was conducted using the Time Series Generator (TiSeG) software [47]. Annual time series were calculated for EVI and the blue (459–479 nm), red (620–670 nm), nir (841–876 nm), mir (1,230–1,250 nm) spectral bands [48]. Bi-annual cumulative precipitation data as averaged over the study region from the Tropical Rainfall Measuring Mission (TRMM, [49]) were used to analyse the relationship of classification error rates to bi-annual rainfall conditions.

Training Database Generation

An object-oriented segmentation was performed on five Landsat scenes to retrieve homogeneous objects based on similar reflectance settings to regionalize in-situ information of the training points on a coarse MODIS pixel size. Using the multi-resolution segmentation function in the Definiens Developer 7.0 software package [50], three segmentation levels were created on pixel level data to locally adapt the optimization procedure for computing homogeneous image segments.

The first segmentation level was generated using a scale parameter of 10. The higher levels were computed on increased shape and compactness values (level 2) and a scale parameter of 25 (level 3). Increasing the scale parameters caused that small scale image objects were merged in homogenous objects to achieve a 232 m pixel size. Each specific botanical field plot was intersected with the surrounded Landsat segments. The segments assigned with a training class label were used as training database for MODIS time series metrics, as visualized in Figure 2. The up-scaling procedure resulted in an increase of the spatial extent of the training database (compare the plot number of the botanical field survey with the sample size applied on the MODIS data in Table 1), a key issue to capture the phenological variability of semi-arid vegetation in the MODIS features within the training process. To avoid the inclusion of natural disturbance by fires in the training database, the MODIS burned area product (MOD45A1, [51]) was used to exclude training areas located on burnt training sites.

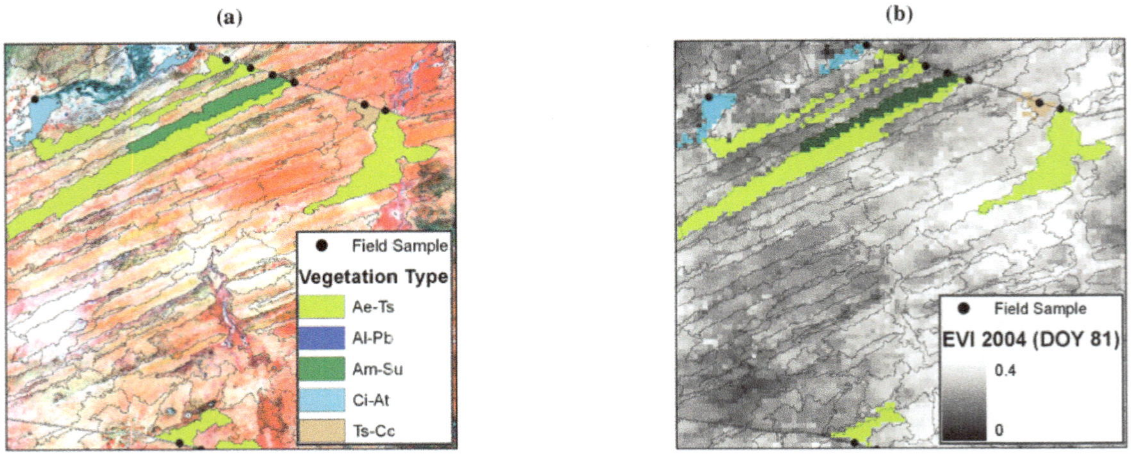

Figure 2: Examples for the generation of training data from *in-situ* to MODIS 232 m pixel size. Botanical field plots were intersected with homogeneous segments retrieved from Landsat imagery [2a]. The training data on the 232 m MODIS pixel is visualized in [2b], displayed on the MODIS image of the 81th day of the year (DOY) 2004. Note the detailed description of the vegetation type legend in Table 1.

Calculation of Time Series Metrics

The feature set for the classification was arranged from two kinds of time series metrics, such as phenological metrics and intensity-related metrics derived from a temporal segmentation of the annual time series. The metrics were computed on an annual basis from 2001 to 2007. The resultant annual statistical features were used to derive inter-annual metrics by computing the long-term statistics for each annual segment metric, as displayed in the flowchart in Figure 3.

Phenological metrics including information on the timing of recurring vegetation cycles (emergence and senescence of canopy) and were generated using the TIMESAT software [52]. An adaptive Savitzky-Golay filter was applied on the upper envelope of the seasonal EVI curve to reduce atmospheric errors and enhance the EVI values to a meaningful phenological curve. The season cut-off parameter was determined to a number of one growing season per year. The start of the growing season (SGS) and the end of the growing season (EGS) were defined as the date for which the EVI value has increased to 25% between minimum and maximum EVI. Based on SGS and EGS, the lenth of the growing season (LGS) and mid position of the growing season (MGS) was derived. Further phenological metrics were extracted that describe the curve characteristics of the seasonal EVI time series such as maximum EVI value, base EVI value, left derivative, right derivative and amplitude. The integrated EVI over the growing season was computed, a widely used proxy for net primary production (NPP).

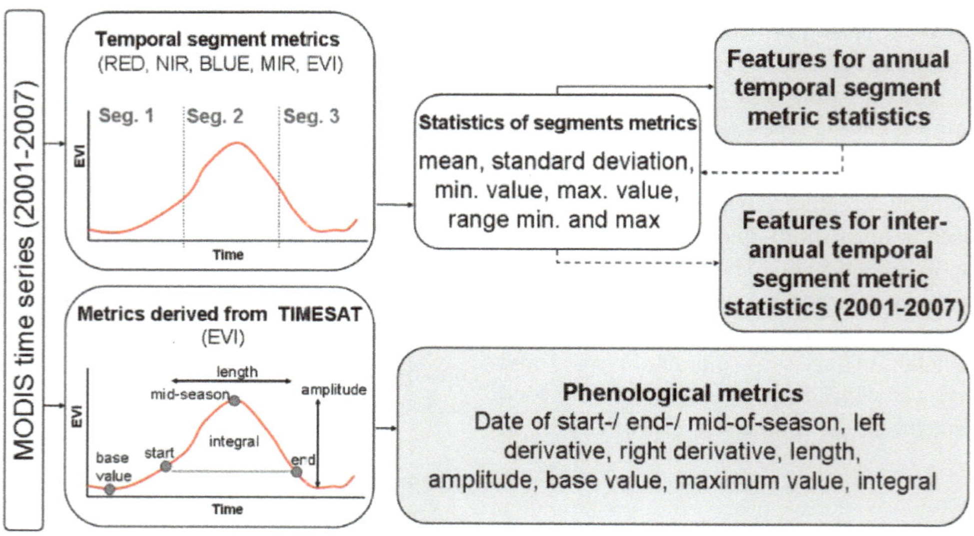

Figure 3: Flowchart of the extraction of intensity-related temporal segment metrics and phenological time series metrics derived from the TIMESAT software [52]. Note the resulting feature sets for the classification in the grey boxes.

Metrics related to reflectance intensity values of the spectral bands of the sensor were used as single date images for the 16-day EVI composites, and additional statistical metrics were computed from the temporally segmented MODIS EVI and reflectance bands. Three temporal segments per year were derived from the annual time series following the main seasonal characteristics of vegetation activity in Namibia. The first segment (Jan.–Apr.) features the main rainy season (mid- to late summer), segment two (May–Jul.) the fall and winter season and segment three (Aug.–Dec.) the hot, dry spring and early summer. Temporal mean, standard deviation, minimum, maximum, and range (between minimum and maximum) were calculated on each temporal segment from the spectral bands of the MOD13Q1 product (EVI, blue, red, NIR, MIR). The feature generation workflow is visualized in Figure 3.

Classifications were performed based on the annual and long-term feature sets. The annual classification sets consist of the 75 MODIS derived annual statistical segment metrics, 10 MODIS-derived phenological date- and NPP-related metrics, and 23 Savitzky-Golay-filtered EVI 16-day composites for the analysed year (DOY 001-353). The total number of the annual feature set was 108. The 75 statistical features from the temporal segmentation for six years of MODIS time series were used to derive 375 inter-annual features.

Separability Analysis

A separability analysis was performed to estimate the suitability of the MODIS time series derivates for classifying vegetation type units. Bhattacharyya distance (B-distance) has been successfully used for time series discriminant analysis [53], land-use classification and pattern recognition in urban areas [54]. The B-distance coefficient indicates the probable proportion of variance reduction and class discrimination for a multi-feature space. The B-distance is implemented in MULTISPEC, a free software package designed for hyperspectral image analysis [55]. The advantage of using B-distance compared with other common distance measures like transform divergence and Jefferies Matusita distance is its large dynamic range, which does not saturate when applied to a large set of features. The classification was performed using a Random Forests. Those non-parametric tree-based classifiers do not require a Gaussian distribution of the data. However, B-distance is a parametric separability measure. Regarding the differing statistical assumptions, a direct combination of both methods is problematic. Here, B-distance and Random Forests were applied independently. Bhattacharyya distance measures were computed on two different feature sets. The first set includes the date- and NPP-related phenological metrics and the annual seasonal segment metrics of the year 2004 as the reference year for the vegetation survey with a total number of 108 features. The second comprises a number of 375 features of long-term temporal segment metrics (2001 to 2007).

Random Forest Classification

Decision trees (DT) are structured in simple binary decisions. They are independent of data distribution and can handle categorical variables. The hierarchical structure further allows a biogeophysical interpretation of the relationship between input features and classes and can be useful if multi-source high dimensional remote sensing data is used. Bagging (bootstrap aggregation) attains to reduce variance of a classification by training a number of weak classifiers with varying bootstrap samples from the training set and subsequently averaging the predictions. Boosting is based on classifications using different takes of weighted training sets to be combined in the resultant prediction [56]. Comparisons of bagging using Random Forest and boosting based on Adaboost achieved best accuracies for Adaboost at the cost of computation time [26,29,57].

Random Forest is a tree-based classifier where multiple trees are produced and combined based on equally weighted majority voting. A randomly selected third of the original training dataset is excluded for training each particular tree. This so-called out-of-bag (OOB) bootstrap sample is randomly permuted among the input features for each tree. With the remaining 2/3 of the training data, trees are grown to their maximal depth using the impurity gini index [22] due to the fact that the random permutation of samples and features antagonizes overfitting. Here, the Random Forest package implemented in the R statistics language [58] was used.

The OOB sample is used to estimate the prediction error for each permutation. Gislason [57] and Breiman [27] showed that the prediction error based on the OOB sample is slightly higher than using an independent test set. The more conservative OOB error was used to estimate classification accuracies. The variable importance function implemented in Random Forest is based on the internal OOB error estimates. The prediction error is computed on every tree based on the OOB bootstrap sample. In a second step, the OOB error is computed by permuting each predictor variable. The difference in the accuracy measures are averaged over the tree ensemble. The accuracy for each variable is normalized by the standard error [58]. In this study, variable importance is examined to evaluate the used features in an ecological context. A detailed overview of Random Forests is given in Breiman [27].

RESULTS

Suitability of MODIS Time Series Metrics for Semi-Arid Vegetation Mapping

The characteristic phenological patterns of each vegetation type is represented by the Enhanced Vegetation Index (EVI) time series, shown in Figure 4, exemplary for the rainy season of the year 2004 with the dry-season transitions. Slight differences in the amplitude of green vegetation become apparent by comparing the maximum EVI values of moderately closed and semi-open shrub- and bushland classes. Semi-open vegetation types reach maximum EVI values between 0.45 and 0.5, while closed vegetation classes reach their maximum at 0.55. Distinct differences are visible in the green-up and senescence between vegetation structural types.

In general, as shown in Figure 4a,Figure 4b, a significantly steep slope during green-up is apparent within the moderately closed shrub- and bushland classes groups *Acacia erioloba - Terminalia sericea* bushlands (Ae-Ts), *Terminalia sericea - Combretum collinum* shrub- and bushlands (Ts-Cc), and *Acacia mellifera - Stipagrostis uniplumis* shrublands (Am-Su). Compared with that, a slow increase in photosynthetic activity is visible in *Pterocarpus angolensis - Burkea africana* woodlands (Pa-Ba) and *Hyphaene petersiana plains* (HP_pl). *Acacia luederitzii - Ptichtolobium biflorum* floodplains of the Omatako

Omuramba (Al-Pb) and *Terminalia prunioides* thickets (Tp_th) have a distinct offset in greening and early senescence. For the rain period of 2003–2004, the time frame of green-up onset varies between end of November and January (day of year 337/001). The semi-open shrub- and bushland vegetation types are also characterized by a slight increase of vegetation activity and a lower peak in amplitude. An appreciable impact of the so-called small rainy season (varying in time and intensity from September to December) is visible in the *Eragrostis rigidior - Urochloa brachyura* grasslands (Er-Ub) expressed by a small peak before the main vegetative activity, which may be typical for vegetation types with a high grass cover.

The results of the separability analysis indicate the temporal and spectral capabilities of the MODIS-MOD13Q1 product to discriminate vegetation type patterns. The Bhattacharyya distance indicates a relative measure of how reliable one class can be statistically separated compared to the remaining classes. A low distance value indicates a low separability and *vise versa*.

Table 2 showing average Bhattacharyya distance measures between Kalahari vegetation type classes, calculated on MODIS intra-annual segment metrics for the season 2003–2004 (lower left values), indicate strong spectral and temporal confusions and thus a more uncertain classification between vegetation types with similar phenology, e.g., *Acacia mellifera - Stipagrostis uniplumis* shrublands, *Acacia erioloba - Terminalia sericea* bushlands, *Terminalia sericea - Combretum collinum* shrub- and bushlands (Ae-Ts vs. Ts-Cc = 0.2, Ae-Ts vs. Am-Su = 0.31). The highest mean inter-class B-distance values reached the Bare areas/ Pans caused by significant delay of the growing period due to late green-up onset caused by a typical flooding situation in the end of the rainy season and a higher surface albedo compared with shrub- and grasslands. The significance of bands within the visible and middle infrared spectral range for discrimination of vegetation community classes due to different soil settings is discussed later in Section 4. Highest B-distance values reach *Combretum imberbe - Acacia tortilis*woodlands (Ci-At) and *Hyphaene petersiana* plains (HP_pl) with 44.2. The reason for that can be seen in contrary phenologies. Ci-At has a stronger slope and early onset and slight slope of decline at the end of the rainy season, whereas HP_pl is characterized by a delay in onset of green-up and fast senescence.

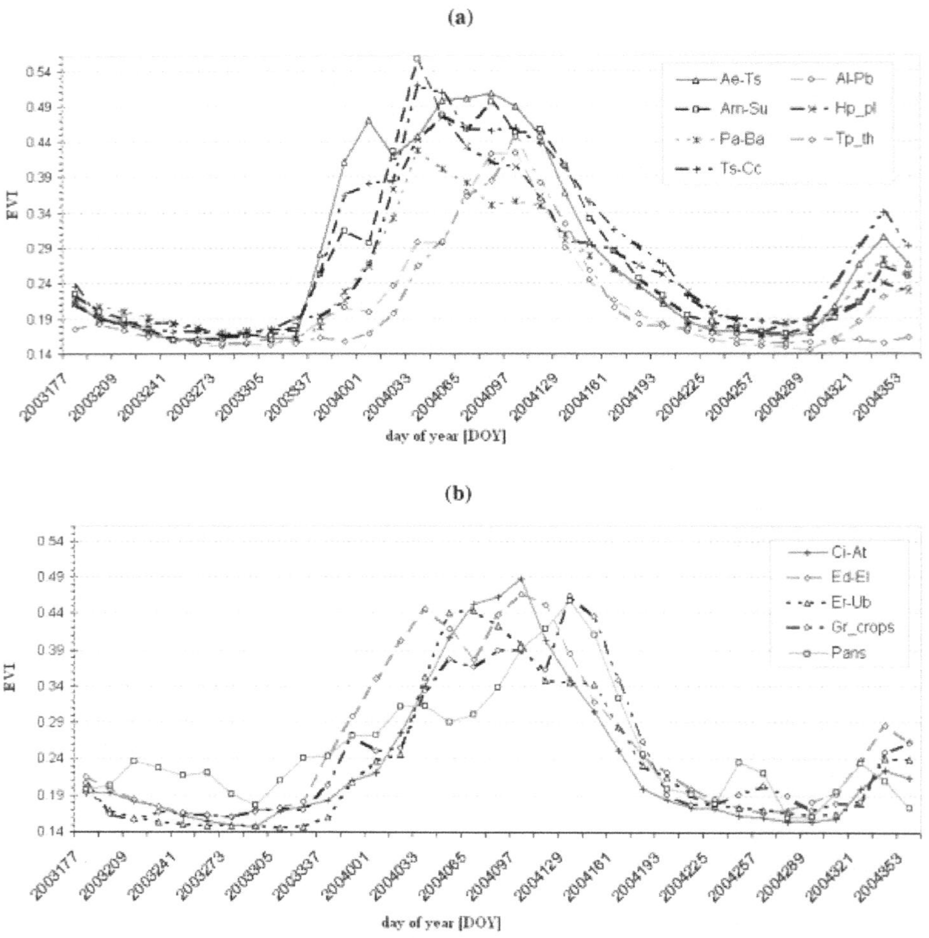

Figure 4: EVI time series of the year 2003–2004 averaged for the vegetation type classes with 4a moderately closed shrub- and bushland vegetation cover and 4b semi-open shrub- and bushland vegetation after Edwards [41] and graminoid crops and pans shown in the MODIS (MOD13Q1, 232 m) Enhanced Vegetation Index (EVI) smothed with a Savitzky-Golay filter. See Table 1 for class labels.

The comparison of the long-term statistics of inter-annual MODIS segment features from 2001 to 2007 (upper right italic values, shown in Table 2) indicate that the inter-class B-distance values could significantly increase by a factor of approximately ten. Lowest average B-distance values reached again *Acacia mellifera - Stipagrostis uniplumis* shrublands, *Acacia erioloba - Terminalia sericea* bushlands, *Terminalia sericea - Combretum collinum* shrub- and bushlands.

Table 2: Matrix showing average Bhattacharyya distance measures between Kalahari vegetation type classes calculated on MODIS intra-annual segment metrics for the season 2003–2004 (lower left) versus long-term inter-annual MODIS segment features (upper right, italic values) from 2001 to 2007. See Table 1 for detailed class labels. Note the bold values benchmarking examples for the lowest and highest B-distance values

Annual MODIS segment features 2004	Inter-annual MODIS segment features 2001-2007											
	Gr_cr	Ae-Ts	Al-Pb	Am-Su	Ci-At	Ed-El	HP_pl	Pa-Ba	Tp_th	Ts_Cc	Er_Ub	Pans
Gr_cr	-	*30.6*	*131*	*31.2*	*191*	*44.8*	*62.3*	*72.2*	*54.5*	*36.9*	*63.8*	*140*
Ae-Ts	3.42	-	*43.7*	***2.87***	*125*	***7.44***	*33.6*	*23.3*	*25.8*	***5.78***	*30*	*151*
Al-Pb	13.4	6.31	-	*37.1*	*77.7*	*63.8*	*167*	*140*	*128*	*81.3*	*103*	*146*
Am-Su	3.49	**0.31**	5.94	-	*124*	***11.8***	*31.8*	*30.5*	*21.8*	***13.5***	*38.7*	*142*
Ci-At	23.8	18.9	10.5	18.8	-	*151*	*284*	*211*	*241*	*153*	*120*	*144*
Ed-El	7.17	**1.47**	6.73	**1.53**	21	-	*30.2*	*37.9*	*31*	***12.3***	*44.1*	*155*
HP_pl	13.3	7.03	**23.1**	6.89	**44.2**	7.5	-	*79.6*	*48.7*	*41*	*116*	*162*
Pa-Ba	7.52	**1.8**	11.6	2.37	**29.7**	3.28	4.64	-	*59.2*	*16.9*	*31.6*	*159*
Tp_th	8.46	6.72	10.5	5.99	**26.6**	9.21	17.3	8.51	-	*40.8*	*78.6*	*143*
Ts_Cc	3.78	**0.2**	6.25	**0.24**	18.3	**1.42**	6.43	**1.61**	6.3	-	*27.4*	*162*
Er_Ub	7.71	5.63	8.18	6.31	8.6	6.73	21.1	11.5	13.6	5.65	-	*146*
Pans	15.3	14.7	17.6	14.5	15.4	16	18.8	17.6	14	14.4	12.1	-

Vegetation Type Mapping

The vegetation type mapping is based on a vegetation survey in Namibia's eastern communal areas and describes the transition zone between Kalahari to the Otavi mountains where calcareous geology is dominating in the NW and the mountain savanna transition zone at the SW border of the study area. The resulting vegetation type map is shown in Figure 5.

In addition to the Omatako Omuramba, a number of smaller omirimba valleys cross the study area in SW-NE direction. In general, the spatial distribution of vegetation types follows the major topographic units. The *Combretum imberbe - Acacia tortilis* woodlands are typical for the main river bed of the Omatako Omuramba, dominated by bright soils, as displayed in Figure 5b.

These open woodlands change to *Enneapogon desvauxii - Eriocephalus luederitzianus* short shrublands on calcareous omirimba and pans (Ed-El), visible in the lower river course where the *Acacia luederitzii - Ptichtolobium biflorum* floodplains of the Omatako Omuramba (Al-Pb) are disappearing. Besides the Omatako river, the smaller omirimba valleys are mainly mapped as Ed-El or *Eragrostis rigidior - Urochloa brachyura* grasslands (Er-Ub), whereas their periphery is mapped as linear patches of *Acacia mellifera - Stipagrostis uniplumis* shrublands (Am-Su), (Figure 5c). *Acacia mellifera - Stipagrostis uniplumis*shrublands (Am-Su) and *Acacia erioloba - Terminalia sericea* bushlands (Ae-Ts) are the dominating vegetation types in the southern part of the study area classified as Thornbush shrubland and Central Kalahari class after Giess [4] and mark the transition zone to the central Kalahari. *Terminalia sericea - Combretum collinum* shrub- and bushlands (Ts-Cc) are dominant on deep Kalahari sands.

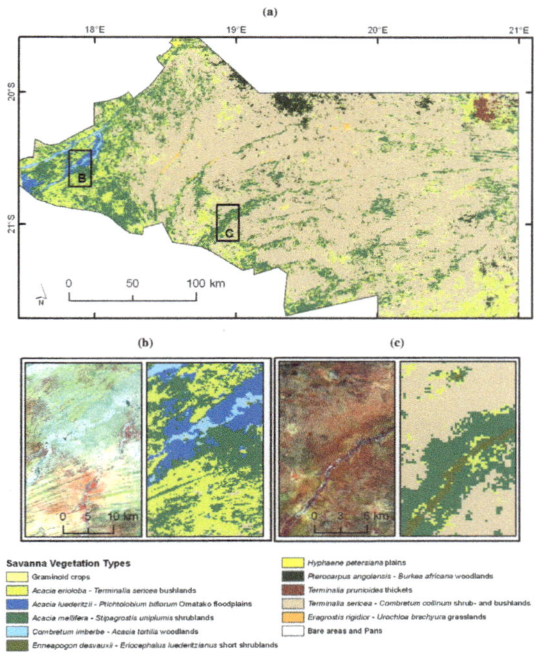

Figure 5: Vegetation type classification derived from inter-annual MODIS time series metrics (2001–2007) based on Random Forest classification. The vegetation type map is shown in 5a. 5b and 5c show the classification result for examples of the Omatako River region (Box B) and an Omarumba valley cut deep into Kalahari sands (Box C), 5b and 5c are compared to Landsat-TM images (RGB-4-3-2).

Due to different soil conditions of the historical longitudinal dune system, e.g., in the eastern part of the study region, *Acacia mellifera - Stipagrostis uniplumis* shrublands were mapped. *Terminalia prunioides* thickets (TP_th) are confined to a patch in the most NE part of the study area, associated with shallow soils on sub-outcropping limestones of the Otavi group [59]. *Hyphaene petersiana* plains (HP_pl) typically occur on calcareous soils and floodplains located at the periphery of the Karstveld. *Pterocarpus angolensis - Burkea africana* woodlands (Pa-Ba) are native in the northern Kalahari woodlands and were detected in the central to northern Kalahari transition. Due to misclassifications, patches of *Pterocarpus angolensis* - *Burkea africana* woodlands *Hyphaene petersiana* plains were detected in the central Kalahari.

Classification Error Assessment

The classification accuracies of the classifications performed on the six annual sets between 2001 and 2007 and the inter-annual set including the complete time series database are notably different.

As described in Section 2.6, the map accuracy was calculated on the out-of-bag data randomly selected in each classification iteration. Every pixel in the training database will be used as reference for the estimation of the classification error.

The range of the kappa statistics for annual classification sets is between 0.87 and 0.91. The inter-annual classification set reaches an increased kappa coefficient of 0.93. The resulting maps contain a higher omission error rate with user's accuracies ranging from 83.79% for the year 2006–2007 to 88.41% for the season 2004–2005. The lowest commission error rate was reached also for the season of 2004–2005 with a producer's accuracy of 96.11%, whereas the season of 2006–2007 scored lowest producer's accuracies of 94.17%. The inter-annual classification set achieved comparatively higher producer's accuracies (97.73%) than user's accuracies (94.86%).

The long-term feature set resulted in good classification accuracies for all classes in a range from 85% to 100%, where bare areas and pans reached best results with 100% user's and producer's accuracies. Semi-open shrub- and bushland vegetation types show increased spectral confusions and thus reached higher commission errors, e.g., *Enneapogon desvauxii* - *Eriocephalus luederitzianus* short shrublands on calcareous omirimba and pans (Ed-El) with 85.03%. Lower producer's accuracies reached *Acacia mellifera* - *Stipagrostis uniplumis* shrublands (Am-Su, 95.09%), *Acacia erioloba* - *Terminalia sericea*bushlands (Ae-Ts, 96.18%), *Enneapogon desvauxii* - *Eriocephalus luederitzianus* short shrublands on calcareous omirimba and pans (Ed-El, 96.60%), and *Terminalia sericea* - *Combretum collinum* shrub- and bushlands (Ts-Cc, 93.48%). More significant differences become apparent if seasonal and inter-seasonal classification sets are compared. Graminoid crops were mapped with user's accuracies between 45% for the season 2006–2007 and 59% in the 2003–2004 season, whereas 93.84% user's and 100% producer's accuracy were estimated based on the long-term feature set. Similar situations were observed for the *Enneapogon desvauxii* - *Eriocephalus luederitzianus* short shrublands and *Eragrostis rigidior* - *Urochloa brachyura* grasslands, as shown in Table 3.

Seasonal and inter-seasonal MODIS time series features were tested for their suitability to map plant communities of different savanna vegetation types in the Kalahari and Kalahari-transition zone. B-distance values indicate a spectral separability that is approximately ten times higher for the inter-seasonal dataset than the seasonal set of MODIS time series metrics. This result is also represented in the classification error rates. Whereas the classification error range varies between the different classes in the annual classification, a more balanced result among the classes was achieved in the long-term set, ending up in a more accurate map with a Kappa coefficient of 0.93.

Table 3: User's-, producer's and overall accuracies and Kappa coefficients for six annual and the inter-annual classifications

class	2001/02		2002/03		2003/04		2004/05		2005/06		2006/07		2001–2007	
	users	*prod.*	users	*prod.*	users	*prod.*	users	*prod.*	users	*prod.*	users	*prod.*	users	*prod.*
Gr_cr	50.76	*91.66*	52.30	*93.15*	59.23	*100.00*	48.46	*98.43*	62.30	*100.0*	45.38	*100.00*	93.84	*100.00*
Ae-Ts	94.10	*92.29*	93.93	*92.64*	95.53	*93.02*	94.64	*91.10*	93.50	*89.49*	93.47	*87.42*	96.35	*96.18*
Al-Pb	95.84	*96.72*	95.97	*96.72*	95.58	*94.72*	97.14	*97.01*	97.14	*95.16*	92.33	*92.45*	98.96	*97.94*
Am-Su	92.67	*89.94*	92.98	*89.52*	94.26	*91.07*	92.93	*91.94*	92.22	*90.34*	90.32	*90.53*	96.63	*95.09*
Ci-At	90.89	*97.06*	91.75	*96.56*	91.23	*93.98*	96.90	*96.90*	95.53	*97.03*	87.62	*94.61*	96.28	*97.96*
Ed-El	75.52	*91.86*	75.00	*92.40*	77.81	*91.32*	77.81	*96.71*	69.71	*92.74*	67.78	*93.67*	85.03	*96.60*
HP_pl	92.72	*97.34*	94.30	*98.02*	90.82	*97.61*	93.67	*95.48*	92.08	*97.32*	91.13	*94.11*	97.78	*96.86*
Pa-Ba	94.19	*97.01*	93.27	*97.11*	91.95	*95.74*	96.04	*97.32*	90.50	*97.86*	*92.48*	*96.42*	*97.22*	*99.86*
Tp_th	81.36	*96.75*	81.81	*98.36*	91.36	*99.01*	96.36	*99.53*	82.27	*98.36*	90.90	*99.00*	95.90	*100.00*
Ts_Cc	93.34	*89.09*	93.11	*89.28*	91.85	*90.03*	92.22	*89.71*	91.37	*87.97*	91.88	*87.22*	95.48	*93.48*
Er_Ub	73.35	*95.70*	75.00	*95.39*	71.71	*96.88*	75.65	*99.13*	73.02	*97.36*	71.38	*96.44*	84.86	*98.85*
Pans	93.33	*98.24*	93.33	*98.24*	95.83	*99.13*	99.16	*100.00*	98.33	*100.00*	90.83	*98.19*	100.00	*100.00*
overall accuracy	**85.67**	***94.47***	**86.07**	***94.78***	**87.26**	***95.21***	**88.41**	***96.11***	**86.50**	***95.30***	**83.79**	***94.17***	**94.86**	***97.73***
Kappa	**0.91**		**0.90**		**0.90**		**0.91**		**0.88**		**0.87**		**0.93**	

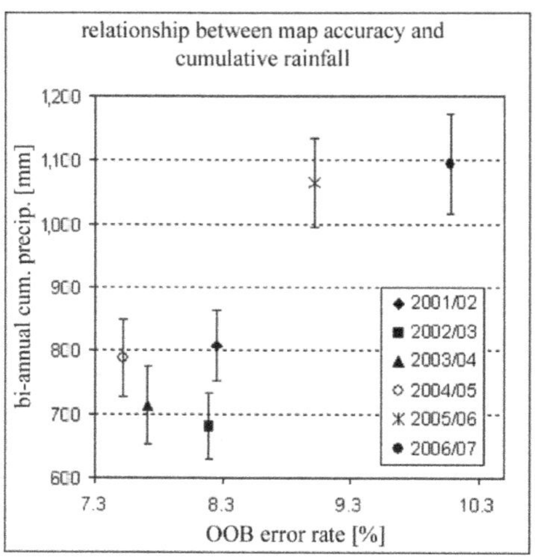

Figure 6: Relationship between the out-of-bag (OOB) error rates for seasonal vegetation type classifications and cumulative bi-annual rainfall (mean and standard deviation). Note the increasing OOB error rates with increasing precipitation.

Precipitation data from the Tropical Rainfall Measuring Mission (TRMM) were used to compare the annual classification error rates with annual precipitation sums. Therefore, bi-annual cumulative rainfall sums were averaged over the study area. Figure 6 shows the comparison of annual OOB error rates and bi-annual cumulative rainfall. Increased OOB error rates occurred in years with high precipitation sums. Compared 00with the previous years, the growing seasons of 2005–2006 ad 2006–2007 with precipitation sums in a range of 1,000 mm to 1,160 mm were wet rainy seasons. These years were mapped with increased error rates from nine to ten percent overall OOB error. In contrast, the seasons of 2002–2003, 2003–2004, and 2004–2005 were characterized by low rainfall below 800 mm. Comparatively low error rates below eight percent were achieved for those seasons.

DISCUSSION

Requirements for Rainfall Amount for Vegetation Type Mapping

To summarize the suitability of multi-temporal MODIS time series features for vegetation type mapping in dry semi-arid savannas, the "best map" could be achieved using inter-annual segment metrics. Significant characteristics from MODIS time series metrics between the mapped vegetation type associations become apparent by including in this case a number of six annual growing periods. The vegetation mapping based on annual satellite time series achieves more reliable results for dryer years.

As shown in Table 1, the major life forms of the vegetation types in the Kalahari are characterized by open to closed shrublands, leading to very similar percent coverage values of the particular life form. This can be challenging in terms of an accurate statistical discrimination when coarse resolution satellite imagery is used. EVI time series are a measure of photosynthetic active vegetation due to the increased reflection of healthy vegetation in the near infrared band. Increased EVI values will be observed due to high rainfall rates. The closer the vegetation cover or the greenness of vegetation, the higher are the observed EVI values. Similar EVI observations for different classes can cause a decreased spectral separability among life form classes. For example, similar EVI values can be observed in a shrub-dominated savanna as in a grass-dominated savanna, if rainfall sums are high enough to produce a full coverage of green vegetation on the MODIS pixel. On a 232 m resolution an observation of a pure pixel for one single homogeneous life form group is improbable in savanna ecosystems due to the heterogeneous vegetation structure. A spectral separation of similar life form compositions, for instance the shrub and herbaceous layer, is therefore difficult, as is reflected by decreased classification accuracies.

The similar intensity level of EVI values in the growing season of different life-form classes may be the reason for decreased classification accuracies in the wet rainy seasons of 2005–2006 and 2006–2007. On the other hand, the dry rainy seasons of 2002–2003, 2003–2004, and 2004–2005 effected an increased significance of the reflectance characteristics of the underlying soil properties. The substantial influence of soil properties in local variations in vegetation phenology is discussed by Zhang *et al.* [60] and thus, the significance of soil-related features for mapping vegetation types in open savannas is discussed in Section 4.2. One reason for the increased accuracy of a long-term feature set can be explained by a characteristic response of each vegetation type class to precipitation conditions, as discussed in Klein *et al.* [61]. The variability of the phenological cycle of different life-form classes of savanna vegetation was analysed by Archibald *et al.* [62]. They showed that woody vegetation is less sensitive to inter-annual variability than grasses. Memory effects of tree species, as discussed below, and differing environmental cues on plant phenology, such as soil moisture for grasses and temperature for some tree and shrub species, can explain those differences. An inclusion of features in the classification process containing information on the inter-annual variability of the photosynthetic activity during the growing seasons, which is a response of spatially and temporally variable rainfall events, can help to separate different life-form classes in savannas.

However, sub-Saharan ecosystems have a high sensitivity to short-term rainfall variability, which can cause significant effects on land-cover dynamics [63]. Hence, vegetation phenology is strongly related to precipitation amounts. Further research has to be conducted to develop a comprehensive understanding of the phenology in dry savannas and their interactions with precipitation patterns.

Spectral and Temporal Requirements for Dry Savanna Vegetation Mapping

Remote sensing applications dealing with time series images can be time consuming in terms of image processing and memory allocation. Variable importance can be useful for a pre-selection of relevant input variables to develop an ecologically-based understanding of remote sensing parameters. Here, the variable importance is used to highlight the most useful time series metrics for mapping in an open savanna ecosystem.

Figure 7 displays the 15 most important features for the classification of the seasonal datasets (2003–2004) for the intensity-related and phenology-related time series metrics, showing the class-wise variable importance score. An analysis of the variable ranking shows that EVI is the most frequently used variable followed by the middle infrared band. Less scored bands are the red and near infrared bands. The list of the 15 high-ranked features indicates a major importance of the phenological information in the EVI metrics. The high-ranked middle infrared and blue bands indicate that reflectance information in the pedological context to a vegetation type is of major importance to distinguish vegetation type classes. The spatial distribution of semi-arid vegetation types is related to subjacent soil properties. Soil type and color are represented in the visible spectral range (soil color) and middle infrared spectral reflectance (soil properties). Regarding the similar life form composition of the vegetation type classes, as displayed in Table 1, the phenological patterns were observed to be very similar. Therefore, the information content of EVI may be too low for accurate vegetation type class discrimination which highlights the importance of the blue and middle infrared band in the variable ranking.

Figure 7 indicates the dominance of intensity-related features from the temporal segmentation (e.g., BLUE 2004 Seg-3 Max, coding the maximum value of segment three in

2004 of the blue band) followed by the occurrence of the EVI 16-day composites (e.g., EVI 2003 DOY-305, coding the EVI value of the day of the year 305 in 2003).

The analysis of the variable importance on a per-class basis indicates the bio-physical relevance of each time series metric for an increase of the accuracy in the tree ensemble. An example is shown for the maximum reflectance of the blue band in the third segment of 2004 (BLUE 2004 Seg-3 Max) assigning the highest relevance for the class bare areas and pans. Due to periodic flooding in pans, the highest variance reduction in the tree ensemble can be achieved using spectral information of the blue band. Similar patterns of increased variable importance of blue reflectance can be observed for the *Eragrostis rigidior - Urochloa brachyura* grasslands and graminoid crops. Both classes are characterized by sparse vegetation cover and assign increased importance values for the features related to soil properties. Similar findings on the response of herbaceaous and woody life form groups were discussed in Archibald *et al.* [62].

Beside the spectral information, the occurrence of date-related features such as the single EVI 16-day composite layers strengthens the hypothesis that dry savanna vegetation types can be statistically discriminated from their phenological patterns. An example is given for the open shrubland classes *Hyphaene petersiana* plains and *Enneapogon desvauxii - Eriocephalus luederitzianus* short shrublands on calcareous omirimba and pans, where EVI features of the beginning rainy season score increased importance values. In general, date-related features of the early start of the growing period have increased relevance for discriminating open savanna vegetation types. An example is given for the Acacia- dominated savanna classes where the EVI 16-day composite of the day of the year 305 (EVI 2003 DOY-305) scores increased positions. As shown in Figure 4 this feature marks the immediate time before the main growing season.

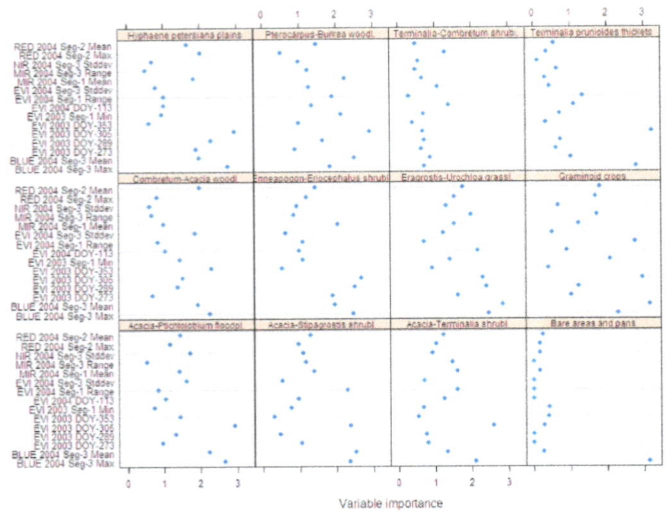

Figure 7: Variable importance visualizing the 15 top-ranked MODIS time series metrics for mapping Kalahari vegetation types based on the seasonal feature set 2003–2004 as decrease of OOB error.

The availability of moisture is a key driver for phenological events of Kalahari Sand species dominated by sandy soils, as discussed by Childes [64]. Changes in soil moisture will affect changes in the species phenology. The degree of syncronisation of phenological events is a function of environmental cues. Leaf-flush in accordance to the first rainfall is an important environmental cue for flush stimulation. Shackleton [65] argue that the initiation of leaves is, beside rainfall, triggered by temperature. This hypothesis confirms the dominance of EVI features of the beginning growing season in the importance ranking (DOY 273, 289, 305, 353). The exposing importance values of the EVI feature coincide with the immediate time before the first rains can be explained by a pre-rain flushing, an effect observed in Australian mesic tropical savannas [66] and for tree and shrub species in Southern African savannas [64]. Examples are shown for the vegetation types dominated by *Terminalia* and *Acacia* species in Figure 7.

The absence of phenological metrics extracted from the TIMESAT package in the top-rank positions indicates a minor relevance of those features for the classification. The reason may be the static concept of defining the threshold of start and end of season. For vegetation types related to pre-rain flush, a threshold of 25% of the seasonal amplitude may be too high to generate meaningful phenological features for the class discrimination. The inclusion of single EVI layers and temporal segment metrics in the feature set seems therefore more useful to classify vegetation types in the Kalahari than using phenological metrics. Further research has to be conducted to analyse the sensitivity of phenological metrics derived from satellite time series regarding environmental cues of semi-arid vegetation.

CONCLUSIONS

This study highlights the importance of combining local scale botanical assessments with coarse scale remote sensing applications. It was demonstrated that even though different vegetation type associations showed similar phenological characteristics, they gave good classification results. Improved accuracies were be achieved by the integration of inter-annual time series metrics of six years, which highlights the importance for studying longer-term inter-annual dynamics in dry savanna ecosystems. The feature ranking indicated a high potential of the spectral bands for the discrimination of vegetation type classes. Beyond EVI, an integration of the the full spectral range from the visible to the middle infrared wavelength range is therefore recommended for mapping landscapes with open vegetation cover.

Results of the separability analyses highlighted the capabilities and limitations for mapping savanna vegetation types. Vegetation types characterized by differing soil properties reached high Batthacharrya distance values and thus enhanced classification accuracies. Lower B-distance measures between closed shrubland and woodland classes due to similar phenology and soil properties indicated less accurate mapping results. In summary, all vegetation types were mapped with lower error rates in drier rainy seasons, which means that long-term phenological observations including the inter-seasonal precipitation variability is useful for an in-depth characterization of semi-arid vegetation.

The Random Forest technique used in this study proved to be robust in terms of classification error, overfitting and feature analysis functionality. The use of novel data-mining methods in combination with bottom-up approaches can therefore improve the understanding of remote sensing applications in biodiversity biogeographic research. Yet, further research has to be conducted in similar ecosystems in a synergetic use of different remotely sensed time series data such as the combination of MODIS products and inter-sensor products with botanical field data. Regarding the highly heterogeneous vegetation structure of savannas, new optical sensor systems, such as the five RapidEye satellites and the Advanced Wide Field Sensor (AWiFS) onboard the IRS satellites, seem promising for accurate mapping of open savanna vegetation structure.

Regarding the optimization of global land-cover products, a process of inter-annual time series metrics beyond one or two years may increase the low class accuracies for semi-arid shrub- and grassland classes. The application of standardized land-cover classification systems is crucial for a bottom-up transfer of botanical information on a coarse remote sensing scale. To expand the vegetation type mapping to the whole Namibia, the most challenging task will be to collect the necessary botanical field data.

A number of more than 10,000 botanical field samples (relevés) are available for Namibia and therefore a highly interdisciplinary project structure is needed to synergize these data on a coarse scale. The translation of data from the botanical survey into the FAO-UN Land Cover Classification System (LCCS) to generate flexible map products will increase the usefulness of land-cover information to a broader user community and will be focussed on in future research.

ACKNOWLEDGEMENTS

This work was conducted within the framework of the Biodiversity Transect Analysis Project (BIOTA Southern Africa) funded by the German Federal Ministry of Education (BMBF, grant no. 01LC0624J). The authors express their thanks to the staff of the National Biodiversity Institute of Namibia (NBRI) for data submission and support during the field work. The authors express their thanks to Michael Gottschling for help in preparing the MODIS data. Furthermore, the authors thank Jens Oldeland for helpful comments on the manuscript, Michael Wiesner for technical support and the two anonymous reviewers for valuable comments.

REFERENCES AND NOTES

1. Strohbach, B. Vegetation Survey of Namibia. *J. Nam. Sci. Soc.* 2001, *49*, 93–118.
2. Burke, A.; Strohbach, B. Vegetation studies in Namibia. *Dinteria* 2000, *26*, 1–24.
3. Volk, O. Die Florengebiete von Südwestafrika. *J. S. W. A. Sci. Soc* 1966, *20*, 25–58.
4. Giess, W. A preliminary vegetation map of South West Africa. *Dinteria* 1971, *4*, 1–114.
5. Volk, O.; Leippert, H. Vegetationsverhältnisse im Windhoeker Bergland, Südwestafrika. *J. S. W. A. Sci. Soc.* 1971, *30*, 5–44.
6. Robinson, E. Phytosociology of the Namib Desert Park, South West Africa. M.Sc. thesis, Department of Botany, University of Natal, Pietermaritzburg, South Africa, 1976.
7. Jürgens, N.; Burke, A.; Sally, M.; Jacobson, K. Desert. In *Vegetation of Southern Africa*; Cowling, R.M., Richardson, D.M., Pierce, S.M., Eds.; Cambridge University Press: Cambridge, UK, 1997.
8. Jankowitz, W.; Venter, H. Die plantgemeenskappe van die Waterberg-platopark. *Madoqua* 1987, *15*, 97–146.
9. le Roux, C.L.G.; Grunow, J.O.; Morris, J.W.; Bredenkamp, G.J.; Scheepers, J.C. A classification of the vegetation of the Etosha National Park. *S. Afr. J. Bot.* 1988, *54*, 1–10.
10. du Plessis, W.P.; Breedenkamp, G.J.; Trollope, W.S.W. Response of herbaceous species to a degradation gradient in the western region of Etosha National Park, Namibia. *Koedoe* 1998, *41*, 19–28.
11. Strohbach, B.; Calitz, A.; Coetzee, M.E. Erosion Hazard Mapping: Modelling vegetative cover. *Agricola* 1996, *8*, 53–59.
12. Sannier, C.A.D.; Taylor, J.C.; du Plessis, W.; Campbell, K. Real-time vegetation monitoring with NOAA-AVHRR in Southern Africa for wildlife management and food security assessment. *Int. J. Remote Sens.* 1998, *19*, 621–639.
13. Colditz, R.; Keil, M.; Strohbach, B.; Gessner, U.; Schmidt, M.; Dech, S. Vegetation structure mapping with remote sensing time series: Capabilities and improvements. In Proceedings of the 32nd International Symposium on Remote Sensing of Environment, San Jose, Costa Rica, June 25–29, 2007.
14. Verburg, P.; van de Steeg, J.; Veldcamp, A.; Villemann, L. From land cover change to land function dynamics: A major challenge to improve land characterization. *J. Environ. Manage.* 2009, *90*, 1327–1335.
15. Herold, M. A joint initiative for harmonization and validation of land cover datasets. *IEEE Trans. Geosci. Remote Sens.*2006, *44*, 1719–1727.
16. Lambin, E. Monitoring forest degradation in tropical regions by remote sensing: some methodological issues. *Global Ecol. Biogeogr.* 1999, *8*, 191–198.
17. Nagendra, H. Using remote sensing to assess biodiversity. *Int. J. Remote Sens.* 2001, *22*, 2377–2400.
18. DeFries, R.; Hansen, M.; Townshend, J.; Sohlberg, R. Global land cover classifications at 8 km spatial resolution: the use of training data derived from Landsat imagery in decision tree classifiers. *Int. J. Remote Sens.* 1998, *19*, 3141–3168.
19. Steenkamp, K.; Wessels, K.; Archibald, S.; von Maltitz, G. Satellite derived phenology of southern Africa for 1985-2000 and functional classification of vegetation based on phenometrics. In Proceedings of the 33rd International Symposium of Remote Sensing of the Environment (ISRSE), Stresa, Italy, May 4–8, 2009.
20. Hansen, M.; DeFries, R.; Townsend, J.; Carrol, M.; Dimeceli, C.; Sohlberg, R. Global percent tree cover at a spatial resolution of 500 meters: first results of the MODIS vegetation contineous fields algorithm. *Earth Interact.* 2003, *7*, 1–15.
21. Gessner, U.; Klein, D.; Conrad, C.; Schmidt, M.; Dech, S. Towards an automated estimation of vegetation cover fractions on multiple scales: Examples of Eastern and Southern Africa. In Proceedings of the 33rd International

Symposium of Remote Sensing of the Environment, Stresa, Italy, May 4–8, 2009.

22. Breiman, L.; Friedmann, J.; Olshen, R.; Stone, C. *Classification and Regression Trees*; CRC Press: Boca Raton, FL, USA, 1984.
23. DeFries, R.; Hansen, M.; Townshend, J. Global discrimination of land cover types from metrics derived from AVHRR pathfinder data. *Remote Sens. Environ.* 1995, *54*, 209–222.
24. Friedl, M.; Brodley, C. Decision tree classification of land cover from remotely sensed data. *Remote Sens. Environ.* 1997, *61*, 399–409.
25. Hansen, M.C.; DeFries, R.; Townshend, J.; Sohlberg, R. Global land cover classification at 1km spatial resolution using a classification tree approach. *Int. J. Remote Sens.* 2000, *21*, 1331–1364.
26. Chan, J.; Huang, C.; DeFries, R. Enahanced algorithm performance for land cover classification from remotely sensed data using bagging and boosting. *IEEE Trans. Geosci. Remote Sens.* 2001, *39*, 693–695.
27. Breiman, L. Random forests. *Mach. Learn.* 2001, *45*, 5–32.
28. Pal, M. Random forest classifier for remote sensing classification. *Int. J. Remote Sens.* 2005, *26*, 217–222.
29. Chan, J.; Paelinckx, D. Evaluation of random forest and adaboost tree-based ensemble classification and spectral band selection for ecotope mapping using airborne hyperspectral imagery. *Remote Sens. Environ.* 2008, *112*, 2999–3011.
30. Watts, J.; Lawrence, R. Merging Random Forest Classification with an object-oriented approach for analysis of Agricultural Lands. *The International Archives of the Photogrammetry, Remote Sensing and Spatial Information Sciences*2008, *37*, 579–582.
31. Running, S.; Loveland, T.; Pierce, L.; Nemany, R.; Hunt, E. A Remote Sensing Based Vegetation Classification Logic for Global Land Cover Analysis. *Remote Sens. Environ.* 1995, *51*, 39–48.
32. Bonan, G.; Levis, S. Landscapes as patches of plant functional types: An integrating concept for climate and ecosystem models. *Global Biochem. Cy.* 2002, *16*, 1–18.
33. Gessner, U.; Conrad, C.; Hüttich, C.; Keil, M.; Schmidt, M.; Schramm, M.; Dech, S. A multi-scale approach for retrieving proportional cover of life forms. In Proceedings of the IEEE International Geoscience and RemoteSensing Symposium, Boston, MA, USA, July 6–11, 2008.
34. Herold, M.; Mayaux, P.; Woodcock, C.; Baccini, A.; Schmullius, C. Some challenges in global land cover mapping: An assessment of agreement and accuracy in existing 1 km datasets. *Remote Sens. Environ.* 2008, *112*, 2538–2556.
35. Jung, M.; Henkel, K.; Herold, M.; Churkina, G. Exploiting synergies of global land cover products for carbon cycle modelling. *Remote Sens. Environ.* 2006, *101*, 534–553.
36. Bicheron, P.; Defourny, P.; Schouten, C.B.L.; Vancutsem, C.; Huc, M.; Bontemps, S.; Leroy, M.; Achard, F.; Herold, M.; Ranera, F.; Arino, O. GLOBCOVER - Products description and validation report, medias France technical documants. Available online at: http://postel.mediasfrance.org/en/DOWNLOAD/Documents/ (accessed May 12, 2009).
37. Mendelsohn, J.; el Obeid, S. *The Communal Lands in Eastern Namibia*; RAISON: Windhoek, Namibia, 2002.
38. King, L. *South African Scenery*; Hafner Publishing Company: New York, NY, USA, 1963.
39. de Pauw, E.; Coezee, M.E.; Calitz, A.J.; Beukes, H.; Vits, C. Production of an Agro-Ecological Zones map of Namibia (first approximation). Part II: Results. *Agricola* 1999, *10*, 33–43.
40. Strohbach, B.; Strohbach, M.; Kutuahuripa, J.; Mouton, H. *A Reconnaissance Survey of the Landscapes, Soils and Vegetation of the Eastern Communal Areas (Otjiozondjupa and Omaheke Regions), Namibia*; National Botanical Research Institute and Agro-Ecological Survey Programme Directorate Agriculture Research and Training, Ministry of Agriculture, Water and Rural Development: Windhoek, Namibia, 2004.
41. Edwards, D. A broad-scale structural classification of vegetation for practical purposes. *Bothalia* 1983, *14*, 705–712.
42. *Global and National Soils and Terrain Digital Databases (SOTER)*; Land and Water Development Division, Food and Agriculture Organisation of the United Nations: Rome, Italy, 1993.
43. Hennekens, S. *TURBO(VEG) Software Package for Input, Processing, and Presentation of Phytosociological Data. User's Guide*; IBN-DLO: Wageningen, The Netherlands, 1996.
44. Hill, M. *TWINSPAN -A FORTRAN Program for Arranging Multivariate Data in an Ordered Two-Way Table by Classification of the Individuals and Attributes*; Cornell University: Ithaca, NY, USA, 1979.
45. Westfall, R.; Theron, G.; van Rooyen, N. Objective classification and analysis of vegetation data. *Plant Ecol.* 1997, *132*, 137–154.
46. Justice, C.; Townshend, J.; Vermote, E.; Masuoka, E.; Wolfe, R.; Saleous, N.; Roy, D.; Morisette, J. An overview of MODIS Land data processing and product status. *Remote Sens. Environ.* 2002, *83*, 3–15.
47. Colditz, R.; Conrad, C.; Wehrmann, T.; Schmidt, M.; Dech, S. TiSeG: Flexible software tool for time-series generation of MODIS data utilizing the quality assessment science data set. *IEEE Trans. Geosci. Remote Sens.* 2008, *46*, 3296–3308.
48. Huete, A.; Didan, K.; Miura, T.; Rodriguez, E.; Gao, X.; Ferreira, L. Overview of the radiometric and biophysical performance of the MODIS vegetation indices. *Remote Sens. Environ.* 2002, *83*, 195–213.
49. Satake, M.; Oshimura, K.; Ishido, Y.; Kawase, S.; Kozu, T. TRMM PR data processing and calibration to be performed by NASDA. In IEEE Conference Proceedings, Geoscience and Remote Sensing Symposium, IGARSS '95, Quantitative Remote Sensing for Science and Applications, Firenze, Italy, July 10–14, 1995.

50. Definiens Developer 7 - Reference Book. Definiens AG: München, Germany, 2007
51. Roy, D.; Boschetti, L.; Justice, C.; Ju, J. The collection 5 MODIS burned area product - global evaluation by comparison with the MODIS active fire product. *Remote Sens. Environ.* 2008, *112*, 3690–3707.
52. Jönsson, P.; Eklundh, L. TIMESAT - A program for analyzing time-series of satellite sensor data. *Comput. Geosci.* 2004, *30*, 833–845.
53. Chaudhuri, G.; Borwankar, J.D.; Rao, P. Bhattacharyya distance-based linear discriminant function for stationary time series. *Comm. Statist.* 1991, *20*, 2195–2205.
54. Herold, M.; Hang, L.X.; Clarke, K. Spatial metrics and image texture for mapping urban land use. *Photogramm. Eng. Remote Sens.* 2003, *69*, 991–1001.
55. Landgrebe, D.; Biehl, L. *An Introduction to MultiSpec*; Purdue University: West Lafayette, IN, USA, 1997.
56. Hastie, T.; Tibshhirani, R.; Friedmann, J. *The Elements of Statistical Learning–Data Mining, Inference, and Prediction*, 2nd ed.; Hastie, T., Tibshirani, R., Friedman, J., Eds.; Springer: New York, NY, USA, 2009.
57. Gislason, P.; Benediktsson, J.; Sveinsson, J. Random Forests for land cover classification. *Pattern Recogn. Lett.* 2006, *27*, 294–300.
58. Liaw, A.; Wiener, M. Classification and regression by random forest. *R News* 2002, *2*, 18–22.
59. Hines, C. An ecological study of the vegetation of eastern Bushmanland (Namibia) and its implications for development. M.Sc. thesis, University of Natal, Pietermaritzburg, South Africa, 1992.
60. Zhang, X.; Friedl, M.; Schaaf, C.; Strahler, A. Monitoring the response of vegetation phenology to precipitation in Africa by coupling MODIS and TRMM instruments. *J. Geophys. Res.* 2005, *110*, 1–14.
61. Klein, D.; Roehrig, J. How does vegetation respond to rainfall variability in a semi-humid West African in comparison to a semi-arid East African environment? In Proceedings of the 2nd Workshop of the EARSeL Special Interest Group on Land Use and Land Cover, Bonn, Germany, September 28–30, 2006; Volume 1, pp. 149–156.
62. Archibald, S.; Scholes, R. Leaf green-up in a semi-arid African savanna - separating tree and grass responses to environmental cues. *J. Veg. Sci.* 2007, *18*, 583–594.
63. Vanacker, V.; Lindermann, M.; Lupo, F.; Flasse, S.; Lambin, E. Impact of short-term fluctuation on interannual land cover change in sub-Saharan Africa. *Global Ecol. Biogeogr.* 2005, *14*, 123–135.
64. Childes, S. Phenology of nine common woody species in semi-arid, deciduous Kalahari Sand vegetation. *Vegetatio*1989, *79*, 151–163.
65. Shackleton, C. Rainfall and topo-edaphic influences on woody community phenology in South African savannas. *Global Ecol. Biogeogr.* 1999, *8*, 125–136.
66. Sekhwela, M.; Yates, D. A phenological study of dominant acacia tree species in areas with different rainfall regimes in the Kalahari of Botswana. *J. Arid Environ.* 2007, *70*, 1–17.

CHAPTER 12

APPLICATION OF ADVANCED REMOTELY-SENSED DATA IN ECOLOGICAL STUDIES OF TERRESTRIAL ANIMAL MOVEMENT

Wiebke Neumann, Sebastian Martinuzzi, Anna B Estes, Anna M Pidgeon, Holger Dettki, Göran Ericsson and Volker C Radeloff

Keywords: Animal trajectories, Movement patterns, Remote sensing, Trade-off resolution, Satellite products, Landsat, LiDAR, MODIS, Animal movement databases

INTRODUCTION

Both remote sensing and animal tracking technology have recently experienced major advances which has the potential to facilitate integrated analyses of environmental and animal movement data in unprecedented detail [1,2] Correspondingly, statistical analyses have advanced to quantify spatial patterns, to account for processes on different scales and for spatial autocorrelation in animal movement data [3]. Improvements in sensor tracking technology, e.g., Global Positioning System (GPS) data, provide animal movement data that capture movement paths (i.e., time series) in ecological landscapes, an improvement over "timeless" position clusters [1]. These advances provide many opportunities, but also challenge the scientific community of movement ecology as acknowledge with special issues in two leading international journals [1,2]. In parallel, remotely-sensed data from different satellite sensors have been available since the 1970s at a wide range of spatial and temporal resolutions, providing a better match to various scales of animal movement. A key challenge to overcome, however, is the time lag between the availability of new remotely-sensed products and their application in ecological research and management. Minimizing this temporal disconnection requires closer collaboration between wildlife ecologists and remote sensing experts to facilitate more rapid implementation of geospatial data processing (*sensu* [4]).

Analyses of animal movement pathways can increase understanding of resource utilization and dispersal dynamics among populations over time, and can ultimately aid animal conservation and management [5]. While habitat analyses tend to overlook the temporal aspects of habitat use, analyses of animal movement pathways (trajectories) integrate each animal location into the larger context of the spatial distribution of the population and changing environmental conditions through time, as well as individuals' constraints, behavior, and survival [5-7]. Animal movement over time is a result of decision-making among behavioral trade-offs, considering animals' internal state, motion, navigation, and external factors [8]. Today, sensor technology can link physiological data such as body temperature and heart rate to movement data, thereby directly linking animals' physiological conditions and movement behavior to landscape features [9]. Movement path analyses (such as state-space models or step selection functions) track animal resource selection over time [10,11], and require increasing application of dynamic environmental covariates to better understand the mechanism for animal movement behavior.

Animals interact with their environment at multiple spatiotemporal scales, resulting in different movement modes [12,13]. Fine-scale temporal movement data reveal functional landscape connectivity by distinguishing movement corridors, barriers, or stop-over-sites, thereby helping to identity areas critical for mammalian and avian movement [14-18]. Fine-scale data also capture residency times and site fidelity, and can identify critical resources such as water or resting sites [19,20]. At broader scales, pathway analysis can distinguish between migration, nomadic behavior, and dispersal [21], which affects

Citation: Wiebke Neumann, Sebastian Martinuzzi, Anna B Estes, Anna M Pidgeon, Holger Dettki, Göran Ericsson and Volker C Radeloff, Application of advanced remotely-sensed data in ecological studies of terrestrial animal movement, DOI: 10.1186/s40462-015-0036-7.

population dynamics, resource utilization, gene flow among populations (e.g. [5]), and the spread of diseases [22].

New remotely-sensed data better capture both the temporal dynamics of landscapes and the three-dimensional structure of habitats. In contrast to field-collected point data, which are typically information-rich but spatially-sparse, remotely-sensed data typically provide spatially continuous data over larger areas, but less information for any given point or pixel due to their relatively coarser resolution. This means that the different spatiotemporal scales at which animal movement occurs require different types of remotely-sensed data to understand the underlying causes of movement (Figure 1 [7,23]). However, there is an inherent trade-off between *fine-but-infrequent* (i.e., fine spatial but coarse temporal grain) versus *coarse-but-frequent* (i.e., coarse spatial but fine temporal grain) data recorded by different satellite sensors (Figure 1; Table 1). More specifically, if the goal is to understand fine-scale movement patterns, fine-but-infrequent data are arguably most useful, because they provide the detail necessary to gain insight into the drivers of movement patterns and because temporal variation of the environment such as land cover or phenology matters less over shorter time spans (see e.g. [13,24-28]). In contrast, to understand fine-scale movement patterns of flying animals, highly frequent information about weather and wind condition is needed because the fine details of e.g., atmospheric turbulence that shape the movement patterns can change very rapidly [29,30]. On the other hand, when broad-scale movement patterns are studied, temporal change in the environment is the key and *coarse-but-frequent* data are better covariates of movement (e.g. [31]).

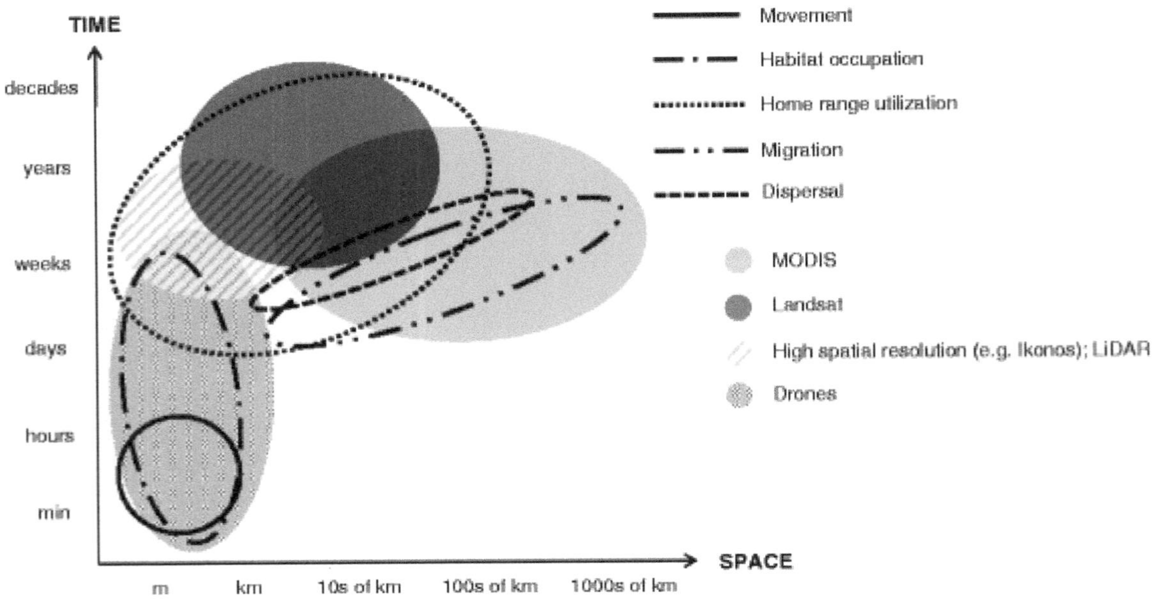

Figure 1: Scales of current resolution in space and time of animal movement and remotely-sensed data.

Table 1: Source and resolution of different remote sensing products

Sensor and satellite	Spatial resolution [m]	Temporal resolution [days]	Swath [km]	Operational since	Link
MODIS (Terra, Aqua)	250, 500, 1000	1-2	2.330	1999 (Terra), 2002 (Aqua)	http://modis.gsfc.nasa.gov/about/design.php
AVHRR***	1100	<1	2.600	1981	http://noaasis.noaa.gov/
VIIRS (Suomi NPP)	750	1-2	3.000	2011	http://npp.gsfc.nasa.gov/
ASTER (Terra)	15, 30, 90	16	60	1999	http://asterweb.jpl.nasa.gov/
ETM+ (Landsat 7)	15, 30, 60	16	183	1999**	http://landsat.gsfc.nasa.gov
TM (Landsat 5)	30, 120	16	185	1984*	http://landsat.gsfc.nasa.gov
Vegetation 1 (SPOT 4)	10, 20	2-3	60	1998	http://www.astrium-geo.com
Vegetation 2 (SPOT 5)	5, 10, 20	2-3	60	2002	http://www.astrium-geo.com
Spot 6	1.5, 6	2	60	2012	http://www.astrium-geo.com
RapidEye satellites	5	5	25	2008	http://www.rapideye.com
IKONOS	0.8, 3.2	3	11	1999	https://www.digitalglobe.com
GeoEye-1	0.4, 1.7	3	15	2008	https://www.digitalglobe.com

QuickBird	0.6, 2.4	3	18	2001	http://www.digitalglobe.com
Worldview 1	0.5	2	18	2007	http://www.digitalglobe.com
Worldview 2	0.5	1	18	2009	http://www.digitalglobe.com

* The very first Landsat launch was in 1972, but is not longer in service. **due to an instrument failure all scenes after May 2003 have data gaps; ***different NOAA satellites. 1978 was a first attempt for AVHRR on a different satellite, which was improved and replaced in 1981 by the AVHRR sensor or the NOAA satellite. Additional missions are planned for 2013–2014, such as Landsat DCM (continuing the Landsat program with pixels sizes of 15, 30, and 100 m), Spot 7 (which combined with Spot 6 will provide satellite imagery at a temporal resolution of 1 day), and Worldview 3, with pixel sizes of 0.3 m and 1.2 m. LiDAR remote sensing data used in research applications are typically acquired from airborne systems, rather than from satellites as those described above. The LiDAR acquisitions specifications, such as laser pulse density and area cover, thus are flexible and depend on the objectives of the particular study. Detailed information about LiDAR data and specifications used in natural resource management are described in [122] and [123]. Aerial photos captured with drones have typically a high spatial resolution (e.g. <10 cm) and they cover small areas (~15-25 km; see [112] for an example of an inexpensive drone designed to monitor forests and biodiversity.

Our goal here was to provide a review for movement ecologists who are less familiar with the variety of remotely-sensed data available and their application in animal movement studies, reducing the present time lag in application of available remotely-sensed products in animal movement studies [4]. Thus, we first review the current applications of remotely-sensed ecological indices that deal with vertebrate movement patterns in terrestrial systems. Second, we introduce newly-available remotely-sensed products and discuss the opportunities that they provide for animal movement studies. We focused on studies that link animal locations and remotely-sensed data in terrestrial ecosystems, published between 2002 and 2013 (Table 2). Specifically, the studies we reviewed investigated how animals interact with their terrestrial environment over time, and are based on trajectories and range shifts by individuals and populations, thereby potentially aiding conservation and management. Current animal movement studies are heavily biased towards larger-bodied animals due to the weight and size of tracking devices. While there is a history of radio-tracking and linking large-bodied seabird movement to environmental features, it has only recently become possible to equip small-bodied terrestrial avian migrants with devices capable of tracking long-range movements [32,33]. Consequently, our review focuses largely on movement studies of large-bodied mammalian species.

Table 2: Sample of ecological animal movement studies that apply remotely-sensed products

Land cover	**Terrain**	**Infrastructure**	**Climate**	**Weather**	**Phenology/ Productivity**	**Species**	**Data origin (when available; multiple layers in bold)**	**Reference**
Small scale (e.g., daily movement, residence time, side fidelity, corridors)								
X	X					Woodland caribou	LANDSAT, DEM[a]	[12]
X	X	X		X		Elk	**Snow water equivalents**, DEM[w], roads[w], habitat[w]	[124]
X	X	X				Bison	Vegetation cover type, geothermal layers[b], DEM	[44]
X	X	X				Roe deer	SPOT, DEM[c], digitized aerial photographs	[50]
X		X				Elk	**LANDSAT** [v]	[34]
X		X		X	X	African elephant	**LANDSAT**, TRMM, **MODIS** (EVI)	[35]
X	X					Moose	Ecoforest maps, DEM	[13]
X	X	X				Brown bear	CORINE, DEM[u], human features[u]	[42]
X	X	X				Gray wolf	LANDSAT[d], DEM[e]	[39]
X						North Island robin	digitized aerial photographs, satellite images	[24]
X	X	X				Grizzly bear	LANDSAT[f], vegetation inventory data, fire-history maps, DEM	[51]
X						Lion	digitized water wholes	[19]

X	X	X			X	African elephant	digitized static features, DEM[g], SPOT (NDVI)	[40]
		X			X	Jaguar	Roads[h], **MODIS** [c](NDVI)	[125]
X		X				American marten	**color-infrared aerial photographs**	[15]
X						Barred Antshrikes, Rufous-naped Wrens	infrared images[g], orthorectified using DEM	[126]
X		X				Black bear	LANDSAT, color orthophotos, stream water layer[c], topographic maps	[52]
					X	Mule deer	**MODIS** [m] (NDVI)	[16]
X						Lion	LiDAR[s]	[110]
Coarse/seasonal scale (e.g., seasonal range change, migration, dispersal)								
	X				X	Wildebeest	DEM, GTOPO30[c], **AVHRR/ NOAA** (NDVI)	[127]
					X	Red deer	**NOAA** (NDVI)	[128]
	X		X		X	Serengeti Wildebeest	DEM, SRTM, LANDSAT, **NOAA**, **SPOT** (NDVI)	[58]
					X	Elk	**MODIS** (NDVI)	[66]
					X	Five migratory bird species[j,k]	**AVHRR/NOAA** (NDVI)	[60]
					X	Mongolian gazelle	MODIS (NDVI)	[129]
	X		X		X	Saiga antelope	DEM, SRTM[i], **MODIS**(NDVI); **WorldClim database**(precipitation)	[68]
X					X	Migratory birds	**LANDSAT**	[14]
X	X		X			Roe deer	EEA-Corine Land cover, CGIAR-DEM/SRTM, NASA-ASTER relative DEM, **MODIS** (snow)	[38]
				X		Great snipes	**NCEP/NCAR** provided by NOAA/OAR/ESRL PSD[r] (wind)	[30]
					X	Four migratory ungulate species[l]	**AVHRR/ NOAA**, **GIMMS** (NDVI)	[31]
	X					Red deer	DEM[t]	[130]
	X			X		Golden Eagle, Turkey vulture	DEM, GTOPO30, **NARR** [o] (wind, temperature)	[29]
				X		Common swift	**NCEP/NCAR** provided by NOAA/OAR/ESRL PSD[r] (wind)	[17]

X		X	X		X	African buffalo	MODIS (EVI, tree cover[p]), TRMM[n], **MODIS** [o] (fire), vegetation map[q]	[41]
X			X		X	Bobolink	**MODIS** (NDVI), **SPOT**[x], **water content estimates** [y]	[59]
Small and coarse scale								
					X	African elephant	**AVHRR/NOAA, SPOT**(NDVI)	[61]
					X	Sheep	**MODIS** (NDVI)	[67]

LAND COVER: vegetation, water, streams, forest age, cutblocks, seral stages; TERRAIN: elevation, slope, aspect, ruggedness, solar radiation, soil wetness; INFRASTRUCTURE: roads, buildings, borders, fences, trails; CLIMATE/WEATHER: wind speed, wind direction, precipitation, temperature, cloud; PHENOLOGY/PRODUCTIVITY: NDVI, EVI, tree cover, fire; [a]British Columbia Ministry of Crown Lands; [b]The Watershed Institute (California State University, Monterey Bay, USA), National Hydrographic Dataset; [c]USGS; [d]Foothills Research Institute Grizzly Bear Research Program; [e]Banff and Kootenay National Park; [f]Alberta Vegetation Inventory; [g]NASA, CARTA program; [h]Selva Maya Zoque y Olmeca database; [i]Surface Radar Topography Mission; [j]Ringing data; [k]Lesser Whitethroat, Whitethroat, Blackcap, Chiffchaff, Willow Warbler; [l]barren-ground caribou, Mongolian gazelle, guanacos, moose; [m]Wyoming View; [n] http://trmm.gsfc.nasa.gov; [o]National Deneter for Environ Predict; [p]Wegmann *et al.*, unpublished data; [q]Mendelsohn 1997, An environmental profile and atlas of Caprivi. Windhoek, Namibia: Gamsberg Macmilian; [r]Boulder, Colorado, USA, http://www.cdc.noaa.gov; [s]Carnegie Airborne Observatory, [t]National Mapping Agency of Norway, [u]Lantmäteriet Sweden, [v]Alberta Sustainable Resource Development, [w]Spatial Analysis Center at Yellowstone National Park, [x]http://bioval.jrc.ec.europa.eu/products/glc2000/products.php, [y] http://neo.sci.gsfc.nasa.gov.

REVIEW

Remotely-sensed products commonly used in ecological animal movement studies

The studies of small-scale movement that we reviewed included data about land cover, infrastructure, and terrain, and less often data on vegetation phenology, climate and weather (Table 2). Surprisingly, we found that even long-term, small-scale movement studies seldom applied multiple years of land cover data, although the satellite data from which they are derived, i.e. Landsat imagery, are available continuously since the 1970s (but see [15,34,35]). Moving beyond a single-date snapshot of a given environmental situation towards incorporation of multiple years of data could help to detect land use change [36] which influences animal movement. Studies that analyzed animal movement at coarser scales (e.g., migration) often used data on vegetation phenology (e.g., the Normalized Difference Vegetation Index (NDVI)), sometimes combined with climate, weather, and terrain data (Table 2). While the use of land cover data from multiple dates was rare, phenology was typically included from multiple dates (Table 2). Unfortunately, we also found that ecological studies frequently lacked details regarding the origin of the satellite data, and the spatial and temporal resolution of the products that had been analyzed (e.g., land cover data; Table 2). This lack of information makes it difficult to determine, for example, if the resolution of the remotely-sensed products matches the resolution of the animal location data and is suitable to address the study question, ultimately impeding the readers' interpretation of results [18].

Land cover data are typically derived from *fine-but-infrequent* remotely-sensed products and researchers can use them to infer animal movement patterns on finer spatial scales (Figure 1). To suggest appropriate conservation actions based on studies of movement, it is important to relate animal movement to land cover and land use, and to understand animals' use of the landscape, such as, the reluctance of dispersing North Island robins *Petroica longipes* to cross pastures, or elephants' *Loxodonta africana* need to be close to water resources during the dry season [24,35]). Ancillary data can capture land use change and infrastructure development and can help to update remotely-sensed maps, thereby creating dynamic landscape data [15]. Of the studies we reviewed, Landsat or aerial photographs were the most common source. Images provided by Google Earth (i.e., satellite data from Landsat, IKONOS, and Quickbird) provide a visualization tool, and have been used to understand habitat use during avian migration for Veery *Catharus fuscescens* [37], but are not suitable for analyses requiring image manipulation.

Digital elevation models (DEM) provide *fine-but-infrequent* information on altitude, aspect, and slope, as well as soil wetness and solar radiation, and have been applied in different ways. DEMs are sometimes combined with *coarse-but-frequent* remotely-sensed atmospheric variables such as wind speed, turbulent kinetic energy, temperature, and cloud height (e.g., 8-day moderate-resolution Imaging Spectroradiometer (MODIS) products, monthly WorldClim data, or 10-day National Oceanic and Atmospheric Administration data (NOAA; i.e., the Advanced Very High Resolution Radiometer (AVHRR); Table 2). DEMs can be a powerful tool and the interpretation of observed patterns can contribute to inference of migratory strategies, for example disentangling species-specific specialization to uplift modes in soaring birds such as Turkey vultures *Cathartes aura* and Golden Eagles *Aquila chrysaetos* [29]. The analysis of digital elevation and climatic data in relation to animal movement data highlights that topographic variability and winter severity affect migratory behavior in ungulates such as roe deer *Capreolus capreolus*, and their opportunistic migration behavior when there are no predictably extreme winter conditions [38].

Human infrastructure influences many wildlife species' habitats. Infrastructure can create barriers and can modify daily and seasonal movement patterns. For example, gray wolves *Canis lupus* cross highways less frequently with increasing human presence [39], human settlements disrupt African elephant *Loxodonta africana* movement [40], and distance to fences affects African buffalo *Syncerus caffer* migrations [41]. In hunted species, infrastructure can also affect temporal movement patterns, e.g., brown bear *Ursus arctos* and moose *Alces alces* avoid roads during the day [42,43]. On the other hand, some species associate preferentially with infrastructure, for example to facilitate movement, as in the case of the American bison *Bison bison* [44]. Digital maps of infrastructure are *fine-but-infrequent* data, and typically derived from land cover data and aerial photos. The use of high spatial resolution imagery and LiDAR (Light Detection and Ranging; pulsed laser signals that measure distances to the earth surface and thereby generate information about the shape of the surface; http://oceanservice.noaa.gov/facts/lidar.html) also provides opportunities to discriminate between different types of human infrastructure and vegetation, such as highways, roads, railroads, pipelines, power-lines, houses, and trees (e.g. [45-48]). In high-resolution satellite imagery, the number of lanes of a given road is easily determined, and LiDAR data can detect roads even under dense forest canopies [49]. Governments and private companies use LiDAR and high spatial resolution data to map human infrastructure for large areas, and the resulting maps can be useful for understanding wildlife movement. Understanding road-crossings, the impact of road density and movement barriers such as wildlife fences, and exploitation or avoidance of road-near habitats, provides insights important for wildlife conservation and management (e.g., elk *Cervus elaphus* [34]; African elephant [35]; roe deer [50]; Grizzly bear *Ursus arctos* [51]; Black bear *Ursus americanus* [52]).

Plant phenology, the annual dynamics of vegetation greenness, can be tracked using NDVI and the Enhanced Vegetation Index (EVI) [53,54], although NDVI is more commonly applied in animal movement studies (see review by [55]). Phenology indices are typically derived from *coarse-but-frequent* satellite imagery, such as 10-day Satellite Pour l'Oservation de la Terre (SPOT) or 16-day MODIS reflectance data, and can be applied to predict individual and population movement and distribution. Yet animal movement ecologists need to consider the different interpretation of traditional phenology and satellite-based landscape phenology; while traditional phenology tracks e.g. flowering and budding of single plant species, satellite-based phenology quantifies dates of greening and browning patterns of multiple plant species at the landscape scale [56,57]. The analysis of variation in broad-scale landscape predictability is useful for gaining insights into long-distance movements in ungulates such as migration, nomadism, and residency [31]. Vegetation proxies such as NDVI also provide opportunities for applications in environmental conservation and paleoecology, and for predictions of past and future population and biodiversity response to climate, phenology, and primary production [55]. In our sample of studies, plant phenology data were commonly used to understand broad-scale migration patterns. For example, new forage growth is important for ungulate and avian long-distance migration (e.g., wildebeest *Connochaetes taurinus* [58]; Bobolink *Dolichonyx oryzivorus* [59]), and environmental conditions en route affect arrival at the breeding grounds of migratory bird species [60]. NDVI is particularly useful for movement studies in highly seasonal ecosystems (e.g., African elephants move more randomly when forage is abundant [61]). The occurrence of fires affects plant phenology and modifies animal movement (e.g., African buffalo [41]). MODIS fire detection data are available from 2001 to present, and may facilitate reconstruction of prior habitat conditions [62]. The combination of vegetation indices (e.g., NDVI) with precipitation data (e.g., TRMM (Tropical Rainfall Measuring Mission)) improves the prediction of the timing and speed of migrations (e.g., zebra *Equus burchelli antiquorum* [63]). For vertical migration, i.e., along an elevational gradient, vegetation proxies combined with DEM can predict seasonal movement that follows vegetation growth at different elevation (e.g., for elephants [64]). For birds, combination of NDVI and weather data such as wind and temperature generates fruitful environmental indices helping to understand global long-term movement patterns (e.g., vultures [65]). Furthermore, new products, such as global data on soil moisture (SMAP, Soil Moisture active Passive, newly launched, http://smap.jpl.nasa.gov/) and the Global Precipitation Measurement (GPM, planned for the near future, http://www.nasa.gov/mission_pages/GPM/main/#.VNtBJy7s7FQ) provide exciting opportunities to improve our understanding of the drivers of animal movement.

Productivity, i.e., seasonal plant growth, can be calculated by merging information from ground biomass measurements and MODIS-derived NDVI, and can predict, e.g., movement patterns of elk at different scales, demonstrating the connection between migration and access to higher-quality forage [66]. With its temporal resolution of typically 10–16 days, NDVI has been used to infer fine-scale movement patterns related to vegetation greenness, where it may represent a static habitat index rather than a phenological index (e.g., forage availability does not affect within-patch movement in sheep *Ovis aries* [67], but affects migration speed across patches (i.e., stop-over-sites) in mule deer *Odocoileus hemionus* [16]).

Climate and weather indices provide information about precipitation (i.e., rainfall and snow from 10-day NOAA, 8-day MODIS, and monthly WorldClim data) and temperature (e.g., 10-day NOAA), and are generally coarse-but-frequent indices. It is important to distinct between climate and weather. *Climate* describes the average pattern of meteorological variables such as temperature, wind, precipitation over longer time in a given area. In contrast, *weather* describes the short-term variations of those meteorological variables. In the context of animal movement, climate affects animals at a broader scale and can, for example, cause seasonal migration. Precipitation and temperature are particularly important determinants of animal movement patterns in seasonal ecosystems (e.g., rainfall and snow influence migration patterns in Saiga antelope *Saiga tatarica tatarica* [68] and roe deer [38]; Table 2). *Weather*, on the other hand, affects short-term movement such as daily movement. In flying animals, dynamics in diurnal weather conditions generate complex environmental conditions that affect animals' movement patterns (e.g., thermal conditions generate inter-specific migration strategies and movement behavior in soaring birds [29,69-71], wind drift such as tail- or crosswinds affect flying patterns in fruit bats *Eidolon helvum* [72], and temperature and turbulent kinetic energy affect bee-eater *Merops apiaster* flight mode [73]. *Climate and weather indices* provide information about precipitation (i.e., rainfall and snow from 10-day NOAA, 8-day MODIS, and monthly WorldClim data) and temperature (e.g., 10-day NOAA), and are generally *coarse-but-frequent* indices.

In summary, inclusion of remotely-sensed data can greatly improve our understanding of the drivers of animal movement and provide novel insights about movement strategies across species and ecosystems. The main environmental indices are available from a broad range of remotely-sensed products that can vary in precision, and accuracy (Table 3). The set of studies that we reviewed reflected the current application of remotely-sensed data in animal movement studies. Single-date Landsat imagery was a common source of land cover information. Terrain data combined with climate and weather data were powerful tools for inferring migration patterns. Yet, movement ecologists should critically weigh which remotely-sensed product best fits their scale of analysis (i.e., *fine-but-infrequent* versus *coarse-but-frequent*), and more importantly, should avoid the expectation that remotely-sensed data are equivalent to field-collected data. On the other hand, field-collected data will never match the spatial continuity of remotely-sensed data collected over large areas. Remotely-sensed and field-collected data have unique strengths, and their integration allows unique insights.

Table 3: Precision and accuracy of main remotely-sensed derived products

Product	Sources	Precision	Accuracy
Land cover	NLCD (National Land Cover Database): developed from Landsat 30-m pixel; available for 1992, 2001, 2006, and 2011. U.S. only.	20 classes	78 and 79% for latest products [117]
	CORINE: developed from Landsat, Spot, and recently higher resolution imagery; available for 1990, 2000, 2006, 2012. Minimum mapping unit/with 25 ha/ 100 m.	44 classes	8% (for 2000 version) [131]
	MODIS Land Cover: available for 2001–2012, 500-m pixel resolution	17 classes	75% [132]
	GlobCover: based on 300-m pixel resolution imagery from MERIS sensor (ENVISAT), available for 2005–06 and 2009.	23 classes	58-79% [133]
Terrain (elevation)	SRTM (Shuttle Radar Topography Mission): 30-m and 90-m pixel resolution	16-bit	Geolocation error = 9.8 meters; absolute height error = 6.9 meters; relative height error = 7.0 meters [134]
	ASTER Global Digital Evalation Map: 30-m pixel	16-bit	Overall accuracy ~17 meters [135]
	TanDem-X (TerraSAR-X add-on for Digital Elevation Measurement): provides elevation data <12-m pixel resolution; launched in 2010.	16-bit	<2 m height accuracy [136]
Climate/ Weather	MODIS Land Surface Temperature/Emissivity: available daily, 8-day, and monthly, at 1-km, 5.6-km resolutions.	8-bit and 16-bit	0.5 K to 1 K [132]
	TRMM (Tropical Rainfall Measuring Mission): 16 times per day, multiple products describing rainfall at 2.4 km and 5-km pixel resolution	na	TRMM Precipitation Radar instrument is able to detect fairly light rain rates down to about .0.7 mm per hour.
	MODIS Cloud Product: daily product of cloud properties at 1-km and 5-km pixel resolution.	na	na
Phenology	MODIS Global Vegetation Phenology product (MCD12Q2): provides estimates of the timing of vegetation phenology at 500-m pixel resolution. MCD12Q2 is produced once a year with 24 months of input data, available from 2001 through 2012	16-bit	Consistent with in-situ measurements [137]
	SPOT Vegetation: provides NDVI global data since 1998 at a 1.15-km pixel resolution	na	na
	MODIS Gross Primary Productivity (GPP) product (MOD17A2): 8-day composite at 1-km spatial resolution	8-bit and 16-bit	Annual estimates of GPP are within 10.4% of average published results [132]
	MODIS Leaf Area Index (LAI) and Fractional Photosynthetically Active Radiation (FPAR): 8-day composite at 1-km spatial resolution	8-bit	Accuracy is 0.66 LAI units RMSE and 0.12 FPAR units RMSE respectively [132]

Note: The Visible Infrared Imaging Radiometer Suite (VIIRS) extends and improves the measurements initiated by the Advanced Very High Resolution Radiometer (AVHRR) and the Moderate Resolution Imaging Spectroradiometer (MODIS). na: no information.

Opportunities for remotely-sensed products in ecological studies of animal movement

Increasing access to remotely-sensed data

High spatial resolution imagery is recorded by sensors such as IKONOS, QuickBird, OrbView-3 and Spot-5, with spatial resolutions from 0.5 - 10 m ([74]; Table 1). High-resolution imagery allows the identification of small habitat features such as infrastructure (e.g. [75]), riparian corridors (e.g. [76,77]), and individual trees (e.g. [78]). Knowledge of specific environmental features is often critical for understanding animals' response to landscapes (see [79]), and improves our insight into species' ecology (e.g. [20,50]). Although the cost of high spatial resolution imagery can be a limitation, cost has declined as more sensors have launched and increased competition [74]. Furthermore, in many areas high-spatial resolution images are available for free, thanks to web-mapping sites such as Google Earth (http://earth.google.com).

However, high-resolution imagery is not available everywhere and imagery archives are limited. In contrast, Landsat satellite imagery is available for free, and archives span 40 years, making Landsat the most commonly used satellite imagery for monitoring land cover [80], and in ecological studies (e.g. [81,82]; Table 1). The launch of the Landsat 8 satellite in February 2013 ([83]; Table 1), together with the planned Sentinel satellites of the European Space Agency [84], promises a new era in which 30-m resolution Landsat-like satellite data are available every two to three days, i.e., with similar temporal resolution as currently available for 500–1,000 m resolution MODIS and VIIRS.

Raw reflectance data (i.e., unclassified data), as an alternative to traditional classified land cover data or calculated indices such as NDVI, experience increasing interest for predicting animal distribution. They overcome classification errors and can provide better predictive ability – even at fine resolution – than more commonly-used remotely-sensed products (see for example in birds [85]).

A computational challenge for many movement ecologists is the integration of these remotely-sensed data sets into animal movement data sets, as remote sensing data often are provided using complex tiling systems in space and time [86]. Further, these data sets often vary in source, format, and projection system, making it difficult if not impossible for anyone others than remote sensing experts to make use of directly. However, new tools are emerging for movement ecologists which greatly automate and simplify the computation and integration [86]. In movement data portals, linking of environmental data to animal movement data is a general trend, making a variety of remotely-sensed products readily available for animal movement ecologists; e.g. through the *Env*-DATA system in *Movebank* ([86]; https://www.movebank.org), as a direct link in *Eurodeer* ([38]; www.eurodeer.org), or under development in *Wireless Remote Animal Monitoring* ([87]; *WRAM*).

Capturing the temporal dynamics of landscapes

The use of high temporal resolution satellite data offers new possibilities for animal movement studies as it means increased opportunities for matching satellite imagery with concurrent high resolution animal location data, and captures many temporal dynamics of landscapes (e.g., phenology, land use change, disturbances). In particular, long-term animal monitoring projects can take advantage of analyzing movement patterns in dynamic landscapes to better understand the impact of a given change and its consequence for management and conservation [88,89].

For studying fine-scale animal movement, an exciting advancement is the automated analysis of dense time series of Landsat images. In the past, studies rarely incorporated more than a few Landsat images due to cost and processing limitations. Since Landsat data became freely available in 2008, new image processing algorithms have been developed that analyze many decades of annual Landsat imagery simultaneously [90,91]. The great advantage of such dense time series (or 'time stacks') is in identifying changes, such as forest clear-cuts that can be detected immediately after the disturbance occurs, but can be missed if images are years apart. Software that analyses forest disturbance from dense time series of LANDSAT imagery includes Vegetation Change Tracker (VCT [90]) and LandTrendr [91]. Dense time series have also been used to monitor desertification [92] and urban development ([93]; Table 1).

For analysis of coarse-scale animal movement, indices of vegetation phenology can be informative [55]. Satellite sensors such as MODIS, AVHRR, VEGETATION, and the recently launched VIIRS provide daily imagery, typically summarized in 8- or 16-day composites that are relatively cloud-free (Table 1). However, composites may contain some erroneous NDVI values, and analyzing dozens of satellite images for a given year is computationally and logistically challenging. Thus algorithms that fit phenology curves to a time series of NDVI data and estimate parameters such as the onset of greenness, or the range of NDVI for each pixel are a major advancement [94,95]. By fitting a curve, the effects of outliers are reduced, reduce data volumes of phenology indices, and offer ecologically meaningful measures of vegetation change over time. One software package for phenology analysis is TIMESAT (http://www.nateko.lu.se/TIMESAT; version 3.0 [94]); MODIS phenology products are also available (https://lpdaac.usgs.gov/products/modis_products_table/). Substantial work has been done to unify data from older and new sensors to establish consistency in time series data [96] and improve phenological assessment through Landsat-MODIS fusion [97].

Although NDVI is the most commonly used index in ecological applications of remotely-sensed data, there is also a broad range of spectral indices that can be derived from satellite data (e.g. chlorophyll, foliage, surface moisture [98,99]) as well as ready-to-use products (e.g. Gross and Net Primary Productivity from MODIS) that offer opportunities for animal movement studies. The spectral fidelity (the number of spectral bands, and bandwidths) of remotely-sensed data is becoming increasingly important, as it is now possible to map individual plant species and nitrogen content of grazing lands. For example, RapidEye sensors provide high-resolution satellite images in which the Red-Edge band is sensitive to chlorophyll content and leaf structure (http://blackbridge.com/rapideye/). RapidEye data are at the cutting-edge of geospatial data, because of their ability to describe foliar nitrogen at impressive resolutions [100], which can help to understand herbivore distribution in grazing systems.

Characterizing the 3d strcuture of habitats

The 3D structural characteristics of vegetation are important determinants of animals' space use [4]. For example, visibility at ground level and canopy cover affect selection of daytime bedding sites (e.g., brown bears [20]). Unfortunately, remotely-sensed products commonly used in animal movement studies, i.e., Landsat and MODIS imagery, are not well suited to quantify vegetation structure, resulting in a resolution gap between the movement data and the remotely-sensed vegetation indices.

LiDAR is a relatively new technology that uses a laser to measure the vertical and horizontal structure of ecosystems ([101]; Table 1), making it possible to characterize animal habitats in novel ways [102]. LiDAR data provide direct information about vegetation height [103] and canopy closure [104], and indirect information about forest age [105], canopy gaps [106], and large trees [107]. The ability of LiDAR to "see through" the forest canopy allows the characterization of topography and understory vegetation, important to many animal species [108]. For these reasons, LiDAR data are great predictors of species habitat use [102], and can track herbivore activity e.g., treefalls by elephants [109]. However, their utility for movement pathway studies still needs to be fully explored (but see [110] for lion *Panthera leo* hunting strategies and for predation risk in roe deer [111]). Most LiDAR data stem from airborne platforms, and are thus costly to obtain. As a result, the application of LiDAR data in ecological movement studies will likely depend on available LiDAR data from other projects for the foreseeable future. However, as more LiDAR platforms become available, costs are likely to decrease, and accessibility of LiDAR products will increase.

Environemntal real time assessment

Finally, scientists and managers are exploring the use of low-cost pre-programmed or remote-controlled drones equipped with video and photo cameras, as alternative ways to survey landscapes, count animal populations, and monitor human activities (see http://www.fort.usgs.gov/RavenA; e.g. [112]). This environmental assessment in real-time provides exciting opportunities for animal movement studies. Weather surveillance radars track arrival of migratory birds and thereby assess quantitative data on bird distribution, helping to identify migratory stopover over broad geographic areas in real-time [14,113].

Model reanalysis products

In addition to the technologies and products outlined before, there is a growing number of efforts to reanalyze remote sensing and weather data to derive consistent, long-term, global or regional environmental products at low to medium (12–1000 km) spatial resolution, but very high (1–24 h) temporal resolution. Examples include the North American Regional Reanalysis (NARR), the NOAA global weather reanalysis (NCEP Reanalysis 2), and the European weather reanalysis (ECMWF) for weather data, or the Oregon University Ocean NPP reanalysis for ocean products. The NASA's MERRA project, for example, provides consistent information on weather variability from 1979 to the present. Finally, although not a reanalysis product per se, Hansen and co-authors [114], recently mapped global tree cover at yearly time steps between 2000 and 2012 at a spatial resolution of 30 meters, with great potential for animal movement studies.

Potential limitations of remotely-sensed products for studies of movement ecology

As much promise as new remotely-sensed data and methods offer, it is important to recognize inherent limitations of these approaches. One important issue is that of errors, uncertainties, and accuracies in remotely-sensed products [115,116]. No land cover classification is perfect, and this can introduce errors in models of animal movement. Closely related to the question of accuracy is the question of the thematic resolution, i.e., the number of land cover classes, which for ecological applications often needs to be high. However, the classification accuracy decreases as more land cover classes are mapped. In the 2006 National Land Cover Dataset (NLCD), for example, forest as a single class has a user accuracy of 93%, but coniferous, deciduous and mixed forests each had accuracies of 81-85% [117]. Some land cover classes are more challenging to map than others; in the NLCD, for example, wetlands user accuracy was <40%. Spectral indices such as NDVI, on the other hand, are mathematical formulas and as such do not have a classification error, but require validation due to e.g. sensor errors or atmospheric contamination [118,119].

Movement ecologists that apply remotely-sensed products to animal movement data also need to consider spatial autocorrelation that are inherent in the data. Advances in statistical tools help to discover spatial patterns and to correct for autocorrelated data, e.g., in mixed models with help of correlation structures [3,120]. Patterns of autocorrelation in animal movement behavior themselves provide valuable information, and are used to better understand the mechanisms of animal movement [121].

The spatial resolution of the satellite data is another critical issue that is pertinent when analyzing fine-scale movement data. First, the size of the pixels, which tend to be 250–1,000 m for those satellites that provide daily observations, are far coarser than GPS-collar-based movement data. Secondly, the smallest observable feature that can reliably be mapped (i.e., the minimum mapping unit) is considerably larger than the pixel size (typically at least four pixels). This means that only Quickbird or Worldview data can detect habitat features of 1–5 m^2, compared to minimum mapping units of 3,600 m^2 for Landsat imagery, and 250,000 m^2 for MODIS data. As movement ecology advances, and integrates more remotely-sensed data, it will be important to keep these inherent limitations of satellite data in mind, and to avoid the temptation to try to obtain more information from the imagery than it contains.

CONCLUSIONS

Animals move in response to environmental changes, and to fulfill their varying needs for environmental resources over days, weeks, and years. New generations of remotely-sensed products with better spatiotemporal resolution provide great opportunities for insights into animal movement, as they characterize environmental patterns such as temporal dynamics of landscapes and three-dimensional structures of habitats. To capture these dynamics and to improve predictive ability, we recommend that researchers analyze time series of remotely-sensed products, as they can capture environmental dynamics that determine movement responses. The integration of landscape dynamism in movement studies becomes particularly important with respect to meeting the conservation and management challenges of future landscape change and corresponding changes in animal movement in space and time. To be able to provide realistic foundations for management and conservation, movement ecologists need to increase their application of dynamic environmental information when analyzing animal movement. With new tools emerging to automate the integration of remotely-sensed data with animal movement data sets, former computational challenges can be overcome, making environmental data more easily accessible for movement ecologists. Yet, the increasing possibilities of integrated analyses of remotely-sensed products and animal movement data presents new challenges to match these types of data logically, bearing in mind the inherent trade-offs posed by the different resolutions. It is important to critically weigh which remotely-sensed product is best suited for a given scale of analyses (e.g., *fine-but-infrequent* versus *coarse-but-frequent* or minimum mapping unit) and how its limitations may affect inferences. Closer collaboration between ecologists and remote sensing experts can facilitate the use of newly available products in movement studies, and minimize the time lag of applying these products in ecological research.

DECLARATIONS

Acknowledgements

We thank the Swedish Research Council FORMAS, the thematic Program "Wildlife and Forestry", the Swedish Environmental Protection Agency, the Swedish University of Agricultural Sciences (SLU), NASA Earth and Space Science Fellowship program, and the NASA Biodiversity program for their support. We are grateful to L. Börger, N.S. Singh, A. Huete, the two referees and the handling editor for their constructive comments on an earlier version of the manuscript.

Authors' contribution

All authors participated in the design of the review and conceptualization of ideas, and writing the manuscript. WN, ABE, and AMP conducted the literature search. All authors read and approved the final manuscript.

REFERENCES

1. Getz WM, Saltz D. A framework for generating and analyzing movement paths on ecological landscapes. Proc Natl Acad Sci U S A. 2008;105:19066–71.
2. Urbano F, Cagnacci F, Calenge C, Dettki H, Cameron A, Neteler M. Wildlife tracking data management: a new vision. Phil Trans R Soc B. 2010;365:2177–85.
3. Aarts G, MacKenzie M, McConnell B, Fedak M, Matthiopoulos J. Estimating space-use and habitat preference from wildlife telemetry data. Eography. 2008;31:140–60.
4. Leyequien E, Verrelst J, Slot M, Schaepman-Strub G, Heitkönig IMA, Skidmore A. Capturing the fugitive: applying remote sensing to terrestrial animal distribution and diversity. Int J Appl Earth Obs. 2007;9:1–20.
5. Morales JM, Moorcroft PR, Matthiopoulos J, Frair JL, Kie JG, Powell RA, et al. Building the bridge between animal movement and population dynamics. Phil Trans R Soc B. 2010;365:228–2301.
6. Lima SL, Zollner PA. Towards a behavioral ecology of ecological landscapes. Trends Ecol Evol. 1996;11:131–5.
7. Schick RS, Loarie SR, Colchero F, Best BD, Boustany A, Conde DA, et al. Understanding movement data and movement process: current and emerging directions. Ecol Lett. 2008;11:1338–50.
8. Nathan R, Getz WM, Revilla E, Holyoak M, Kadmon R, Saltz D, et al. A movement ecology paradigm for unifying organismal movement research. Proc Natl Acad Sci U S A. 2008;105:19052–9.
9. Signer C, Ruf T, Schober F, Fluch G, Paumann T, Arnolds W. A versatile telemetry system for continuous measurement of heart rate, body temperature and locomotor activity in free-ranging ruminants. Method Ecol Evol. 2010;1:75–85.
10. Patterson TA, Thomas L, Wilcox C, Ovaskainen O, Matthiopoulos J. State-space models of individual animal movement. Trends Ecol Evol. 2008;23:87–94.
11. Thurfjell H, Ciuti S, Boyce MS. Applications of step-selection functions in ecology and conservation. Mov Ecol. 2014;2:4.
12. Johnson CJ, Parker KL, Heard DC, Gillengham MP. A multiscale behavioral approach to understanding the movements of woodland caribou. Ecol Appl. 2002;12:1840–60.
13. Leblond M, Dussault C, Quellet J-P. What drives fine-scale movements of large herbivores? A case study using moose. Ecography. 2010;33:1102–12.
14. Buler JJ, Moore FR. Migrant-habitat relationships during stopover along an ecological barrier: extrinsic constraints and conservation implications. J Ornithol. 2011;152(Suppl1):S101–12.
15. Cushman SA, Raphael MG, Ruggiero LF, Shirk AS, Wasserman TN, O'Doherty EC. Limiting factors and landscape connectivity: the American marten in the Rocky Mountains. Landsc Ecol. 2011;26:1137–49.
16. Sawyer H, Kauffman MJ. Stopover ecology of a migratory ungulate. J Anim Ecol. 2011;80:1078–87.
17. Åkesson S, Klaassen R, Holmgren J, Fox JW, Hedenström A. Migration routes and strategies in a highly aerial migrant, the Common swift Apus apus, revealed by light-level geolocators. PLoS One. 2012;7:e41195.
18. Zeller KA, McGarigal K, Whiteley AR. Estimating landscape resistance to movement: a review. Landsc Ecol. 2012;27:777–97.
19. Valeix M, Loveridge AJ, Davidson Z, Madzikanda H, Fritz H, MacDonald DW. How key habitat features

influence large terrestrial carnivore movements: waterholes and African lions in a semi-arid savanna of north-west Zimbabwe. Landsc Ecol. 2010;25:337–51.

20. Ordiz A, Stoen O-G, Delibes M, Swenson JE. Predators or prey? Spatio-temporal of human-derived risk by brown bears. Oecologia. 2011;166:59–67.

21. Börger L, Fryxell JM. Quantifying individual differences in dispersal using net square displacement. In: Clobert J, Baquette M, Benton T, Bullock J, editors. Dispersal Ecology and Evolution. USA: Oxford Press; 2012. p. 222–30.

22. Takekawa JY, Newman SH, Xiao X, Prosser DJ, Spragens KA, Palm EC, et al. Migration of waterfowl in the east Asian flyway and spatial relationship to HPAI H5N1 outbreaks. Avian Dis. 2010;54:466–76.

23. Moorcraft PR. Mechanistic approaches to understanding and predicting mammalian space use: recent advances, future directions. J Mammal. 2012;93:903–16.

24. Richard Y, Armstrong DP. Cost distance modeling of landscape connectivity and gap-crossing ability using radio-tracking data. J Appl Ecol. 2010;47:603–10.

25. Rivrud IM, Loe LE, Mysterud A. How does local weather predict red deer home range size at different temporal scales? J Anim Ecol. 2010;79:1280–95.

26. Van Beest FM, Rivrud IM, Loe LE, Milner JM, Mysterud A. What determines variation in home range size across spatiotemporal scales in a large browsing herbivore? J Anim Ecol. 2011;80:771–85.

27. Bischof R, Loe LE, Meisingset EL, Zimmermann B, van Moorter B, Mysterud AD. A Migratory Northern Ungulate in the Pursuit of Spring: Jumping or Surfing the Green Wave? Am Nat. 2012;180:407–24.

28. Morellet N, Bonenfant C, Börger L, Ossi F, Cagnacci F, Heurich M, et al. Seasonality, weather and climate affect home range size in roe deer across a wide latitudinal gradient within Europe. J Anim Ecol. 2013;82:1326–39.

29. Bohrer G, Brandes D, Mandel JT, Bildstein KL, Miller TA, Lanzone M, et al. Estimating updraft velocity components over large spatial scales: contrasting migration strategies of golden eagles and turkey vultures. Ecol Lett. 2012;15:96–103.

30. Klaassen RHG, Alerstam T, Carlsson P, Fox JW, Lindström Å. Great flights by great snipes: long and fast non-stop migration over benign habitats. Biol Lett. 2011;7:833–5.

31. Mueller T, Olson KA, Dressler G, Leimgruber P, Fuller T, Nicolson C, et al. How landscape dynamics link individual- to population-level movement patterns: a multispecies comparison of ungulate relocation data. Glob Ecol Biogeogr. 2011;20:683–94.

32. Wakefield ED, Phillips RA, Matthiopoulos J. Quantifying habitat use and preferences of pelagic seabirds using individual movement data: a review. Mar Ecol Prog Ser. 2009;391:165–82.

33. McKinnon EA, Fraser KC, Stutchbury BJM. New discoveries in landbird migration using geolocators, and a flight plan for the future. The Auk. 2013;130:211–22.

34. Frair JL, Merrill EH, Beyer HL, Morales JM. Thresholds in landscape connectivity and mortality risks in response to growing road networks. J Appl Ecol. 2008;45:1504–13.

35. Loarie SR, Van Arde RJ, Pimm SL. Fences and artificial water affect African savannah elephant movement patterns. Biol Conserv. 2009;142:3086–98.

36. Estes AB, Kuemmerle T, Kushnir H, Radeloff VC, Shugart HH. Land-cover change and human population trends in the greater Serengeti ecosystem from 1984–2003. Biol Conserv. 2012;147:255–63.

37. Heckscher CM, Taylor SM, Fox JW, Afanasyev V. Veery (Catharus fuscescens) wintering locations, migratory connectivity, and a revision of its winter range using geolocator technology. The Auk. 2011;128:531–42.

38. Cagnacci F, Focardi S, Heurich M, Stache A, Hewison AJM, Morellet N, et al. Partial migration in roe deer: migratory and resident tactics are end points of a behavioural gradient determined by ecological factors. Oikos. 2011;120:1790–802.

39. Musiani M, Anwar SM, McDermid GJ, Hebblewhite M, Marceau DJ. How humans shape wolf behavior in Banff and Kootenay National Parks. Can Ecol Model. 2010;221:2374–87.

40. Boettiger AN, Wittemyer G, Starfield R, Volrath F, Douglas-Hamilton I, Getz WM. Inferring ecological and behavioral drivers of African elephant movement using a linear filtering approach. Ecology. 2011;92:1648–57.

41. Naidoo R, Du Preez P, Stuart-Hill G, Jago M, Wegmann M. Home on the range: factors explaining partial migration of African buffalo in a tropical environment. PLoS One. 2012;7:e36527.

42. Martin J, Basille M, Van Moorter B, Kindberg J, Allaine D, Swenson JE. Coping with human disturbance: spatial and temporal tactics of the brown bear (Ursus arctos). Can J Zool. 2010;88:875–83.

43. Neumann W, Ericsson G, Dettki H, Radeloff VC. Behavioral response to infrastructure of wildlife adapted to natural disturbances. Land Urban Plan. 2013;114:9–27.

44. Bruggeman JE, Garrott RA, White PJ, Watson FGR, Wallen R. Covariates affecting spatial variability in bison travel behavior in Yellowstone National Park. Ecol Appl. 2007;17:1411–23.

45. Neubert M, Hecht R, Gedrange C, Trommler M, Herold H, Krüger T, et al. Extraction of railroad objects from very high resolution helicopter-borne LiDAR and ortho-image data. Int Arch Photogramm Remote Sens Spat Inf Sci. 2008;38:25–30.

46. Pacifi F, Chini M, Emery WJ. A neural network approach using multi-scale textural metrics from very high-resolution panchromatic imagery for urban land-use classification. Remote Sens Environ. 2009;6:1276–92.

47. Hubbard TD, Braun ML, Westbrook RE, Gallagher PE. High-resolution lidar data for infrastructure corridors. In: Hubbard TD, Koehler RD, Combellick RA, editors. High-resolution lidar data for Alaska infrastructure corridors: Alaska Division of Geological & Geophysical

Surveys Raw Data File 2011-3L. Alaska: Anchorage Quadrangle; 2011.

48. Bohrer G, Zhu KP, Jones RL, Curtis PS. Optimizing wind power generation while minimizing wildlife impacts in an urban area. PLoS One. 2013;8:e56036.

49. White RA, Dietterick BL, Mastin T, Strohman R. Forest roads mapped using LiDAR is steep forested terrain. Remote Sens. 2010;2:1120–41.

50. Coulon A, Morellet N, Goulard M, Cargnelutti B, Angibault J-M, Hewison AJM. Inferring the effects of landscape structure on roe deer (Capreolus capreolus) movements using a step selection function. Landsc Ecol. 2008;23:603–14.

51. Roever CL, Boyce MS, Stenhouse GB. Grizzly bear movements relative to roads: application of step selection functions. Ecography. 2010;33:1113–22.

52. Lewis JS, Rachlow JL, Horne JS, Garton EO, Wakkinen WL, Hayden J, et al. Identifying habitat characteristics to predict highway crossing areas for black bears within a human-modified landscape. Landsc Urban Plan. 2011;101:99–107.

53. Huete A, Didan K, Miura T, Rodriguez EP, Gao X, Ferreira LG. Overview of the radiometric and biophysical performance of the MODIS vegetation indices. Remote Sens Environ. 2002;83:195–213.

54. Zhang X, Friedl MA, Schaaf CB, Strahler AH, Hodges JCF, Gao F, et al. Monitoring vegetation phenology using MODIS. Remote Sens Environ. 2003;84:471–5.

55. Pettorelli N, Ryan S, Mueller T, Bunnefeld N, Jedrzejewska B, Lima M, et al. The unforeseen successes of the Normalized Difference Vegetation Index (NDVI) in animal ecology. Climate Res. 2011;46:15–27.

56. Schwartz MD, Reed BC, White MA. Assessing satellite-derived start-of-season measure in the conterminous USA. Int J Climatol. 2002;22:1793–805.

57. Zhang X, Goldberg MD. Monitoring fall foliage coloration dynamics using time-series satellite data. Remote Sens Environ. 2011;115:382–91.

58. Boone RB, Thirgood SJ, Hopcraft JGC. Serengeti wildebeest migratory patterns modeled from rainfall and new vegetation growth. Ecology. 2006;87:1987–94.

59. Renfrew RB, Kim D, Perlut N, Smith J, Fox J, Marra PP. Phenological matching across hemispheres in a long-distance migratory bird. Divers Distrib. 2013;19:1008–19.

60. Tøttrup AP, Thorup K, Rainio K, Yosef R, Lehikoinen E, Rahbek C. Avain migrants adjust migration in response to environmental conditions en route. Biol Lett. 2008;4:685–8.

61. Wittemyer G, Polansky L, Douglas-Hamilton I, Getz WM. Disentangling the effects of forage, social rank, and risk of movement autocorrelation of elephants using Fourier and wavelet analyses. Proc Natl Acad Sci U S A. 2008;105:19108–13.

62. Justice CO, Giglio L, Roy D, Boschetti L, Csiszar I, Davies D, et al. MODIS-Derived Global Fire Products. In: Ramachandran B, Justice CO, Abrams MJ, editors. Land Remote Sensing and Global Environmental Change, Remote Sensing and Digital Image Processing. New York: Springer; 2011. p. 661–79.

63. Bartlam-Brooks HLA, Beck PSA, Bohrer G, Harris S. In search of greener pastures: Using satellite images to predict the effects of environmental change on zebra migration. J Geophys Res-Biogeo. 2013;118:1427–37.

64. Bohrer G, Beck PS, Ngene SM, Skidmore AK, Douglas-Hamilton I. Elephant movement closely tracks precipitation-driven vegetation dynamics in a Kenyan forest-savanna landscape. Mov Ecology. 2014;2:2.

65. Dodge S, Bohrer G, Bildstein K, Davidson SC, Weinzierl R, Bechard MJ, et al. Environmental drivers of variability in the movement ecology of turkey vultures (*Cathartes aura*) in North and South America. Philos T Roy Soc B. 2014;369:20130195.

66. Hebblewhite M, Merrril E, McDermid G. A multi-scale test of the forage maturation hypothesis in a partially migratory ungulate population. Ecol Monogr. 2008;78:141–66.

67. Ares JO, Bertiller M. Modeling high-frequency position data of large herbivores with a phase-state model. Ecol Model. 2010;221:2323–9.

68. Singh NJ, Grachev IA, Bekenov AB, Milner-Gullard EJ. Tracking greenness across a latitudinal gradient in central Asia – the migration of the saiga antelope. Divers Distrib. 2010;16:663–75.

69. Duerr AE, Miller TA, Lanzone M, Brandes D, Cooper J, O'Malley K, et al. Testing an emerging paradigm in migration ecology shows surprising differences in efficiency between flight modes. PLoS One. 2012;7:e35548.

70. Mellone U, Klaassen RHG, Garcia-Ripolles C, Liminana R, Lopez-Lopez P, Pavon D, et al. Interspecific comparison of the performance of soaring migrants in relation to morphology, meteorological conditions and migration strategies. PLoS One. 2012;7:e39833.

71. Nathan R, Spiegel O, Fortmann-Roe S, Harel R, Wikelski M, Getz WM. Using tri-axial acceleration data to identify behavioral modes of free-ranging animals: general concepts and tools illustrated for griffon vultures. J Exp Biol. 2012;215:986–96.

72. Sapir N, Horvitz N, Dechmann DKN, Fahr J, Wikelski M. Commuting fruit bats beneficially modulate their flight in relation to wind. Proc R Soc B. 2014;281:20140018.

73. Sapir N, Horvitz N, Wikelski M, Avissar R, Mahrer Y, Nathan R. Migration by soaring or flapping: numerical atmospheric simulations reveal that turbulence kinetic energy dictates bee-eater flight mode. Proc R Soc B. 2011;278:3380–6.

74. Wang K, Franklin SE, Guo X, Cattet M. Remote sensing of ecology, biodiversity and conservation: a review from the perspective of remote sensing specialists. Sensors. 2010;10:9647–67.

75. Weng Q. Remote sensing of impervious surfaces in the urban area: Requirements, methods and trends. Remote Sens Environ. 2012;117:34–49.

76. Johansen K, Phinn S, Lowry J, Douglas M. Quantifying indicators of riparian condition in Australian tropical

savannas: integrating high spatial resolution imagery and field survey data. Int J Remote Sens. 2008;29:7003–28.

77. Aksoy S, Akcay HG, Wassenaar T. Automatic maping of linear woody vegetation features in agricultural landscapes using very high resolution imagery. IEEE Trans Geoscience Remote Sens. 2010;48:511–22.

78. Boggs GS. Assessments of SPOT 5 and QuickBird remote sensing sensed imagery for mapping tree cover in savannas. Int J Appl Earth Obs. 2010;12:217–24.

79. Gottschalk TK, Aue B, Hotes S, Ekschmitt K. Influence of gran size on species-habitat models. Ecol Model. 2011;222:3403–12.

80. Cohen W, Goward S. Landsat's role in ecological applications of remote sensing. Bioscience. 2004;54:535–45.

81. Woodcoock CE, Allen R, Anderson M, Belward A, Bindschadler R, Cohen W, et al. Free access to LANDSAT imagery. Science. 2008;320:1011.

82. Newton AC, Hill RA, Echeverria C, Golicher D, Benayas JMR, Cayuela L, et al. Remote sensing and the future of landscape ecology. Prog Phys Geogr. 2009;33:528–46.

83. Wulder MA, White JC, Masek JG, Dwyer J, Roy DP. Continuity of Landsat observations: Short terms consideration. Remote Sens Environ. 2011;115:747–51.

84. Aschbacher J, Milagro-Perez MP. The European Earth monitoring (GMES) programme: Status and perspectives. Remote Sens Environ. 2012;120:3–8.

85. Shirley SM, Yang Z, Hutchinson RA, Alexander JD, McGarigal K, Betts MG. Species distribution modelling for the people: unclassified landsat TM imagery predicts bird occurrence at fine resolutions. Divers Distrib. 2013;19:855–66.

86. Dodge S, Bohrer G, Weinzierl R, Davidson SC, Kays R, Douglas D, et al. The environmental-data automated track annotation (Env-DATA) system: linking animal tracks with environmental data. Mov Ecol. 2013;1:3.

87. Dettki H, Ericsson G, Giles T, Norrsken-Ericsson M. Wireless Remote Animal Monitoring (WRAM) - A new international database e-infrastructure for telemetry sensor data from fish and wildlife. In: The European Society of Telemetry, editor. Proceedings 2012: Convention for Telemetry, Test Instrumentation and Telecontrol. 2013. p. 247–56.

88. Thirgood S, Mosser A, Tham S, Hopcraft G, Mwangomo E, Mlengeya T, et al. Can parks protect migratory ungulates? The case of the Serengeti wildebeest. Anim Conserv. 2004;7:113–20.

89. Graham MD, Douglas-Hamilton I, Adams WM, Lee PC. The movement of African elephants in a human-dominated land-use mosaic. Anim Conserv. 2009;12:445–55.

90. Huang C, Goward SN, Masek JG, Thomas N, Zhu Z, Vogelmann JE. An automated approach for reconstructing recent forest disturbance history using dense Landsat time series stacks. Remote Sens Environ. 2010;114:183–98.

91. Kennedy RE, Yang ZG, Cohen WB. Detecting trends in forest disturbance and recovery using yearly Landsat time series: 1. LandTrendr - Temporal segmentation algorithms. Remote Sens Environ. 2010;114:2897–910.

92. Hostert P, Röder A, Hill J. Coupling spectral unmixing trend analysis for monitoring long-term vegetation dynamics in Mediterranean rangelands. Remote Sens Environ. 2003;87:183–97.

93. Schneider A, Friedl M, Potere D. Monitoring urban areas globally using MODIS 500 m data: New methods and datasets based on "urban ecoregions". Remote Sens Environ. 2010;114:1733–46.

94. Eklundh L, Jönsson P. TIMESAT 3.0 - Software Manual. Lund University. 2010. http://www.nateko.lu.se/TIMESAT/timesat.asp

95. Tan B, Morisette JT, Wolfe RE, Gao F, Ederer GA, Nightingale J, et al. An enhanced TIMESAT algorithm for estimating vegetation phenology metrics from MODIS data. IEE J Sel Top Appl Earth Obs Remote Sens. 2011;4:361–71.

96. Beck HE, McVicar TR, van Dijk AIJM, Schellekens J, de Jeu RAM, Bruijnzeel LA. Global evaluation of four AVHRR–NDVI data sets: Intercomparison and assessment against Landsat imagery. Remote Sens Environ. 2011;115:2547–63.

97. Roy DP, Ju J, Lewis P, Schaaf C, Gao F, Hansen M, et al. Multi-temproal MODIS-LANDSAT data fusion for relative radiometric normalization, gap filling, and prediction of Landsat data. Remote Sens Environ. 2008;112:3112–30.

98. Gitelson AA, Merzlyak MN. Signature analysis of leaf reflectance spectra: algorithm development for remote sensing of chlorophyll. J Plant Physiol. 1996;148:494–500.

99. Ganguly S, Nemani RR, Zhang G, Hashimoto H, Milesi C, Michaelis A, et al. Generating global leaf area index from landsat: algorithm formulation and demonstration. Remote Sens Environ. 2012;122:185–202.

100. Ramoelo A, Skidmore AK, Cho MA, Schlerf M, Mathieu R, Heitkönig IMA. Regional estimation of savanna grass nitrogen using the red-edge band of the spaceborne RapidEye sensor. Int J Appl Earth Obs. 2012;19:151–62.

101. Bergen KM, Goetz S, Dubayah RO, Henebry GM, Hunsacker CT, Imhoff ML, et al. Remote sensing of vegetation 3D structure for biodiversity and habitat: review and implications for LiDAR-RaDAR spaceborne missions. J Geophys Res. 2009;114:G00E06.

102. Vierling KT, Vierling LA, Gould W, Martinuzzi S, Clawges R. Lidar: Shedding new light on habitat characterization and modeling. Front Ecol Environ. 2008;6:90–8.

103. Dubayah RO, Sheldon SL, Clark DB, Hofton MA, Blair JB, Hurtt GC, et al. Estimation of tropical forest height and biomass dynamics using lidar remote sensing at La Selva, Costa Rica. J Geophys Res. 2010;115:G00E09.

104. Thomas V, Treitz P, McCaughey JH, Morrison I. Mapping stand-level forest biophysical variables for a mixed-wood boreal forest using lidar: an examination of scanning density. Can J For Res. 2006;36:34–46.

105. Falkowski MJ, Evans JS, Martinuzzi S, Gessler PE, Hudak AT. Characterizing forest succession with Lidar data: an evaluation for the Inland Northwest USA. Remote Sens Environ. 2009;113:946–56.

106. Gaulton R, Malthus TJ. LiDAR mapping of canopy gaps in continuous cover forests: A comparison of canopy height model and point cloud based techniques. Int J Remote Sens. 2010;31:1193–211.

107. Hyde P, Dubayah R, Peterson B, Blair JB, Hofton M, Hunsacker C, et al. Mapping forest structure for wildlife habitat analysis using waveform lidar; validation of montane ecosystems. Remote Sens Environ. 2005;96:427–37.

108. Martinuzzi S, Vierling L, Gould W, Falkowski M, Evans L, Hudak A, et al. Mapping snags and understory shrubs for a LiDAR-based assessment of wildlife habitat suitability. Remote Sens Environ. 2009;113:2533–46.

109. Asner GP, Levick SR. Landscape-scale effects of herbivores on treefall in African savannas. Ecol Lett. 2012;15:1211–7.

110. Loarie SR, Tambling CJ, Asner GP. Lion hunting behavior and vegetation structure in an African savanna. Anim Beh. 2013;85:899–906.

111. Lone K, Loe LE, Gobakken T, Linnell JDC, Odden J, Remmen J, et al. Living and dying in a multi-predator landscape of fear: roe deer are squeezed by contrasting pattern of predation risk imposed by lynx and humans. Oikos. 2014;123:641–51.

112. Koh LP, Wich SA. Dawn of drone ecology: low-cost autonomous aerial vehicles for conservation. Trop Conserv Sci. 2012;5:121–32.

113. Baratchi M, Meratnia N, Havinga PJM, Skidmore AK, Toxopeus BAG. Sensing solutions for collection spatio-temporal data for wildlife monitoring applications: a review. Sensors. 2013;13:6054–88.

114. Hansen MC, Potapov PV, Moore R, Hancher M, Turubanova SA, Tyukavina A, et al. High-resolution global maps of 21st-century forest cover change. Science. 2013;342:850–3.

115. Foody GM, Atkinson PM. Uncertainty in Remote Sensing and GIS. England: John Wiley & Sons Ltd; 2002.

116. Olofsson P, Foody GM, Stehman SV, Woodcock CE. Making better use of accuracy data in land change studies: Estimating accuracy and area and quantifying uncertainty using stratified estimation. Remote Sens Environ. 2013;129:122–31.

117. Wickham JD, Stehman SV, Gass L, Dewitz J, Fry JA, Wade TG. Accuracy assessment of NLCD 2006 land cover and impervious surface. Remote Sens Environ. 2013;130:294–304.

118. Berry PAM, Garlick JD, Smith RG. Near-global validation of the SRTM DEM using satellite radar altimetry. Remote Sens Environ. 2007;106:17–27.

119. Masuoka E, Roy D, Wolfe R, Morisette J, Sinno S, Teague M, et al. MODIS Land Data Products: Generation, Quality Assurance and Validation. In: Ramachandran B, Justice C, Abrams M, editors. Land Remote Sensing and Global Environmental Change. New York: Springer; 2011. p. 509–31.

120. Pinheiro J, Bates D. Mixed-effects models in S and S-PLUS. New York: Springer; 2000.

121. Boyce MS, Pitt J, Northrup JM, Morehouse AT, Knoff KH, Cristescu B, et al. Temporal autocorrelation functions for movement rates from global positioning system radiotelemetry data. Phil Trans R Soc B Biol Sci. 2010;365:2213–9.

122. Hudak AT, Evans JS, Stuart Smith AM. LiDAR Utility for Natural Resource Managers. Remote Sens. 2009;1:934–51.

123. Evans JS, Hudak AT, Faux R, Smith AMS. Discrete Return Lidar in Natural Resources: Recommendations for Project Planning, Data Processing, and Deliverables. Remote Sens. 2009;1:776–94.

124. Fortin D, Beyer HL, Boyce MS, Smith DW, Duchesne T, Mao JS. Wolves influence elk movements: behavior shapes a trophic cascade in Yellowstone National Park. Ecology. 2005;86:1320–30.

125. Colchero F, Conde DA, Manterola C, Chávez C, Rivera A, Ceballos G. Jaguars on the move: modeling movement to mitigate fragmentation from road expansion in the Mayan Forest. Anim Conserv. 2011;14:158–66.

126. Gillies CS, Beyer HL, St Clair CC. Fine-scale movement decisions of tropical forest birds in a fragmented landscape. Ecol Appl. 2011;21:944–54.

127. Musiega DE, Kazadi SN. Simulating the East African wildebeest migration patterns using GIS and remote sensing. Afr J Ecol. 2004;42:355–62.

128. Pettorelli N, Mysterud A, Yoccoz NG, Langvatn R, Stenseth NC. Importance of climatological downscaling and plant phenology for red deer in heterogeneous landscapes. Proc R Soc B. 2005;272:2357–64.

129. Olson KA, Mueller T, Bolortsetseg S, Leimgruber P, Fagan WF, Fuller TK. A mega-herd of more than 200,000 Mongolian gazelles Procapra gutturosa: a consequence of habitat quality. Oryx. 2009;43:149–53.

130. Mysterud A, Loe LE, Zimmermann B, Bischof R, Veiberg V, Meisingset E. Partial migration in expanding red deer populations at northern latitudes – a role for density dependence? Oikos. 2011;120:1817–25.

131. EEA Technical Report. The thematic accuracy of Corine land cover 2000. European Environment Agency. 2006;7. http://www.eea.europa.eu. Accessed 19 Jan 2015.

132. NASA. MODIS land validation. http://landval.gsfc.nasa.gov/. Accessed 19 Jan 2015.

133. Bicheron P, Defourny P, Brockmann C, Schouten L, Vancutsem C, Huc M, et al. Globcover – Products description and validation report. Medias-France. 2008. http://due.esrin.esa.int/files/GLOBCOVER_Products_Description_Validation_Report_I2.1.pdf. Accessed 19 Jan 2015.

134. Rodríguez E, Morris CS, Belz JE. A global assessment of the SRTM performance. Photogramm Eng Rem S. 2006;72:249–60.

135. Tachikawa T, Kaku M, Iwasaki A, Gesch D, Oimoen M, Zhang Z, et al. ASTER Global Digital Elevation Model Version 2 – Summary of Validation Results. NASA Land Processes Distributed Active Archive Center & Joint Japan US ASTER Science Team. 2011. https://www.jspacesystems.or.jp/ersdac/GDEM/ver2Validation/Summary_GDEM2_validation_report_final.pdf. Accessed 19 Jan 2015.

136. Microwave and Radar Institute. TanDEM-X - A New High Resolution Interferometric SAR Mission. TanDEM-X Science Home. 2015. http://www.dlr.de/hr/en/desktopdefault.aspx/tabid-2317/3669_read-5488/. Assessed 21 Jan 2015.

137. Zhang X, Friedl MA, Schaaf CB. Global vegetation phenology from Moderate Resolution Imaging Spectroradiometer (MODIS): evaluation of global patterns and comparison with in situ measurements. J Geophys Res. 2006;111:G04017.

CHAPTER 13

SPATIAL ENHANCEMENT OF MODIS-BASED IMAGES OF LEAF AREA INDEX: APPLICATION TO THE BOREAL FOREST REGION OF NORTHERN ALBERTA, CANADA

Quazi K. Hassan, and Charles P.-A. Bourque

Keywords: Boreal forest, data fusion, enhanced vegetation index, leaf area index, MODIS

INTRODUCTION

Leaf area is an important factor in terrestrial ecosystems and represents the area involved in: (i) the plant-to-air and air-to-plant transfer of gases and particulate matter, (ii) energy transfer and exchange, (iii) eco-physiological processes of evapotranspiration, photosynthesis, and net primary production, and (iv) forest landscape growth and development. LAI is commonly defined as ½ of the total green area of forests, accounting for all sides of the leaves, per unit horizontal ground-surface area [1] and represents the vertical integration of leaf area along the length of forest canopies. LAI is dynamic and varies according to tree species composition, stand developmental stage, site conditions, time of year, and management regime [2].

Determination of spatiotemporal dynamics of LAI across large areas (e.g., covering hundreds to thousands of hectares) is only possible with remote-sensing-based techniques (e.g., [3-6]). These techniques typically rely on establishing empirical relationships between normalised difference vegetation index (NDVI) and LAI [7-11]. However, NDVI tends to saturate over dense vegetation (LAI > 3–4; [12]) or during the height of the growing season [13], despite LAI being dynamic during this time. Since 2000, Moderate Resolution Imaging Spectroradiometer (MODIS)-based 8-day composites of LAI have been made available to the scientific community at 1-km resolution by the National Aeronautics and Space Administration (NASA), USA, at no cost to the user. These products are primarily generated by solving a 3-dimensional radiative-transfer model using atmospherically-corrected MODIS spectral surface reflectance and biome identification [3,14,15]. In understanding ecosystem dynamics at large spatial scales, the scientific community would greatly benefit from simple techniques that could be used to improve the spatial resolution of existing products of LAI.

Enhanced vegetation index (EVI, at 250-m resolution [4]) brings with it two clear advantages over NDVI concerning the depiction of LAI, namely: (i) EVI is sensitive to variations in LAI (NDVI, in contrast, responds more to plant greenness [4]); and (ii) EVI, unlike NDVI, does not saturate at peak foliage concentrations [4,13,16]. Preliminary results from Huete et al. [4] reveal the presence of a linear relationship between EVI and LAI for a number of biomes. Here we attempt to confirm Huete et al.'s observation of linearity for the boreal forest of northern Alberta, Canada, for the April–October period of 2005–2008. Evidence of a strong linear correlation between EVI and LAI would provide basis for augmenting coarse-resolution images of LAI with data fusion methods. The data fusion technique by Hassan et al. [17], which we implement in this study, was first developed to enhance images of growing degree days (GDD; a temperature-related index developed from MODIS-based land surface temperature [13,17]) from 1-km to 250-m resolution using images of EVI (at 250-m resolution) as basis for fusion. Data fusion was possible in this study because of the strong linear correlation between GDD and EVI (r2 = 0.87; [13]). The method by Hassan et al. [17] differs from other data fusion techniques due to its dependence on the relationship between

Citation: Quazi K. Hassan 1, and Charles P.-A. Bourque, Spatial Enhancement of MODIS-based Images of Leaf Area Index: Application to the Boreal Forest Region of Northern Alberta, Canada, doi:10.3390/rs2010278.

ecologically-correlated variables rather than on spectral similarities between reflectance bands [e.g., 18,19].

Following the development of spatially-enhanced images of LAI for the Athabasca Oil Sands Region of northern Alberta (Figure 1) for June 2009, we validate the capabilities of Hassan et al.'s method [17] by comparing point-estimates of LAI extracted from the enhanced images with corresponding field-based estimates.

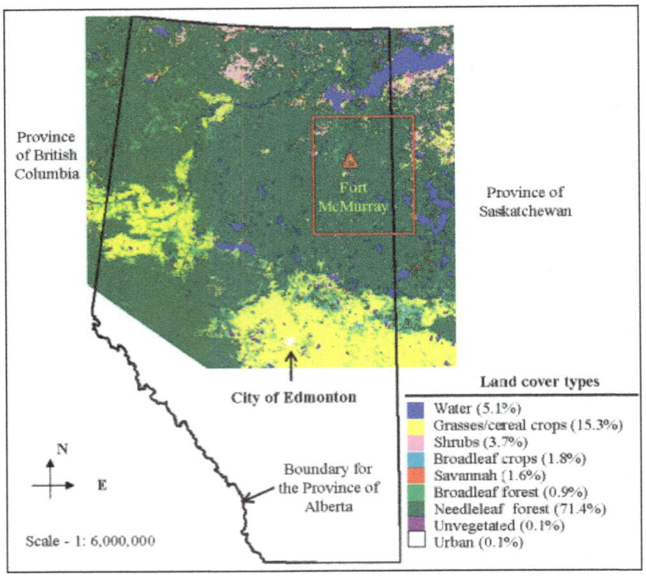

Figure 1: Extent of study area. A landcover map, derived from an annual composite of 2004 MODIS images, appears in the background. The black polygon outlines the Province of Alberta. The red box identifies the Athabasca Oil Sands Region, an area where field-based estimates of LAI were acquired to validate estimates of LAI generated from MODIS data.

METHODS

General Description of the Study Area and Data Requirements

Figure 1 shows the extent of the study area and landcover for northern Alberta (MOD12Q1 v. 4; annual 1-km resolution composite of 2004 MODIS images of landcover). The study area extends to the northern boundary of the Province of Alberta in the north, ~200 km inside the Province of Saskatchewan to the east, just south of the City of Edmonton to the south, and ~25 km inside the Province of British Columbia to the west (Figure 1). The majority of the study area is characterised by (i) a boreal plains, forest-dominated landscape consisting of 71.4% needleleaf and 0.9% broadleaf forests, and (ii) a colddry climate [20]. Tree species common to the area include trembling aspen (Populus tremuloides), balsam poplar (Populus balsamifera), white spruce (Picea glauca), black spruce (Picea mariana), and jack pine (Pinus banksiana). In poorly drained areas (i.e., fens and bogs), tamarack (Larix laricina) and black spruce often dominate. For Fort McMurray, near the centre of the Athabasca Oil Sands Region (56°39.0'N latitude, 111°13.2'W longitude; Figure 1), the mean annual temperature and annual precipitation for the 1971-2000 normal period are 0.7 °C and 455.5 mm, respectively. The January and July mean temperatures for the same period are −18.8 °C and 16.8 °C [21].

MODIS-based products used in this study included (i) fifty two 8-day and twenty six 16-day composites of LAI and EVI (i.e., MOD15A2 v. 5 and MOD13Q1 v. 5 images at 1-km and 250-m resolution, respectively), for the April–October period of 2005–2008 for assessing the presence of correlation between LAI and EVI, and (ii) a second set of eight LAI and four EVI composites for June 2009 to validate spatially-enhanced estimates of LAI (250-m resolution). The first set of images was collected over the April–October period because (i) this period coincided with the growing season, and (ii) LAI varied most over this time [20]. To validate MODIS-based LAI for June 2009 after image enhancement, field-based LAI for June 2009 were acquired at 75 sites across the Athabasca Oil Sands Region of northern Alberta (Figure 1; see Section 2.4).

Image Pre-Processing

LAI and EVI images for the 2005–2008 and June 2009 periods were re-projected to UTM (NAD83), Zone 12. In order to generate images for the entire study area, it required that two adjacent scenes of both LAI and EVI images be mosiacked. Based on a preliminary cell-by-cell evaluation of average LAI (at 1-km resolution) and EVI (at 250 m) for the growing season of 2006 (Figure 2), it was observed that acceptable correlation existed within 80% of the data for needleleaf forests ($r^2 = 0.57$).

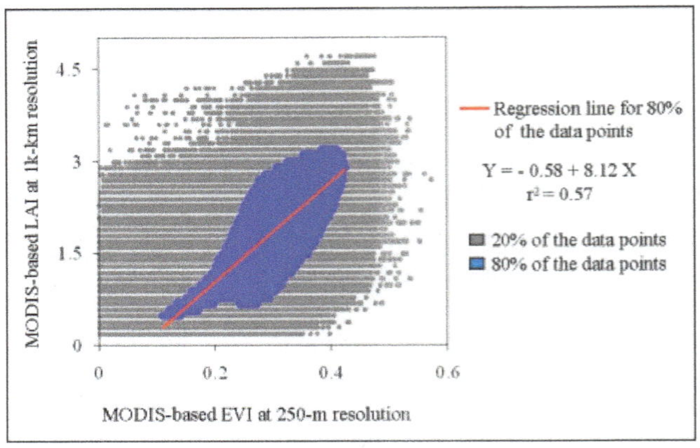

Figure 2: Scatterplot of cell-to-cell comparison of growing season (2006) averages for LAI and EVI for needleleaf forests; F-statistic = 2,120.31 and $p < 0.0001$.

High noise to signal ratio in the data (Figure 2) was most probably associated with (i) inconsistency in temporal and spatial resolutions in the MODIS-based products used in this study (8–16 day averages over 250-m to 1-km resolution), and (ii) spatial variation in atmospheric transmissivity and other extrinsic factors that affect LAI calculations based on satellite data. To reinforce the signal between LAI and EVI, values were averaged across the entire study area (irrespective of biome) for each 8-day period. Similar averaging schemes have been used by other researchers, e.g., [16], to reduce variability inherent in satellite data. Temporal behaviour of 2005-2008 mean values was then used to determine the degree of correlation between "study area" LAI and EVI (see Section 3, for more details).

Data Fusion

Data fusion, after Hassan et al. [17], involves the generation of an artificial image (AI), which takes into account statistical properties of EVI within a 3 cell × 3 cell moving window, i.e.,:

$$AI = \frac{EVI_{ins}}{EVI_{mean}}, \quad (1)$$

where EVI_{ins} is the instantaneous point-value of EVI and EVI_{mean} is the mean of all EVI values inside the moving window. In theory, AI can be viewed as an index that describes the relation of an instantaneous value of EVI to the mean value of EVI of surrounding pixels, and functions as a weight in the calculation of LAI at 250-m resolution, i.e.,:

$$LAI_{250\ m} = AI \times LAI_{1\ km}, \quad (2)$$

where $LAI_{1\ km}$ represents the initial NASA-produced MODIS image of LAI at 1-km resolution.

Field-based LAI Data Acquisition and Processing

Validation of MODIS-based LAI was based on a June-2009 field-collection of LAI at 75 sites across the Athabasca Oil Sands Region of northern Alberta (Figure 1). To address variation in light conditions during the measurement period, LAI was determined with both the Tracing Radiation and Architecture of Canopies (TRAC) sensor and LAI-2000 Plant Canopy Analyzer because of their differences in lighting requirements. The TRAC is optimised for direct light and the LAI-2000, for diffused light conditions. Final selection of LAI from the two sensors is based on recorded sky conditions and the time when measurements were taken. For partially cloudy skies in late morning (09:30 AM LST) to mid-afternoon (05:30 PM LST), an average LAI is calculated from the individual LAI measurements determined with the TRAC and LAI-2000 sensors. At low sun elevations and increased diffused light conditions during early morning (sunrise to 09:30 AM LST) and late afternoon (05:30 PM LST to sundown), final LAI is based on the LAI-2000 sensor reading, regardless of sky conditions.

In general, light measurements were taken along transects of 250 m, unless site conditions (e.g., changes in stand composition, physical barriers) prevented the use of long transects. Both sensors required that light measurements be taken in open light conditions (in fields, alongside roads) prior to measuring light conditions in tall vegetation to define a light reference use in the calculation of LAI.

Calculation of LAI is performed after each measurement session with the assistance of sensor-specific processing software available with the purchase of the equipment.

RESULTS AND DISCUSSION

Figure 3 shows the temporal behaviour of LAI and EVI for the April–October period of 2005–2008. It revealed that both LAI and EVI followed near parallel tracks with a distinctive peak each year at around 20 July [day of year (DOY): 201; Figure 3a–e]. Polynomial fits to both LAI and EVI revealed strong correlations with DOY, i.e., for LAI, r^2 -values ranged from 0.92–0.95 for individual years and for EVI, r^2 -values ranged from 0.97–0.99; for all cases considered, p-values were

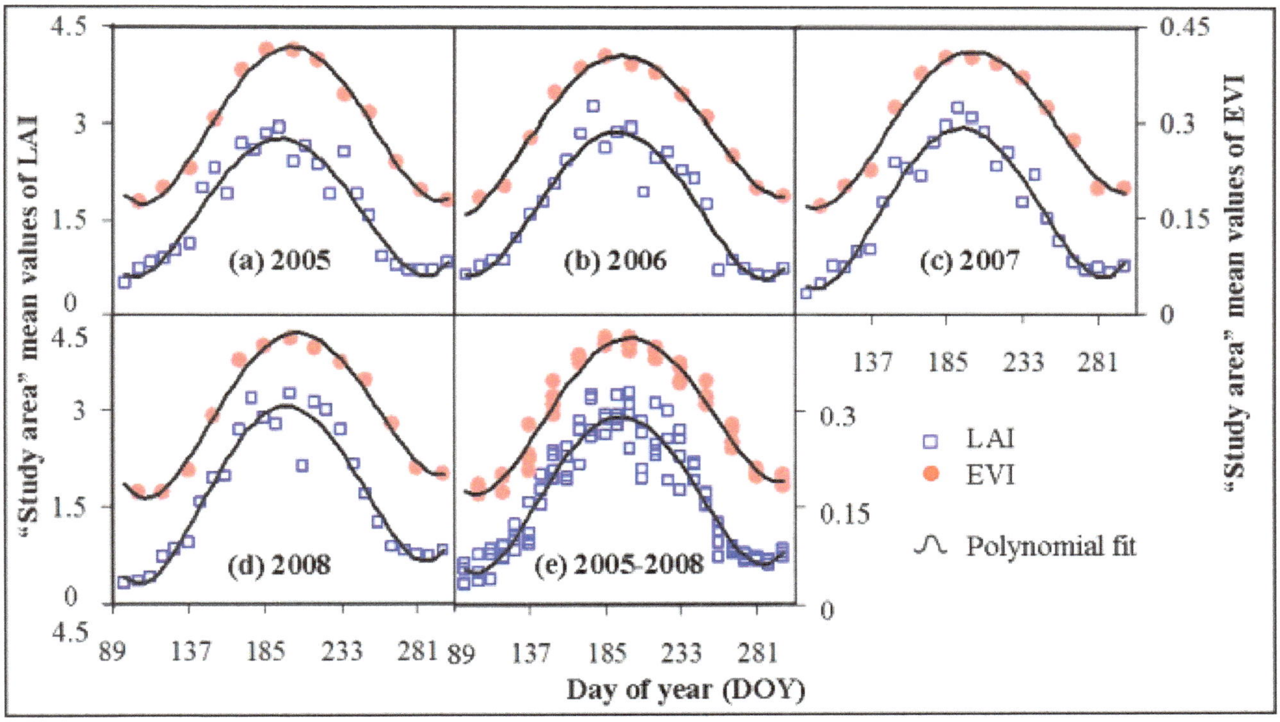

Figure 3: Temporal behaviour of LAI and EVI for the April–October period (DOY 97–297) of (a) 2005, (b) 2006, (c) 2007 and (d) 2008, and for (e) all years (2005–2008).

Table 1: Polynomial fit (i.e., $y = a_0x^4 - a_1x^3 + a_2x^2 - a_3x + a_4$) and associated statistics; x and y denote DOY and either LAI or EVI. In all cases considered, p-values were <0.0001

Variable	Year	a_0	a_1	a_2	a_3	a_4	r^2	F-stat
LAI	2005	3e-08	2e-05	0.0065	0.70	26.82	0.93	70.03
EVI		3e-09	3e-06	0.0007	0.08	3.40	0.99	189.24
LAI	2006	3e-08	2e-05	0.0063	0.67	25.50	0. 92	59.93
EVI		2e-09	2e-06	0.0004	0.04	1.53	0. 98	149.15
LAI	2007	4e-08	3e-05	0.0075	0.81	31.00	0. 95	100.82
EVI		3e-09	2e-06	0.0006	0.06	2.59	0. 98	122.07
LAI	2008	4e-08	3e-05	0.0084	0.94	36.71	0. 94	76.98
EVI		3e-09	3e-06	0.0007	0.08	3.60	0. 98	110.84
LAI	All years	3e-08	3e-05	0.0072	0.78	30.00	0. 92	279.71
EVI		3e-09	2e-06	0.0006	0.07	2.78	0. 97	351.15

Figure 4 shows the relation between the "study area" (Figure 1) mean values of LAI and EVI. Strong linear relationships are observed, producing r^2 -values ranging from 0.85–0.95 and p-values <0.0001. Results for northern Alberta confirm Huete et al.'s [4] observation of linearity in other biomes. However, more investigation is needed for widespread application across a greater number of biomes globally.

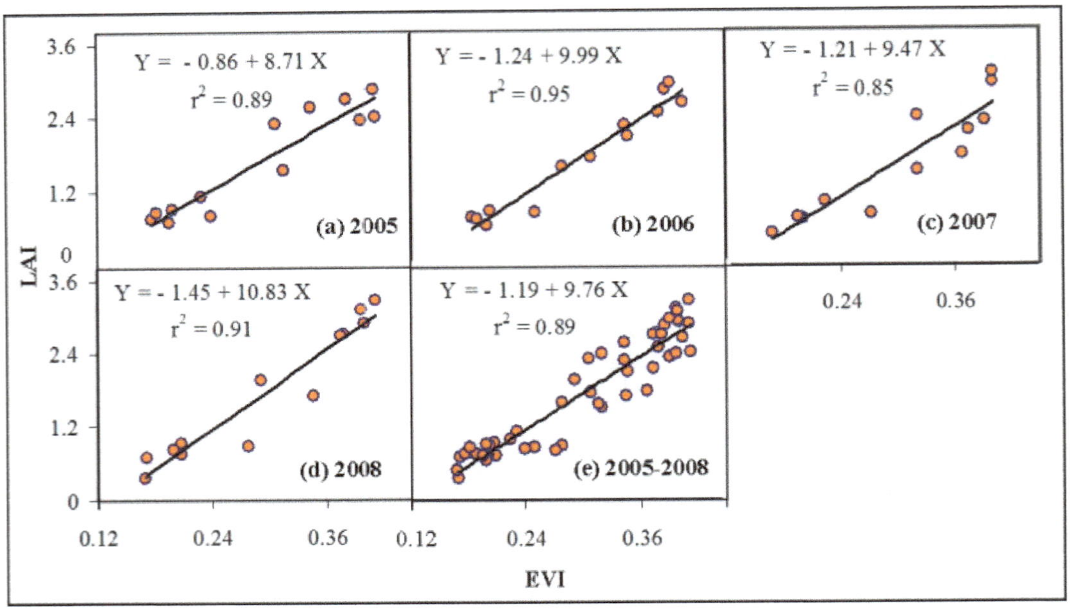

Figure 4: Data comparison of LAI and EVI for the April–October period (DOY 97–297) of (a) 2005 (F-statistics = 87.69), (b) 2006 (F-statistics = 225.08), (c) 2007 (F-statistics = 63.68) and (d) 2008 (F-statistics = 119.65), and for (e) all years (2005–2008; F-statistics = 416.98). For all cases considered, the p-values were<0.0001.

In this study, we assume that the 1-km resolution images of LAI have been previously validated by NASA [3,14,15]. As a result, examination of the level of agreement between mean values of LAI at 250-m resolution and the original values at 1-km resolution serves as an appropriate means of validation (Figure 5). For each of the landcover types identified in Figure 1, Figure 5 provides some sample calculations for six arbitrary 1-km pixels. It revealed that the mean LAI at 250-m resolution compared to the original LAI at 1-km resolution were nearly similar for two landcover categories, i.e., (i) 3.6 vs. 3.7 for needleleaf forests [Figure 5a], and (ii) 4.6 vs. 4.7 for grasses/cereal crops [Figure 5b], and exactly the same for the remaining four categories, i.e., (i) 1.7 for shrubs [Figure 5c]; (ii) 3.0 for broadleaf crops [Figure 5d]; (iii) 2.7 for savannah [Figure 5e]; and (iv) 5.4 for broadleaf forests [Figure 5f]. A study-area-wide comparison of the original 1-km and enhanced 250-m image of LAI for an 8-day period in June 2009 (DOY 169–176), further illustrates the extent the two products match at 1-km resolution ($r2 = 0.87$, with a slope = 1.02; Figure 6). Eighty percent of the data points are distributed along the 1:1 line (blue symbols in Figure 6). Mean absolute difference between LAI at 1-km and 250-m resolution (Figure 6) is roughly 0.59.

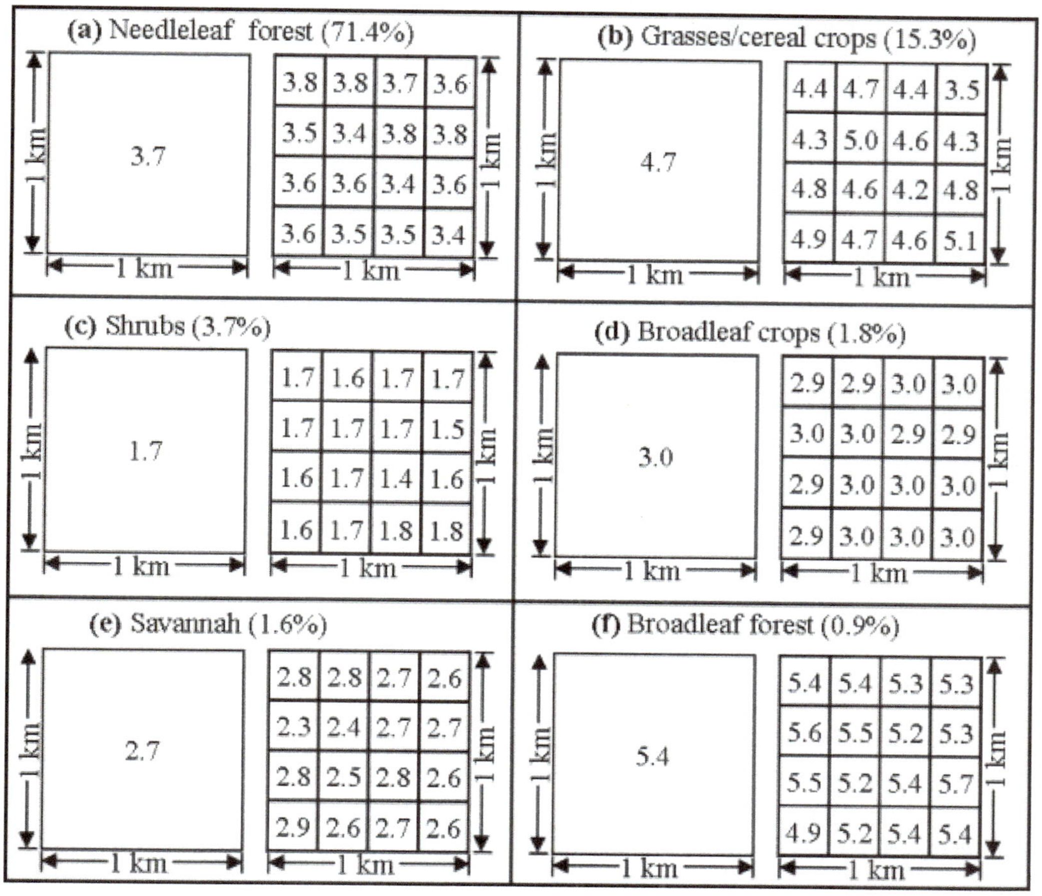

Figure 5: Comparison of LAI values at 1-km resolution without data fusion (left panel) and at 250-m resolution with data fusion (right panel) for various landcover types identified in Figure 1, including (a) needleleaf forests (comprising 71.4% of the study area), (b) grasses/cereal crops (15.3%), (c) shrubs (3.7%), (d) broadleaf crops (1.8%), (e) savannah (1.6%), and (f) broadleaf forests (0.9%).

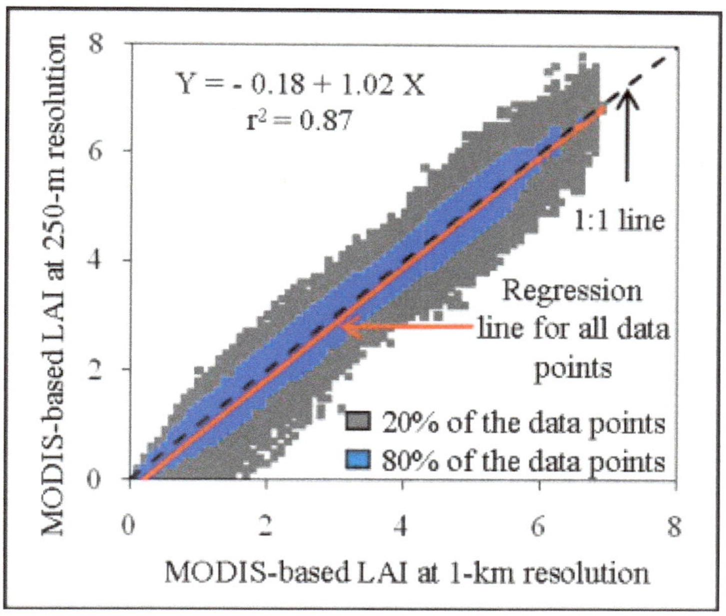

Figure 6: Study-area-wide comparison of LAI from the original 1-km and enhanced 250-m image for an 8-day period in June 2009 (DOY 169–176); F-statistic=10164 and $p < 0.0001$.

Figure 7 provides an example display of the spatial distribution of LAI at 250-m resolution for an 8-day period in June 2009 (DOY 169–176) based on the application of data fusion to the Athabasca Oil Sands Region (Figure 1). Greatest LAI (5–8) is observed to occur in the river valley, where an abundance of hardwood trees are found. Lowest LAI (0–1) occurs on a major plain formation (east of the road network; Figure 7), characterised by coarse sandy soils and a predominance of low-productivity jack pine stands in the dryer areas and an abundance of wetlands (fens and bogs) in local depressions [22].

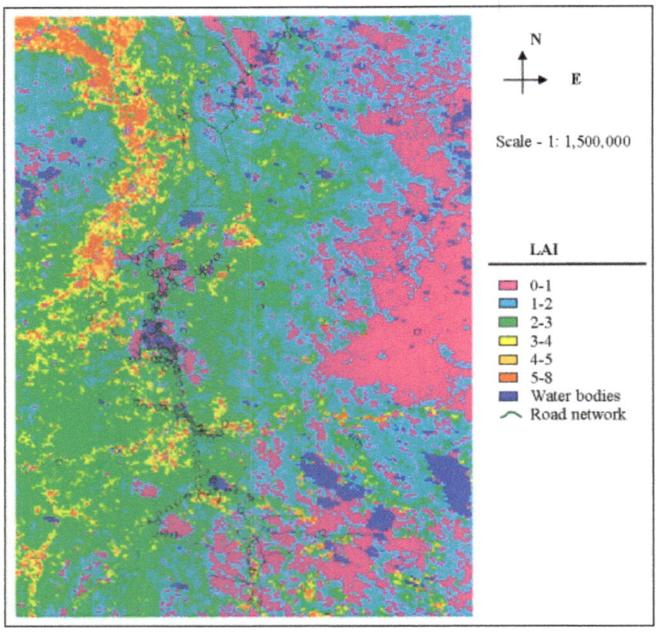

Figure 7: Spatial distribution of LAI for an 8-day period in June 2009 (DOY 169–176) at 250-m resolution for the Athabasca Oil Sands Region (red box in Figure 1). Open circles in the centre, along the road network (green lines) and elsewhere in the image, denote the sites where field-based measurements of LAI were acquired.

To provide further validation of Hassan et al.'s method [17], we have also compared the LAI-values acquired from the field with those developed by data fusion (Figure 8). It revealed the presence of a strong linear relationship between estimates for 72 of the field sites (i.e., $r^2 = 0.71$ and slope = 0.70; Figure 8). LAI for three sites near open water were removed from the analysis, because at 1-km resolution the MODIS sensor was unable to discriminate the LAI-site characteristics from the adjacent water body—a feature that could not be corrected with data fusion. Of the 72 MODIS-based LAI used in the analysis 71% of the values fell within ±20% of the field-based values. Twenty percent deviation between satellite-based and field-based estimates is common when ecological systems are concerned [23,24]. This discrepancy is most likely a characteristic of the fact:

1. Field-based measurements are collected at a particular time of a day, whereas MODIS-based estimates are generated from 8- and 16-day composites, and
2. Field-based measurements are collected by sensing the canopy from below, and as a result the measure of LAI cannot account for the understorey component. In contrast, RS-based methods measure LAI from sensing the top of the canopy. Under sub-optimal overstorey densities, understorey canopy effects are most likely to influence sensor readings and subsequent calculation of LAI.

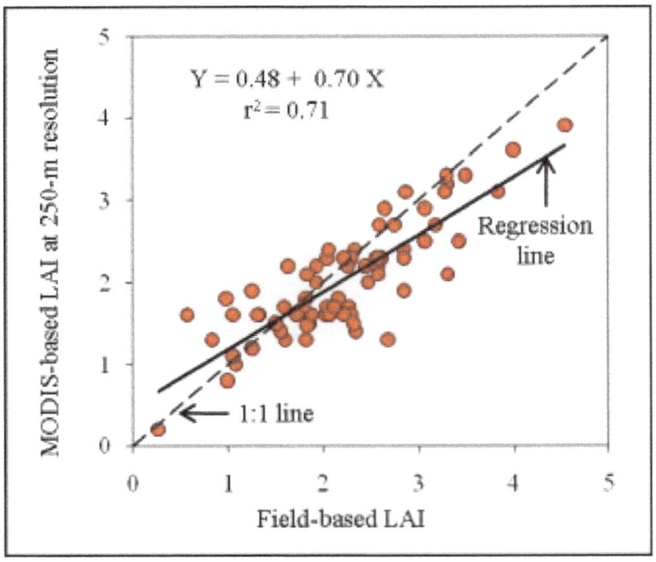

Figure 8: Comparison of 72 field-based measurements with their corresponding MODIS-based LAI for June 2009 at 250-m resolution; F-statistic = 171.23 and p-value < 0.0001.

CONCLUDING REMARKS

In this paper we demonstrate a relatively strong linear correlation between LAI and EVI. Because of this linear relationship, we can use an existing data fusion technique to enhance existing LAI images (at 1-km resolution) with EVI (at 250-m resolution). Reasonable agreement was obtained when pointestimates from spatially-enhanced LAI images for June 2009 were compared with their field-based equivalents collected at the locations where the field measurements were taken. Having the ability to enhance existing LAI products is critical to our understanding of earth-system processes at moderate spatial resolution over large areas.

ACKNOWLEDGEMENTS

The authors would like to acknowledge the Cumulative Environmental Management Association of Alberta for providing funding support to: Q. Hassan for developing LAI maps for northern Alberta from MODIS data and their subsequent validation, and C. Bourque for the acquisition of field-based measurements used in validation. The authors would also like to acknowledge NASA for providing the MODIS data free of charge, and the four anonymous reviewers for providing input for improving the quality of the manuscript.

REFERENCES AND NOTES

1. Chen, J.M.; Black, T.A. Defining leaf area index for non-flat leaves. Plant Cell Environ. 1992, 15, 421-429.
2. Jonckheere, I.; Fleck, S.; Nackaerts, K.; Muys, B.; Coppin, P.; Weiss, M.; Baret, F. Review of methods for in situ leaf area index determination, Part I. Theories, sensors and hemispherical photography. Agric. For. Meteorol. 2004, 121, 19-35.
3. Myneni, R.B.; Nemani, R.R.; Running, S.W. Estimation of global leaf area index and absorbed PAR using radiative transfer models. IEEE Trans. Geosci. Remote Sens. 1997, 35, 1380-1393.

4. Huete, A.; Didan, K.; Miura, T.; Rodriguez, E.P.; Gao, X.; Ferreira, L.G. Overview of the radiometric and biophysical performance of the MODIS vegetation indices. Remote Sens. Environ. 2002, 83, 195-213.
5. Ahl, D.E.; Gower, S.T.; Burrows, S.N.; Shabanov, N.V.; Myneni, R.B.; Knyazikhin, Y. Monitoring spring canopy phenology of a deciduous broadleaf forest using MODIS. Remote Sens. Environ. 2006, 104, 88-95.
6. Chen, P.-Y.; Fedosejevs, G.; Tiscareno-Lopez, M.; Arnold, J.G. Assessment of MODIS-EVI, MODIS-NDVI and VEGETATION-NDVI composite data using agricultural measurements: An example at corn fields in western Mexico. Environ. Monitor. Assess. 2006, 119, 69-82.
7. Gupta, R.K.; Prasad, T.S.; Vijayan, D. Relationship between LAI and NDVI for IRS LISS and Landsat TM bands. Advan. Space Res. 2000, 26, 1047-1050.
8. Johnson, L.F. Temporal stability of an NDVI-LAI relationship in a Napa Valley vineyard. Aus. J. Grape Wine Res. 2003, 9, 96-101.
9. Wang, Q.; Adiku, S.; Tenhunen, J.; Granier, A. On the relationship of NDVI with leaf area index in a deciduous forest site. Remote Sens. Environ. 2005, 94, 244-255.
10. Davi, H.; Soudani, K.; Deckx, T.; Dufrene, E.; Le Dantec, V.; François, C. Estimation of forest leaf area index from SPOT imagery using NDVI distribution over forest stands. Int. J. Remote Sens. 2006, 27, 885-902.
11. Fan, L.; Gao, Y.; Brück, H.; Bernhofer, C. Investigating the relationship between NDVI and LAI in semi-arid grassland in Inner Mongolia using in-situ measurements. Theo. Appl. Climatology 2009, 95, 151-156.
12. Sellers, P.J. Canopy resistance, photosynthesis and transpiration. Int. J. Remote Sens. 1985, 6, 1335-1372.
13. Hassan, Q.K.; Bourque, C.P.-A.; Meng, F.-R.; Richards, W. Spatial mapping of growing degree days: An application of MODIS-based surface temperatures and enhanced vegetation index. J. Appl. Remote Sens. 2007, 1, 013511:1-013511:12.
14. Knyazikhin, Y.; Martonchik, J.V.; Myneni, R.B.; Diner, D.J.; Running, S.W. Synergistic algorithm for estimating vegetation canopy leaf area index and fraction of absorbed photosynthetically active radiation from MODIS and MISR data. J. Geophys. Res. 1998, 103, 32257-32276.
15. Knyazikhin, Y.; Martonchik, J.V.; Diner, D.J.; Myneni, R.B.; Verstraete, M.; Pinty, B.; Gobron, N. Estimation of vegetation canopy leaf area index and fraction of absorbed photosynthetically active radiation from atmosphere-corrected MISR data. J. Geophys. Res. 1998, 103, 32239-32256.
16. Huete, A.R.; Didan, K.; Shimabukuro, Y.E.; Ratana, P.; Saleska, S.R.; Hutyra, L.R.; Yang, W.; Nemani, R.R.; Myneni, R.B. Amazon rainforests green-up with sunlight in dry season. Geophys. Res. Lett. 2006, 33, L06405:1-L06405:4.
17. Hassan, Q.K.; Bourque, C.P.-A.; Meng, F.-R. Application of Landsat-7 ETM+ and MODIS products in mapping seasonal accumulation of growing degree days at an enhanced resolution. J. Appl. Remote Sens. 2007, 1, 013539:1-013539:10.
18. Gao, F.; Masek, J.; Schwaller, M.; Hall, F. On the blending of the Landsat and MODIS surface reflectance: predicting daily Landsat surface reflectance. IEEE Trans. Geosci. Remote Sens. 2006, 44, 2207-2218.
19. Zhang, Y. Understanding image fusion. Photogramm. Eng. Remote Sens. 2004, 70, 657-661.
20. Hassan, Q.K.; Bourque, C.P.-A. Spatiotemporal dynamics of MODIS-based leaf area index in the boreal forest of northern Alberta. In Proceedings of the Canadian Symposium on Remote Sensing, Lethbridge, Alberta, Canada, June 2009.
21. Environment Canada. Available online: http://climate.weatheroffice.ec.gc.ca (accessed on 30 November 2009).
22. Natural Regions Committee. Natural Regions and Subregions of Alberta; Downing, D.J., Pettapiece, W.W., Eds.; Government of Alberta: Edmonton, Alberta, Canada; Pub. No. T/852, 2006.
23. Botkin, D.B. Forest Dynamics: An Ecological Model; Oxford University Press: Cary, NC, USA, 1993.
24. Botkin, D.B.; Simpson, L.G. The First Statistically Valid Estimate of Biomass for a Large Region. Biogeochemistry 1990, 9, 161-174.

CHAPTER 14

THE ENVIRONMENTAL-DATA AUTOMATED TRACK ANNOTATION (ENV-DATA) SYSTEM: LINKING ANIMAL TRACKS WITH ENVIRONMENTAL DATA

Somayeh Dodge, Gil Bohrer, Rolf Weinzierl, Sarah C Davidson, Roland Kays, David Douglas, Sebastian Cruz, Jiawei Han, David Brandes and Martin Wikelski

Keywords: Animal movement, Migration, Movebank, Movement ecology, Remote sensing, Track annotation, Weather

BACKGROUND

The movement of an organism is influenced not only by its internal state and biological factors driving its movement, but also external factors—the environment and underlying context [1]. Environmental conditions may trigger certain movement patterns or invoke a particular behavioral response, and thus determine local movements or long-distance migrations [2, 3, 4]. Animals can optimize their energy expenditure during movement by selecting for locations and times when the conditions are supportive for movement. For example, raptors in their southward fall migration select a preferential mode of uplift that best fits their flight capacity [5, 6, 7]. Spatial and temporal variability in environmental conditions may affect all types of movement and any scale, from local to global. Some of the most challenging movements to study include large-scale movement, such as migrations and movements that cross broad geographic areas and traverse diverse environments and landscapes. The added challenge in these studies is that environmental conditions cannot be measured locally as part of the study, because they are needed over a very extensive area. Long-distance migrants include some of our most endangered species, and thus it is critical to address questions at the core of movement ecology, such as "when do animals start migrating?", "which strategies should animals adopt while migrating?" and "do movement rules change in a changing environment, and if so, how?" [8].

Today, with the rapid improvement and miniaturization of tracking technology, movement ecology has entered a new data-rich era, with tremendous growth in animal tracking data at previously unseen spatial and temporal resolution. Complementing this are large arrays of online remote sensing datasets describing the earth system and informing models that forecast the future environment. Combining these datasets is an active area of research, addressing a variety of questions to gain a better understanding of the interaction between animal movement and the environment. Manual annotation of animal tracking data (i.e. adding information to locations by an expert) and simulations of the environment along movement tracks have been successfully used to discover meaningful interactions between movement and external variables [2, 5, 7, 9, 10, 11, 12, 13, 14, 15, 16, 17, 18, 19, 20, 21, 22, 23, 24, 25]. Improving access to these environmental data will increase our understanding of their broad effects on our planet, motiving the development of RNCEP, a data organization and visualization package for R for working with data from National Centers for Environmental Prediction (NCEP) / National Center for Atmospheric Research (NCAR) Reanalysis data [26].

Manual annotation is not practical for large global environmental datasets owing to several technical and logistical challenges. The remote sensing datasets needed to study how environmental conditions influence animal movements are provided using complex tiling system in

Citation: Somayeh Dodge, Gil Bohrer, Rolf Weinzierl, Sarah C Davidson, Roland Kays, David Douglas, Sebastian Cruz, Jiawei Han, David Brandes and Martin Wikelski, The environmental-data automated track annotation (*Env-DATA*) system: linking animal tracks with environmental data, DOI: 10.1186/2051-3933-1-3.

space and time that need to be aggregated to cover the entire movement track. Environmental and animal movement data are usually collected in different spatial and temporal scales and it is therefore necessary to choose appropriate scales for the annotation process. Likewise, an appropriate interpolation technique must be applied in order to integrate data at different resolutions. Moreover, environmental data are diverse in source, format, and projection system. It is essential to apply appropriate data transformation techniques in order to integrate such heterogeneous datasets. Accordingly, effective storage, indexing, and retrieval strategies must be applied to handle large volume of environmental datasets. These challenges limit many potential non-technical users from accessing these data and applying annotation in a manual mode. Nonetheless, compiling combined movement-environment datasets would be highly beneficial for movement studies. Such vast datasets are well suited for sophisticated, context-aware data mining and pattern recognition techniques that allow researchers to discover patterns of movement in response to changes in the environment [27, 28]. Hence, an integrated system capable of managing and analyzing movement tracks of animals linked to large remote sensing, climatic, and land use datasets will greatly facilitate the next generation of research into movement ecology.

This paper contributes to movement ecology research by describing a new open system, Environmental-Data Automated Track Annotation (*Env-DATA*), that automates annotation of movement trajectories with ambient atmospheric observations and underlying landscape information. The aim of the system is to provide efficient movement track annotation and knowledge discovery methods to allow scientists to examine relationships between observed animal movements and a breadth of information about environmental conditions. The *Env-DATA* system utilizes large computational servers to co-register the animal tracks with environmental data without requiring the user be an expert in the processing of such data. The system facilitates the investigation of biological research questions about movement behavior of animals, including threatened and endangered species that are of concern due to the impact of climate and environmental changes. *Env-DATA* will facilitate discovery of unique information about niche selection and habitat, movement patterns and landscape connectivity of moving animals, and how these may be affected by variability and long-term changes in climate and landscape. Such knowledge is crucial for planning and management of protected areas and for forecasting population status and habitat needs in future conditions of climate and land use changes.

METHODS

The movebank *Env-DATA* system

This paper primarily focuses on the architecture and technical characteristics of the *Env-DATA* Track Annotation service. The *Env-DATA* system expands the capabilities of Movebank, an animal movement data portal (http://www.movebank.org). Movebank is a free, online database of animal tracking data, which provides biologists and animal movement researchers with a secure online archive to store, manage, process, and share animal movement data [29, 30]. Figure 1 illustrates the main components of the Movebank *Env-DATA* system. The *Env-DATA* system extensions within Movebank include three main services: (1) the Track Annotation Service, (2) the Track Simulation Service, and (3) the Knowledge Discovery and Visualization Service.

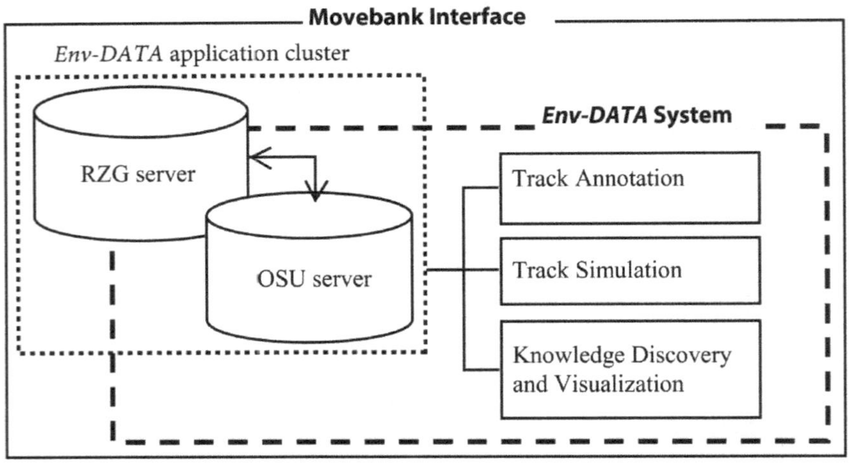

Figure 1: Movebank *Env-DATA* System. RZG: Computing Center Garching, Germany; OSU: The Ohio State University Supercomputer Center. The gray box highlights the *Env-DATA* system components within Movebank.

The Track Simulation and the Knowledge Discovery and Visualization services are developed as external libraries using the Java(TM) Platform and the R programming language. These libraries can read tracking data directly from Movebank (provided the authenticated user has the necessary access rights), and read-write annotated data that were prepared by the Track Annotation Service. They were developed to serve as a convenient access layer to *Env-DATA* and can be used to transfer data to any other programming or data analysis environment. They can, for example, directly communicate and exchange data with common analysis, visualization and simulation tools like R, MATLAB and Google Earth, and niche analysis software such as Maxent [31]. Examples for such applications using R and Google Earth are described in the case study section below. In addition, several open movement analysis packages providing tools for knowledge discovery, data mining, modeling, and visualization have been developed by others, including the Move R-package [32]—a movement track analysis and visualization package— and MoveMine [33]—a track segmentation and classification package.

To ensure its relevance and effectiveness, Movebank services and the *Env-DATA* system were designed and tested in collaboration with several wildlife research partners from the U.S. Fish and Wildlife Service (FWS), the U.S. National Park Service (NPS), and the U.S. Geological Survey (USGS), who contributed to the design of *Env-DATA* to ensure its applicability and relevance to contemporary conservation and wildlife management [34, 35, 36].

System infrastructure

The system runs on a Linux application cluster and manages data flow using a MySQL database. Two technically identical installations of the *Env-DATA* application cluster are located at the Computing Center Garching (RZG) of the Max Planck Society in Garching, Germany, and at the Ohio Supercomputer Center (OSC) of The Ohio State University in Columbus (OH, USA). They serve as storage and processing systems for the environmental data cached directly from their original data sources. Each cluster (i.e. at RZG and OSC) serves different sets of environmental variables depending on their proximity to the original data source. The Movebank application server, which serves the animal tracking data, is also located at the RZG (Figures 1 and 2).

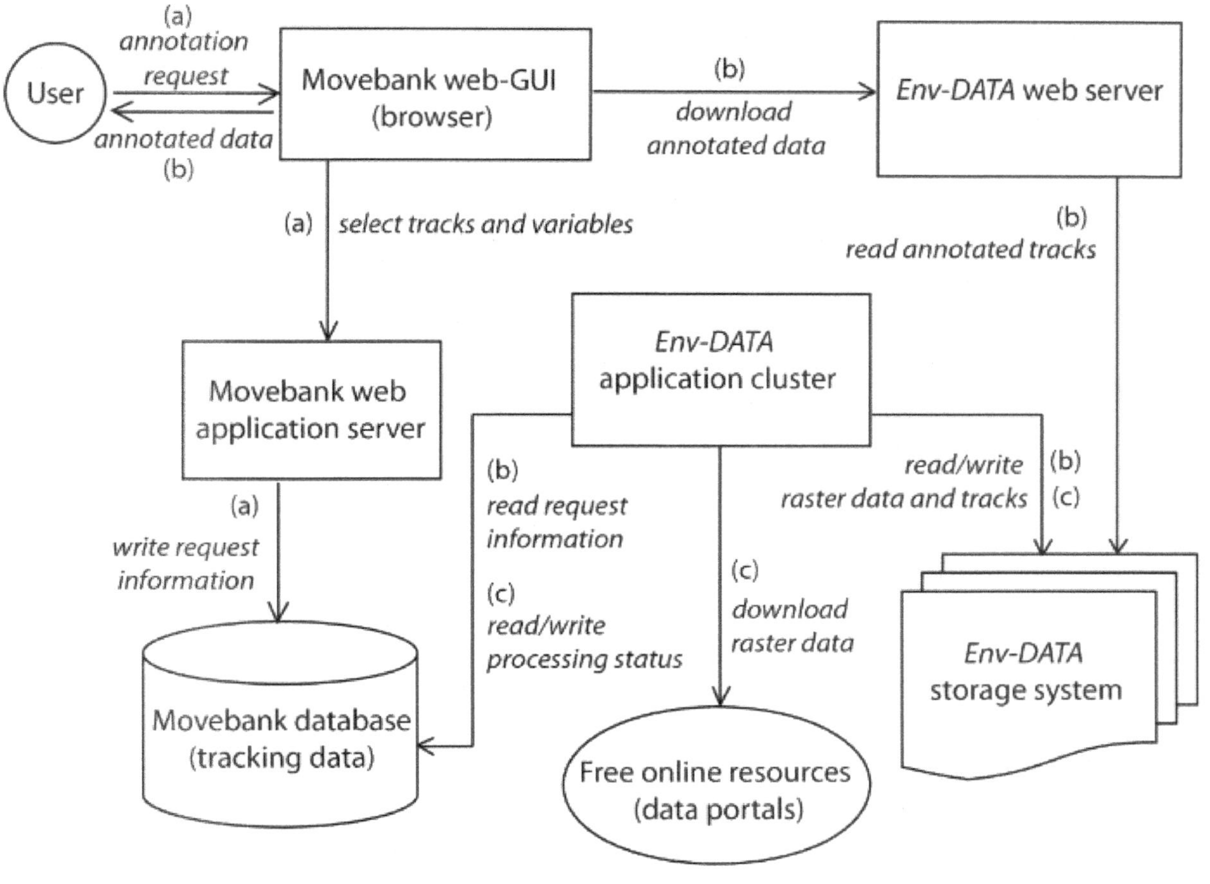

Figure 2: *Env-DATA* Track Annotation Service Components. The figure illustrates the workflow of an annotation request through the different servers and components of the system. Steps shown indicate the following: (a) selection and submission of a data annotation request by the User, (b) read annotation request information, process Track Annotation in the *Env-DATA* application cluster, storage of annotation results in the *Env-DATA* storage system, and delivery of results to User through the *Env-DATA* web server, and (c) environmental data acquisition and storage in the *Env-DATA* storage system through the *Env-DATA* application cluster. RZG: Computing Center Garching, Germany; OSU: The Ohio State University Supercomputer Center.

Environmental data

Our goal is to provide access to the most relevant global gridded datasets of environmental data. Datasets that are available in this first release of *Env-DATA* include NASA's MODIS vegetation, ocean, ice and fire data products; NCEP Global and North American Regional Reanalyses (NARR); the European Centre for Medium-Range Weather Forecasts (ECMWF) moderate resolution reanalysis; Oregon State University Ocean Net Primary Productivity (NPP); NOAA's Ocean Surface Current Analyses (OSCAR) ocean currents and sea surface temperatures; NASA's Tropical Rainfall Measuring Mission (TRMM) precipitation; normalized difference vegetation index from the AVHRR sensor; ESA's GlobeCover land use and land cover; NASA's ASTERGDEM 30 m topography; and the Columbia University Human Geography dataset. The raw data are obtained in different formats, such as NetCDF, GRIB, HDF, GeoTIFF, and ASCII. See Table 1 for more details about the datasets. *Env-DATA* also offers derived variables that are particularly suited to aerial movement studies, such as tail-wind support and uplift availability (Table 1).

Table 1: Available environmental datasets for the trajectory annotation service

Datasets	Data Description	Data Source	Projection system/ Grid	Temporal Coverage	Geographic coverage (Latitude/ Longitude)	Temporal resolution	Spatial resolution	Data Format
Tropical Rainfall Measuring Mission (TRMM) [37]	Tropical precipitation	NASA http://trmm.gsfc.nasa.gov/	Regular lat/lon grid	1998- present	50°N–50°S 180°E–180°W	3-hour	0.25°	Unformatted binary
AVHRR land NDVI [38]	Normalized difference vegetation index from the AVHRR (low resolution) sensor	USGS (USA only) http://phenology.cr.usgs.gov/get_data_1km.php NASA (global) http://glcf.umiacs.umd.edu/data/gimms/	Albert's Equal Area grids	1989–present, 1982–present	CONUS, 90° N–90°S 180° E–180°W	1-week, 2-week	1 km (USA), 8 km (global)	Unformatted binary
NCEP Global Reanalysis 2 [39]	Global weather reanalysis	NOAA http://www.esrl.noaa.gov/psd/data/gridded/data.ncep.reanalysis2.html	Regular (non-Gaussian) grid	1948–present	90°N–90°S 180°E–180°W	6-hour	2.5° (208 km)	NetCDF
North American Regional Reanalysis (NARR) [40]	Regional (North America only) weather reanalysis	NOAA http://www.emc.ncep.noaa.gov/mmb/rreanl/	Lambert Conformal, Conic Grids	1979–present	90°N–1°N 0°–170W°	3-hour	32 km (at 40°N)	GRIB
ECMWF Reanalysis [41]	Global weather reanalysis	ECMWF http://www.ecmwf.int/	Regular grid	1979–present	89.463°N–89.463°S 180°E–180°W	6-hour	0.7°	GRIB
MODIS Land	Earth-surface, reflectivity and vegetation variables	NASA https://lpdaac.usgs.gov/	Geographic/ Sinusoidal grid	2002–2012	90°N–90°S 180°E–180°W	Daily, 8-day, 16-day, monthly	5.6 km (0.05°)	HDF- EOS
MODIS Ocean	Ocean surface, color, and productivity variables	NASA http://oceancolor.gsfc.nasa.gov/	Cylindrical Equidistant				4 km, 9 km	HDF- EOS
MODIS Snow	Snow and ice variables	NASA http://modis-snow-ice.gsfc.nasa.gov/	Cylindrical Equidistant				1 Km, 4 Km	HDF- EOS
Ocean productivity [42]	Ocean net primary productivity (NPP) reanalysis	http://www.science.oregonstate.edu/ocean.productivity/	Equidistant Cylindrical projection, lat/lon grid	1997–2009	90°N–90°S 180°E–180°W	8-day, monthly	Grid sizes 1080x2160 (1/6 degree) 2160x4320 (1/12 degree)	HDF
ASTER GDEM	Very high-resolution topography	USGS http://asterweb.jpl.nasa.gov/gdem.asp	Regular grid, (WGS84 ellipsoid)		83°N–83°S 180°E–180°W		1 arc-second	GeoTIFF
SRTM [43]	High resolution topography	NASA http://www.cgiar-csi.org/data/srtm-90m-digital-elevation-database-v4-1	Regular grid, (WGS84 ellipsoid)		60°N–60°S 180°E–180°W		3 arc-second	HGT
GlobCover	Land cover and land-use type	ESA http://dup.esrin.esa.it/prjs/prjs68.php	Plate-Carrée projection (WGS84 ellipsoid)	2009	90°N–65°S 180°E–180°W		20 arc-seconds	HDF
Socioeconomic data (Population Density Grid)	Human geography	http://sedac.ciesin.columbia.edu/gpw/global.jsp	Regular grid (WGS84 ellipsoid)	1990–2010	85°N–58°S 180°E–180°W	5 years	30 arc-second (1km)	ASCII
Ocean Surface Current Reanalysis (OSCAR)	Ocean surface currents	NASA http://www.oscar.noaa.gov/	Regular grid	1993–present	60°N–60°S 180°E–180°W	5-day, monthly	1 degree, 1/3 degree	NetCDF
ETOPO1	Ice surface and bedrock	NASA http://www.ngdc.noaa.gov/mgg/global/global.html	Regular grid (WGS84 ellipsoid)	1940–2008	90°N–90°S 180°E–180°W		1 arc-minute	NetCDF
Distance to the Nearest Coast	Distance to the nearest coast	NASA http://oceancolor.gsfc.nasa.gov/DOCS/DistFromCoast/	Regular grid		90°N–90°S 180°E–180°W		0.04° 0.01°	Text file, GeoTiff
Derived wind variables for flight	Tail-wind support and cross wind [36]; Thermal and orographic uplift [5]	Calculated derived variables, based on ECMWF or NCEP data	Regular grid	1979–present	89.463°N–89.463°S 180°E–180°W	6-hour	0.7°	ASCII
Derived topographic variables	Slope and aspect [25]; Rugosity [44]	Calculated derived variables, based on ASTERGDEM	Regular grid		83°N–83°S 180°E–180°W		1 arc-second	ASCII

Track annotation service

The term "path annotation", borrowed from computer science, is used when additional data about important variables encountered through a particular path are added to the dataset describing an object's trajectory. In the context of animal movement, path (track) annotation includes environmental variables co-located in time and space with the moving organism's coordinates [7]. The *Env-DATA* Track Annotation Service is the fundamental extension of the Movebank portal that attributes environmental data to each tracking location (in space and time) along a movement path. The service consists of several components as illustrated in Figure 2 and described below.

The *Env-DATA* application cluster resides on the RZG and OSC servers and is the main core of the system (Figures 1 and 2). Data flow required for the annotation service is handled through the Movebank web application server and *Env-DATA* web server using MySQL. The annotation service is triggered by a request from the User using the Movebank web-GUI (arrows (a) in Figure 2) and is processed at the *Env-DATA*application cluster. The annotation results are then stored in the *Env-DATA* storage system and an email notification, including an http download link, is sent to the user through the *Env-DATA* web server when data are available for download (arrows (b) in Figure 2). The *Env-DATA* web server is a dedicated machine that runs a Tomcat web application server and provides an http interface to the service running on the *Env-DATA*application cluster. The *Env-DATA* web server functionality is limited to accepting annotation requests, storing them in the database, and delivering the results. All data processing is performed on the *Env-DATA* application cluster, which is a family of Linux compute nodes.

The annotation service is conducted offline because of the large volumes of data involved. Latency of the service depends on the time required to download the necessary environmental data. The annotation workflow involves several steps, described below.

Annotation request

Users may request two types of annotation: (1) annotation of a gridded geographic area or (2) annotation of a set of trajectories. For trajectory annotations, the *output* spatial and temporal resolutions and extent are determined by the system according to the spatial and temporal constraints of the *input* trajectories.

A trajectory annotation request starts with the selection of one or more animal tracks from an existing user-created study in Movebank. After selecting the trajectories the user is asked to choose a subset of environmental variables (for example, window (1) in Figure 3). There are two methods (represented by separate tabs) for browsing and selecting variables (window (1) in Figure 3): (1) select variables by source, with environmental variables organized according to their original satellite missions or dataset portal (e.g. MODIS or ECMWF, see Table 1 for complete list); or (2) select variables by type, where the variables are organized according to their geophysical composition (e.g. weather, topography, earth surface and vegetation, ocean). Both methods lead the user through a hierarchical classification (e.g. weather → temperature → surface temperature → NCEP air temp at 10 m) to the point where specific variables can be selected and added to the annotation request. Summary information about each variable and links to source documentation are available from the interface. Next, the user chooses an interpolation method for each environmental variable (window (2) in Figure 3). The annotation service provides three interpolation methods: nearest neighbor, bilinear, and inverse weighted distance, as explained in "Data Interpolation" below.

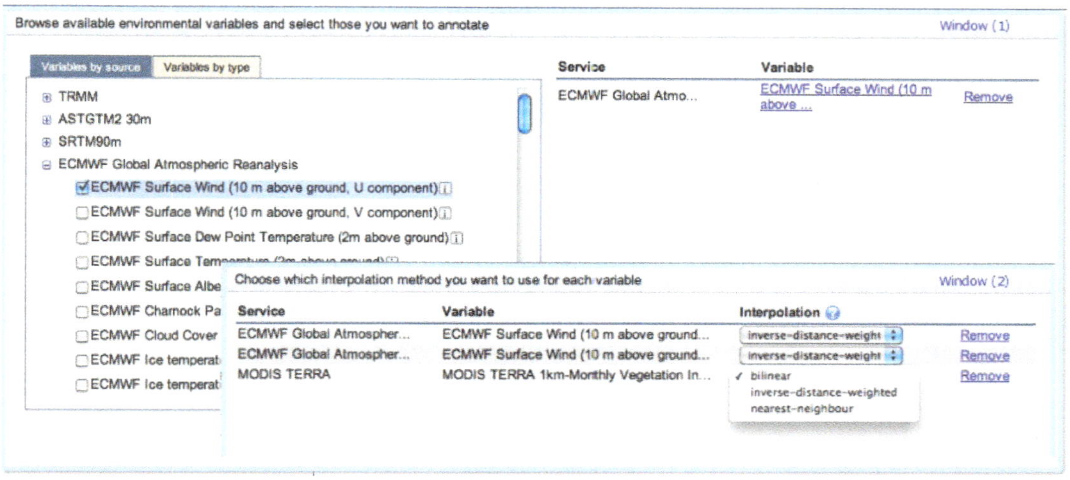

Figure 3: An example for the graphical user interface (GUI) that serves the annotation system users. The figure illustrates an annotation request for the data in the variable "surface wind (10m above ground, U component)" from the global weather reanalysis dataset ECMWF (see Table 1 for more details), and selection of interpolation methods for each requested variable.

For annotations of geographic areas, the user specifies the coordinates of four corners of a region of interest, a coordinate reference system (e.g. WGS84 lon/lat), required timestamps (e.g. as "yyyy-MM-dd HH:mm:ss", comma separated), and a spatial resolution for the target grid (in number of pixels). Data will be interpolated if the requested resolution is higher than the native resolution of the dataset. The result is a bundle of mapped data that can be read for numerical analysis (for example by a niche model, such as Maxent), or used as GeoTiff images or as KML image overlays in Google Earth.

After selecting the desired information for either annotation method, the user provides an e-mail address to which a link to the completed data file will be mailed and submits the request.

Data acquisition

As the combined volume of globally available environmental data is on the order of petabytes, it is not feasible to locally mirror all of the source environmental data. Instead, the *Env-DATA* application cluster applies a caching strategy to retain the most frequently accessed data and download any other data upon request if it is not already locally stored at the *Env-DATA* storage system (arrows (c) in Figure 2). New data requests (provided as a list of locations, times and variables) are translated to lists of needed data sources, sorted according to data service, variable, timestamp and scene (i.e. a raster tile). Multiple data sources are listed when the location in the movement path requires interpolation between scenes and/or in time, or when derived variables (such as thermal uplift) require a combination of several input environmental variables. The data-sources list is compared with the stored metadata table and data that are not stored locally are requested from their provider using an ftp/http, or OPeNDAP interface. The system ranks each scene according to the frequency at which it has been accessed since download. The least accessed scenes are deleted when space is needed for new data.

Data retrieval, indexing, and transformation strategies

The environmental data are acquired in a variety of data formats (e.g. GeoTIFF, NetCDF, HDF, GRIB). Prior to the annotation, *Env-DATA* extracts the required variables from the original data structure using available Java libraries associated with the data formats. For each variable an n-dimensional grid is defined ($1 \leq n \leq 4$, possible dimensions are x, y, z, t) which spans the complete domain of definition of the variable. Each point in the grid is assigned a tile index and tile indices are mapped to the names of files that contain the data for a tile. When a set of points is submitted for annotation by the *Env-DATA* system, each point is transformed to the coordinate reference system of the annotation variable using proper projection techniques. A set of neighboring grid points is then determined and the names of the files containing data for these grid points are identified. For those files not already stored in the *Env-DATA* storage system, an asynchronous job is started to download them from the original source. When the system runs out of disc space for storing new files, a garbage collection job deletes files following a least recently used (LRU)-based algorithm. Track annotation and interpolation starts when all required files have been downloaded. Data files are read in blocks and blocks are cached in an in-memory LRU cache.

The block structure is chosen to match the physical structure of the underlying file in order to optimize read performance. In order to optimize cache use, data points are sorted according to the files they require for annotation and the within-file block structure.

Separate processing steps (e.g. pre-processing, data download, garbage collection and track annotation) are performed in parallel on a pool of compute nodes at the *Env-DATA* application cluster, while resource access is coordinated by a bespoke locking system implemented on MySQL.

Data interpolation

Once all necessary data sources are locally available, the environmental data are interpolated along all trajectory points. Prior to the interpolation, the trajectories are first transformed to the native grid of the environmental variables (e.g. the Sinusoidal grid for MODIS or Lambert Conformal for NARR), if required. After the transformation, for each point along a trajectory, the interpolation first is applied in space, then in time. For each trajectory point $p_i(x_i, y_i, t_i)$, four adjacent pixels (or more if necessary) are located in the global grid of environmental data in space (Figure 4a) at two temporal timestamps before and after t_i (Figure 4b). The values of the four neighbor pixels (i.e. $v_1 \ldots v_4$ and $v_1' \ldots v_4'$) at two timestamps t and t' are then extracted and used to compute an interpolated value for the trajectory point p_i.

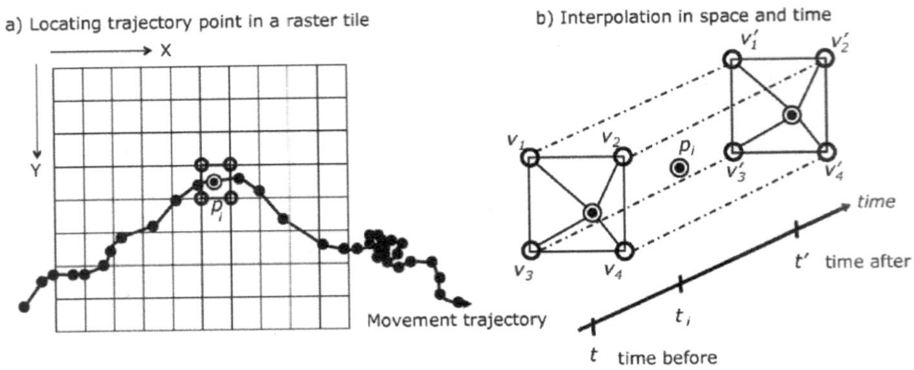

Figure 4: Interpolation in space and time. (a) The variable data for track-point P_i is first interpolated in space (using one of several interpolation methods) based on the data from the available points in the environmental dataset native grid around P_i. (**b**) Similar spatial interpolations are conducted at the two nearest available points in time, the nearest before and nearest after the timestamp of the track-point P_i. Then, the two interpolated spatial values are interpolated in time to the timestamp of P_i.

The system allows three types of interpolation in space: *nearest neighbor*, *bilinear*, and *inverse weighted distance*[45, 46], and two types in time: *nearest neighbor* and *inverse weighted distance*. The user can apply different methods of interpolation to space and to time. However, the available interpolation strategy differs according to the type of data. For instance, for categorical data, such as land-use type, only the nearest neighbor interpolation can be applied, whereas for continuous-numeric data types either a nearest neighbor, bilinear (in regular grids), or inverse weighted distance interpolation can be selected based on the resolution of data [45, 46]. Spatial distances used by the nearest neighbor and inverse weighted distance interpolations are calculated as great circle distances on the WGS84 ellipsoid.

Results delivery

The annotated trajectories are delivered to the user via http download in comma-separated values (csv) format, and gridded geographic areas as GeoTiff or KML images. The user receives an e-mail with a link when the download is ready. The annotated data are stored in the *Env-DATA* storage system at OSU and delivered to the user through the *Env-DATA* web server (cf. arrow (b) in Figure 2). Metadata for the output is uniform across different annotation requests and movement data types, which simplifies linking the output with further analysis tools. A user-group community in Movebank provides a place to exchange user-developed codes and statistical and analytical methods, suggest knowledge discovery and data mining methods, or comment on existing tools and research challenges.

CASE STUDY: GALAPAGOS ALBATROSS (*PHOEBASTRIA IRRORATA*)

This case study illustrates an application of the *Env-DATA* System to examine environmental factors associated with the movements of nine Galapagos Albatrosses (*Phoebastria irrorata*), tracked from June to September 2008.

Methods – movement data collection and annotation

The original tracking data were collected from birds breeding at two sites on Isla Española, Punta Cevallos (1.39° S, 89.62° W) and Punta Suarez (1.38° S, 89.75° W), as well as a small island close to the Ecuadorian mainland, Isla de la Plata (1.58° S, 81.15° W). The birds were tracked throughout the entire breeding season. GPS loggers were deployed on 28 adult albatrosses at the beginning of incubation period. The loggers weighed 22 g with a 9-month battery life and were designed and produced by e-obs GmbH (Munich, Germany). Units were programmed to record GPS locations every 90 minutes. Units were attached to taped bundles of dorsal feathers and secured with cable ties. Data from GPS units were downloaded remotely via an ultra-high frequency (UHF) radio link to a stationary base station that was installed in the vicinity of equipped bird nests at each separate colony. A base station consists of a UHF antenna, a receiver with flash memory and two 6 volt 12 amp-hour batteries. The base station begins to upload data when any unit (or equipped bird) comes within 1000 m of the station. This system has the advantage of allowing data to be retrieved automatically without physically recovering the tag, reducing handling-induced stress to the birds and labor required to collect data in the field. Every four weeks a researcher would go to sites to retrieve data from base stations, change batteries and perform general maintenance duties. On Isla Española, tags were deployed on 23 June 2008 at Punta Cevallos and 31 May 2008 at Punta Suarez, and data were retrieved from base stations on 23 June, 21 July, 12 August and 18 September 2008. On Isla de la Plata, tags were deployed on 7–24 June 2008, data were downloaded on 7 July, 4 August, 1 September, and 6 October of 2008, and loggers were recovered on 7 October.

For the purpose of this case study, we used tracking data from nine albatrosses that made extensive movements for almost the entire period from June to September 2008. The tracks were segmented to transit flights from/to the Galapagos Islands and the Peruvian coastal foraging segments. The transit flights are segmented according the flight speed obtained from GPS points (speed >5 m/s) between longitudes 90°W and 82.5°W. The Peruvian coastal foraging area is defined as areas between longitudes 82.5°W and 75°W. Using *Env-DATA*, the nine albatross tracks were annotated with wind speed (m/s) and wind direction (degrees from North) computed from u- and v-wind components obtained from the NCEP Reanalysis 2 dataset, and Ocean Net Primary Production (NPP) data from Oregon State University (Table 1). For the annotation 6-hour, 2.5° NCEP Reanalysis 2, and 8-day, 2160x4320 ocean NPP datasets are used (cf. Table 1).

RESULTS AND DISCUSSION

Figures 5, 6, 7, and 8 illustrate the visual exploration tools from the Knowledge Discovery and Visualization service package of *Env-DATA*. R code for the generation of these plots is provided in the supporting material (Additional File 1). Figure 5a shows the nine Galapagos Albatross trajectories annotated with 8-day ocean NPP (cf. Function 1, Additional File 1) and Figure 5b shows a gridded geographical area of MODIS chlorophyll-a concentration during one month in the nesting season in 2008, provided as a KML file and plotted using Google Earth. The albatrosses flew to the Peruvian coast to forage where ocean productivity was high. Figure 6 facilitates the comparison of available NPP versus NPP use along flight tracks of the nine albatrosses (the 3D plot is generated using Functions 3, Additional File 1). Figure 6a–b shows the distribution of ocean NPP available at the coastal area of Peru—between 82.5°W, 12°S and 75°W, 4°S—and around the Galapagos Islands—between 95°W, 2.5°S and 90°W, 2.5°N. To illustrate the availability of ocean NPP (Figure 6a–b, Figure 6e), the 8-day NPP datasets are averaged over the period of June–September 2008 in the native grid. Figure 6c–d shows probability density histograms of annotated ocean NPP values along track segments during foraging flight along the Peruvian coast and along flight tracks around the Galapagos, respectively. In addition, Figure 6e provides a 3D visualization of the available NPP versus NPP use along the actual flight tracks (plotted using Function 3, Additional File 1). The NPP histogram constructed from locations along the foraging flight (what was used) illustrates a bias toward high NPP values (Figure 6c), even when compared to the availability near the coast (Figure 6a), which suggests that albatrosses might carefully choose times and locations to forage where NPP is high. However, the similarity between the NPP use and availability histograms around the islands suggests that they do not show preferences for high NPPs along transit flights between the foraging areas and the nesting grounds.

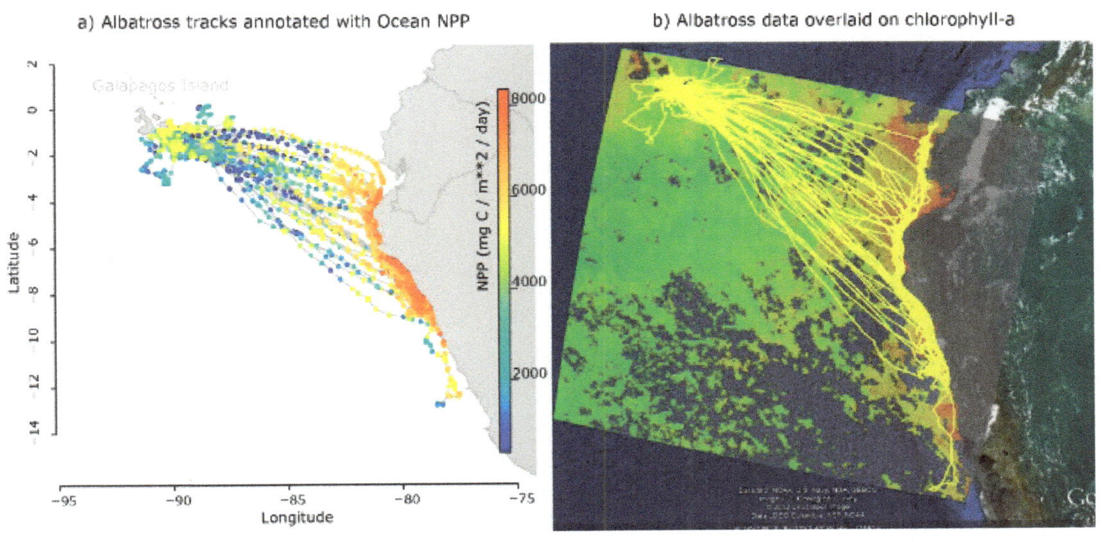

Figure 5: Nine annotated albatross trajectories. The tracks of nine adult albatrosses, overall containing 8286 GPS locations, during the breeding season in June to September 2008, (**a**) color coded with annotated values of 8-day ocean NPP (see Table 1 for more information on this variable), (**b**) the same tracks (yellow lines) plotted on the geographic area annotation using the monthly MODIS-ocean chlorophyll-a variable (Table 1) for the month of July 2008. We used the KML data format and combined the annotated area with a Google-Earth satellite image of the region using the program Matlab and its "Google Earth Toolbox".

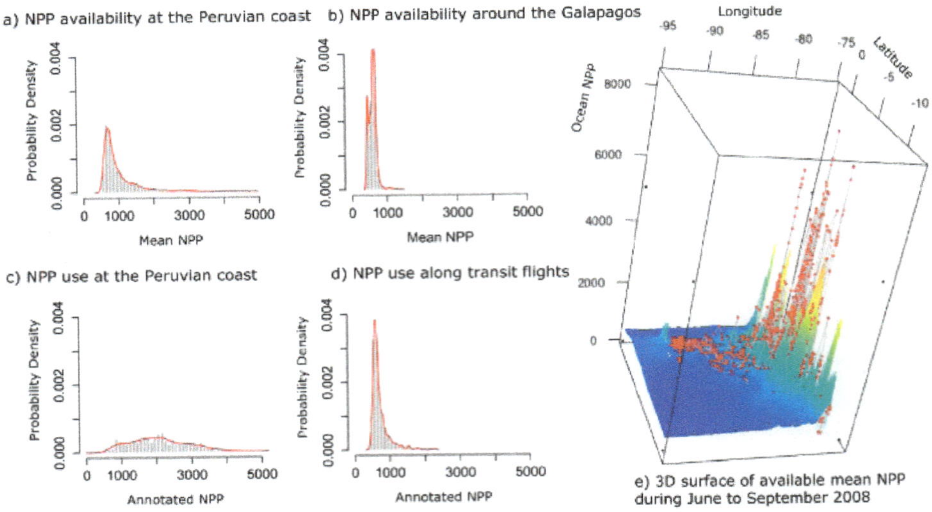

Figure 6: Probability density histograms and 3D surface plot of Ocean NPP. Available net primary ocean production (NPP, mg C/m^2/day) compared to NPP along the tracks of nine Galapagos albatrosses during June to September 2008. Red lines fitted on NPP histograms (left) highlight probability density distributions of NPP use versus NPP availability. Red points connected with gray lines on a 3D surface (right) illustrate the annotated albatross tracks overlaid on the averaged ocean NPP during June to September 2008.

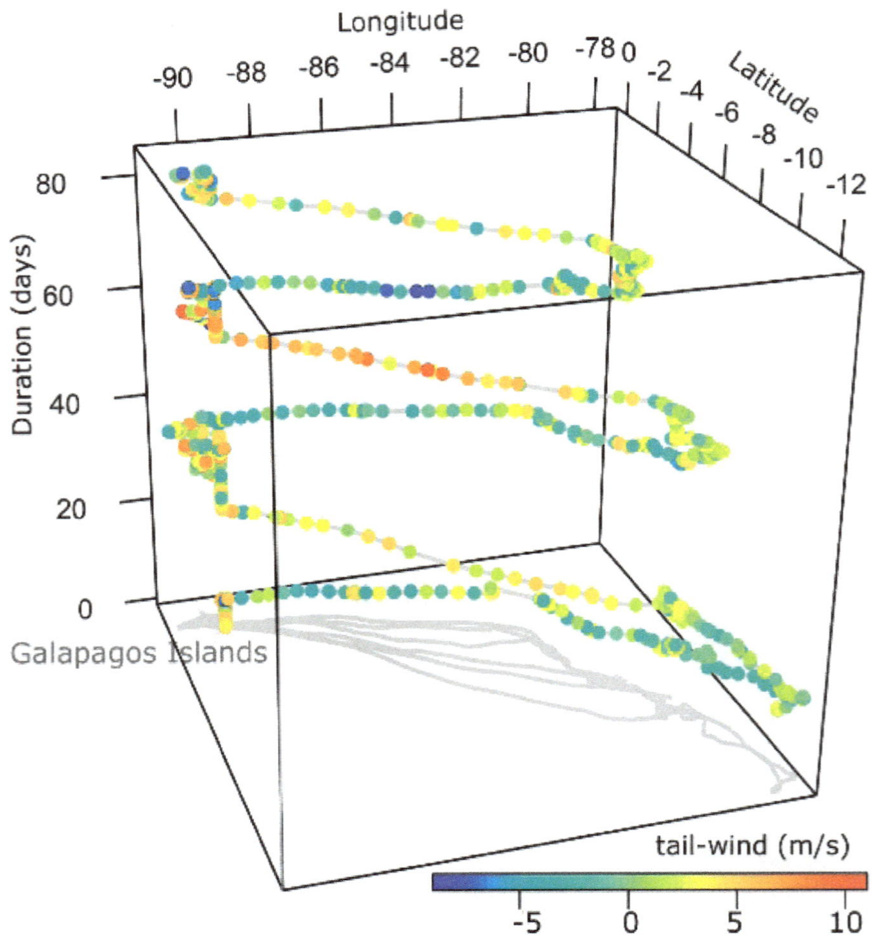

Figure 7: Space-time-cube illustration of an albatross' flights annotated by tail-wind support. The track contains 1326 GPS locations of one individual albatross from 23 June to 15 September 2008. The albatross' outbound flights towards the Peruvian coast are hampered by head winds while the return flights are facilitated by tail-wind assistance.

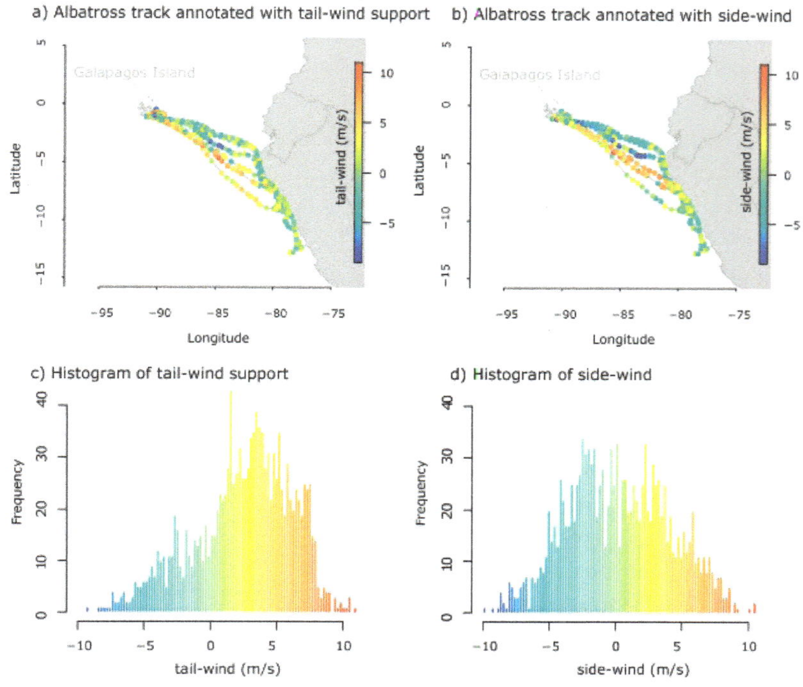

Figure 8: Map (top) and histogram (bottom) illustration of an albatross' flights annotated by tail-wind support and side-wind (cross wind). The track contains 1326 GPS locations of one individual albatross from 23 June to 15 September 2008.

These findings are in accordance with previous studies showing that parents make long-range trips to preferred, productive areas [47].

We use a space-time-cube visualization to illustrate the role of wind on the albatross' flight patterns (see Figure 7, Function 2, Additional File 1). Recent studies suggest that wind was an important factor in determining migration and short-term flight patterns of pelagic birds [4, 24, 36, 48]. Here, we show how *Env-DATA* annotation can assist an investigation of wind dependencies and flow assistance. Figures 7 and 8 show a single albatross trajectory annotated by tail-wind support and side-wind (cross wind), two derived variables (Table 1) computed using wind direction and wind speed and movement direction (flight heading) of the albatross along its flight path, based on the formulation from [24, 36]. The space-time-cube illustrates how wind assistance facilitates the albatross' flights toward the Galapagos Islands (orange to red colors represent higher wind assistance), while the flights to the coast are often challenged by head wind (aquamarine to blue colors represent wind resistance). The flight pattern in Figures 7 and 8 is characteristic to most other flight tracks in our albatross dataset. As seen in Figure 7 and Figure 8a–b, the albatross repeatedly takes a more northern route to the coast relying mostly on side winds, and then moves south (presumably foraging) before returning to the Galapagos Islands using a tail-wind assisted route (cf. Figure 8a and Figure 8c). The observed clock-wise pattern is in accordance with previous findings [4, 48]. Weimerskirch et. al [48] found that albatrosses prefer tail or side winds and therefore use predictable weather systems to fly in large looping tracks; when going south movements are in a clockwise direction. This enables albatrosses to achieve high speeds while expending little energy. The travel direction towards continental South America and back to the Galapagos undertaken by waved albatross means they almost always have side-winds (cf. Figure 8b and Figure 8d).

CONCLUSIONS

We presented the Environmental-Data Automated Track Annotation (*Env-DATA*) System, an openly available portal within Movebank (http://www.movebank.org), and illustrated how the system assists the discovery of environmental conditions associated with animal movements. As compared to the existing RNCEP package, that provides access to, organization, and visualization of atmospheric NCEP/NCAR datasets, the *Env-DATA*annotation service streamlines the co-registration of animal tracking data with a diverse range of environmental variables obtained from satellite remote sensing products and global reanalysis models including the MODIS ecological, ocean, land cover and land use data sets, the NCEP/NCAR and ECMWF weather reanalysis datasets, high-resolution Digital Elevation Models (DEMs), and ecological and human-socioeconomic reanalyses (e.g. the Population Density Grid). This project has overcome the numerous technical and methodological challenges in order to enable processing of a large array of remote sensing, weather and geographical data for the analysis of animal movement tracks:

- optimizing storage and retrieval times for a very large dataset of environmental variables from multiple data providers,
- applying effective interpolation techniques in order to maintain the link between animal tracks and their embedding environment in space and time,
- applying suitable spatiotemporal indexing strategies for data retrieval, and
- maintaining a large database of remote sensing data.

In addition, our system is intended as a general tool that can be used by researchers at all levels of technical ability for a wide range of animal movement data types and research questions. Thus additional challenges were:

- establishing linkages between heterogeneous environmental and movement data, obtained from various sources, collected in different spatial and temporal resolutions and scales; and
- developing a user-friendly interface within Movebank to allow users to browse, access documentation about variables and source datasets, and select and request annotated data.

In addition to the main annotation service, we are currently developing the Knowledge Discovery and Visualization Service and the Track Simulation Service within the *Env-DATA* System. In future releases, we aim to exploit deterministic and probabilistic computational GIS methods, spatiotemporal data mining techniques, and well-known statistical approaches, with the underlying goal to discover patterns and structures among the movements of animals.

Availability of supporting data

The albatross dataset supporting the results of this article is accessible through DOI: 10.5441/001/1.3hp3s250and can also be viewed at http://www.movebank.org in the study "Galapagos Albatrosses". The R scripts used for generating Figures 5–8 in this article are provided in Additional File 1.

AUTHOR'S CONTRIBUTION

SD drafted the manuscript, prepared the figures, analyzed the case study data, and participated in development of the system. GB conceived the study, coordinated the project. RW implemented the system and participated in the design of the study. SCD participated in the coordination of the study. SD, RK, DD, JH, DB, and MW participated in the design and coordination of the study. SD, GB, RW, DD and SCD contributed to the writing. SC provided the albatross data and contributed to the case study. All authors participated in editing the manuscript.

DECLARATIONS

Acknowledgment

We thank Keith Bildstein, Ran Nathan, Kamran Safi, Nir Sapir, John Takekawa, Zhenhui Li, Richard Lanctot, Steve Lewis, Stephen Yezerinac, Carol McIntyre, Erica Craig, Jean-Francois Therrien, Josh Adams, Mark Fuller, and Ron Therrien for ideas and collaboration that helped shape and test the system. We also thank Martin Storhas and Matthias Berger for their work on developing the user interface for the *Env-DATA* System. This research was funded by NASA under grant #NNX11AP61G. Super computer service at The Ohio Supercomputer Center was provided under OSC grant #PAS0626-1. GB was supported in part by the U.S. National Science Foundation grant IOS-1145952. Any opinions, findings, and conclusions or recommendations expressed in this material are those of the author(s) and do not necessarily reflect the views of the National Science Foundation. Use of trade names does not reflect endorsement by the U.S. Government. Galapagos Albatross data used for the case study were collected with permission and support of the Galapagos National Park Service.

REFERENCES

1. Nathan R, Getz WM, Revilla E, Holyoak M, Kadmon R, Saltz D, Smouse PE: A movement ecology paradigm for unifying organismal movement research. Proc Natl Acad Sci U S A. 2008, 105: 19052-19059. 10.1073/pnas.0800375105.
2. Shamoun-Baranes J, Bouten W, Van Loon EE: Integrating meteorology into research on migration. Integr Comp Biol. 2010, 50: 280-292. 10.1093/icb/icq011.
3. Gordon DM: Variation and Change in Behavioral Ecology. Ecology. 1991, 72: 1196-1203. 10.2307/1941093.
4. Chapman JW, Klaassen RHG, Drake VA, Fossette S, Hays GC, Metcalfe JD, Reynolds AM, Reynolds DR, Alerstam T: Animal orientation strategies for movement in flows. Curr Biol. 2011, 21: R861-R870. 10.1016/j.cub.2011.08.014.
5. Bohrer G, Brandes D, Mandel JT, Bildstein KL, Miller TA, Lanzone M, Katzner T, Maisonneuve C, Tremblay JA: Estimating updraft velocity components over large spatial scales: contrasting migration strategies of golden eagles and turkey vultures. Ecol Lett. 2012, 15: 96-103. 10.1111/j.1461-0248.2011.01713.x.
6. Mandel JT, Bildstein KL, Bohrer G, Winkler DW: Movement ecology of migration in turkey vultures. Proc Natl Acad Sci U S A. 2008, 105: 19102-19107. 10.1073/pnas.0801789105.
7. Mandel JT, Bohrer G, Winkler DW, Barber DR, Houston CS, Bildstein KL: Migration path annotation: cross-continental study of migration-flight response to environmental conditions. Ecological applications: a publication of the Ecological Society of America. 2011, 21: 2258-2268. 10.1890/10-1651.1.
8. Bowlin MS, Bisson I-A, Shamoun-Baranes J, Reichard JD, Sapir N, Marra PP, Kunz TH, Wilcove DS, Hedenström A, Guglielmo CG, Åkesson S, Ramenofsky M, Wikelski M: Grand challenges in migration biology. Integr Comp Biol. 2010, 50: 261-279. 10.1093/icb/icq013.
9. Vanak AT, Thaker M, Slotow R: Do fences create an edge-effect on the movement patterns of a highly mobile mega-herbivore?. Biol Conserv. 2010, 143: 2631-2637. 10.1016/j.biocon.2010.07.005.
10. Sapir N, Wikelski M, Avissar R, Nathan R: Timing and flight mode of departure in migrating European bee-eaters in relation to multi-scale meteorological processes. Behav Ecol Sociobiol. 2011, 65: 1353-1365. 10.1007/s00265-011-1146-x.
11. Sapir N, Horvitz N, Wikelski M, Avissar R, Mahrer Y, Nathan R: Migration by soaring or flapping: numerical atmospheric simulations reveal that turbulence kinetic energy dictates bee-eater flight mode. Proceedings Biological sciences / The Royal Society. 2011, 278: 3380-3386. 10.1098/rspb.2011.0358.

12. Nathan R, Spiegel O, Fortmann-Roe S, Harel R, Wikelski M, Getz WM: Using tri-axial acceleration data to identify behavioral modes of free-ranging animals: general concepts and tools illustrated for griffon vultures. J Exp Biol. 2012, 215 (Pt 6): 986-996.

13. Fryxell JM, Hazell M, Börger L, Dalziel BD, Haydon DT, Morales JM, McIntosh T, Rosatte RC: Multiple movement modes by large herbivores at multiple spatiotemporal scales. Proc Natl Acad Sci U S A. 2008, 105: 19114-19119. 10.1073/pnas.0801737105.

14. Dalziel BD, Morales JM, Fryxell JM: Fitting probability distributions to animal movement trajectories: using artificial neural networks to link distance, resources, and memory. Am Nat. 2008, 172: 248-258. 10.1086/589448.

15. Boettiger AN, Wittemyer G, Starfield R, Volrath F, Douglas-Hamilton I, Getz WM: Inferring ecological and behavioral drivers of African elephant movement using a linear filtering approach. Ecology. 2011, 92: 1648-1657. 10.1890/10-0106.1.

16. Getz W, Saltz D: A framework for generating and analyzing movement paths on ecological landscapes. Proceedings of the National Academy of Sciences. 2008, 105 (49): 19066-19071. 10.1073/pnas.0801732105.

17. Adams J, Flora S: Correlating seabird movements with ocean winds: linking satellite telemetry with ocean scatterometry. Mar Biol. 2009, 157: 915-929.

18. Shamoun-Baranes J, Bouten W, Camphuysen CJ, Baaij E: Riding the tide: intriguing observations of gulls resting at sea during breeding. IBIS: The International Journal of Avian Science. 2011, 153: 411-415.

19. Shamoun-Baranes J, Van Loon EE, Alon D, Alpert P, Yom-Tov Y, Leshem Y: Is there a connection between weather at departure sites, onset of migration and timing of soaring-bird autumn migration in Israel?. Glob Ecol Biogeogr. 2006, 15: 541-552. 10.1111/j.1466-8238.2006.00261.x.

20. Shamoun-Baranes J, Leshem Y, Yom-Tov Y, Liechti O: Differential use of thermal convection by soaring birds over central Israel. Condor. 2003, 105: 208-218. 10.1650/0010-5422(2003)105[0208:DUOTCB]2.0.CO;2.

21. Shamoun-Baranes J, Baharad A, Alpert P, Berthold P, Yom-Tov Y, Dvir Y, Leshem Y: The effect of wind, season and latitude on the migration speed of white storks *Ciconia ciconia*, along the eastern migration route. J Avian Biol. 2003, 34: 97-104. 10.1034/j.1600-048X.2003.03079.x.

22. Spaar R, Bruderer B: Soaring migration of Steppe Eagles *Aquila nipalensis* in Southern Israel: Flight behaviour under various wind and thermal conditions. J Avian Biol. 1996, 27: 289-301. 10.2307/3677260.

23. Klaassen M, Beekman JH, Kontiokorpi J, Mulder RJW, Nolet BA: Migrating swans profit from favourable changes in wind conditions at low altitude. J Ornithol. 2004, 145: 142-151. 10.1007/s10336-004-0025-x.

24. Kemp MU, Shamoun-Baranes J, Van Loon EE, McLaren JD, Dokter AM, Bouten W: Quantifying flow-assistance and implications for movement research. J Theor Biol. 2012, 308: 56-67.

25. Katzner TE, Brandes D, Miller T, Lanzone M, Maisonneuve C, Tremblay JA, Mulvihill R, Merovich GT: Topography drives migratory flight altitude of golden eagles: implications for on-shore wind energy development. J Appl Ecol. 2012, 49: 1178-1186. 10.1111/j.1365-2664.2012.02185.x.

26. Kemp MU, van Loon E, Shamoun-Baranes J, Bouten W: RNCEP: global weather and climate data at your fingertips. Methods in Ecology and Evolution. 2012, 3: 65-70. 10.1111/j.2041-210X.2011.00138.x.

27. Dodge S, Weibel R, Lautenschütz A-K: Towards a taxonomy of movement patterns. Inf Vis. 2008, 7: 240-252. 10.1057/palgrave.ivs.9500182.

28. Buchin M, Dodge S, Speckmann B: Context-Aware Similarity of Trajectories. Lecture Notes in Computer Science, Proceedings of the 7th International Conference GIScience 2012. 2012, 7478: 43-56.

29. Kranstauber B, Cameron A, Weinzerl R, Fountain T, Tilak S, Wikelski M, Kays R: The Movebank data model for animal tracking. Environ Model Software. 2011, 26: 834-835. 10.1016/j.envsoft.2010.12.005.

30. Wikelski M, Kays R: Movebank: archive, analysis and sharing of animal movement data. World Wide Web electronic publication, http://www.movebank.org, access in 2013

31. Phillips SJ, Anderson RP, Schapire RE: Maximum entropy modeling of species geographic distributions. Ecol Model. 2006, 190: 231-259. 10.1016/j.ecolmodel.2005.03.026.

32. The Move R-package.http://cran.r-project.org/web/packages/move/index.htmlhttp://computational-ecology.com/main-move.html,

33. Li Z, Han J, Ji M, Tang L, Yu Y, Ding B, Lee J-G, Kays R: MoveMine: Mining Moving Object Data for Discovery of Animal Movement Patterns. ACM Transactions on Intelligent Systems and Technology (ACM TIST). 2011, 2: 37-

34. Hawkes LA, Balachandran S, Batbayar N, Butler PJ, Chua B, Douglas DC, Frappell PB, Milsom WK, Newman SH, Prosser DJ, Sathiyaselvam P, Scott GR, Takekawa JY, Wikelski M, Witt MJ, Yan B, Bishop CM: The paradox of extreme high-altitude migration in bar-headed geese *Anser indicus*. Proceedings Biological sciences / The Royal Society. 2013, 280: 10.1098/rspb.2012.2114.

35. Battley PF, Warnock N, Tibbitts TL, Gill RE, Piersma T, Hassell CJ, Douglas DC, Mulcahy DM, Gartrell BD, Schuckard R, Melville DS, Riegen AC: Contrasting extreme long-distance migration patterns in bar-tailed godwits *Limosa lapponica*. J Avian Biol. 2012, 43: 21-32. 10.1111/j.1600-048X.2011.05473.x.

36. Safi K, Kranstauber B, Kays R, Weinzerl R, Griffin L, Cruz S, Takekawa J, Wikelski M, Bohrer G: Flying with the wind. Movement Ecology. in press

37. Kummerow C, Barnes W, Kozu T, Shiue J, Simpson J: The Tropical Rainfall Measuring Mission (TRMM) Sensor Package. J Atmos Oceanic Tech. 1998, 15: 809-817. 10.1175/1520-0426(1998)015<0809:TTRMMT>2.0.CO;2.

38. Tucker C, Pinzon J, Brown M, Slayback D, Pak E, Mahoney R, Vermote E, El Saleous N: An extended AVHRR 8-km NDVI dataset compatible with MODIS and SPOT vegetation NDVI data. Int J Remote Sens. 2005, 26: 4485-4498. 10.1080/01431160500168686.

39. Kanamitsu M, Ebisuzaki W, Woollen J, Yang S-K, Hnilo JJ, Fiorino M, Potter GL: NCEP–DOE AMIP-II Reanalysis (R-2). Bull Am Meteorol Soc. 2002, 83: 1631-1643. 10.1175/BAMS-83-11-1631.

40. Mesinger F, DiMego G, Kalnay E, Mitchell K, Shafran PC, Ebisuzaki W, Jović D, Woollen J, Rogers E, Berbery EH, Ek MB, Fan Y, Grumbine R, Higgins W, Li H, Lin Y, Manikin G, Parrish D, Shi W: North American Regional Reanalysis. Bull Am Meteorol Soc. 2006, 87: 343-360. 10.1175/BAMS-87-3-343.

41. Dee DP, Uppala SM, Simmons AJ, Berrisford P, Poli P, Kobayashi S, Andrae U, Balmaseda MA, Balsamo G, Bauer P, Bechtold P, Beljaars ACM, Van de Berg L, Bidlot J, Bormann N, Delsol C, Dragani R, Fuentes M, Geer AJ, Haimberger L, Healy SB, Hersbach H, Hólm EV, Isaksen L, Kållberg P, Köhler M, Matricardi M, McNally AP, Monge-Sanz BM, Morcrette JJ: The ERA-Interim reanalysis: configuration and performance of the data assimilation system. Q J Roy Meteorol Soc. 2011, 137: 553-597. 10.1002/qj.828.

42. Behrenfeld M, Falkowski PG: A consumer's guide to phytoplankton primary productivity models. Limnol Oceanogr. 1997, 42: 1479-1491. 10.4319/lo.1997.42.7.1479.

43. Reuter HI, Nelson A, Jarvis A: An evaluation of void-filling interpolation methods for SRTM data. Int J Geogr Inf Sci. 2007, 21: 983-1008. 10.1080/13658810601169899.

44. Jenness JS: Calculating landscape surface area from digital elevation models. Wildl Soc Bull. 2004, 32: 829-839. 10.2193/0091-7648(2004)032[0829:CLSAFD]2.0.CO;2.

45. Li J, Heap AD: A review of comparative studies of spatial interpolation methods in environmental sciences: Performance and impact factors. Ecological Informatics. 2011, 6: 228-241. 10.1016/j.ecoinf.2010.12.003.

46. Lam NS-N: Spatial Interpolation Methods: A Review. Cartogr Geogr Inf Sci. 1983, 10: 22-

47. Anderson DJ, Huyvaert KP, Wood DR, Gillikin CL, Frost BJ, Mouritsen H: At-sea distribution of waved albatrosses and the Galápagos Marine Reserve. Biol Conserv. 2003, 110: 367-373. 10.1016/S0006-3207(02)00238-0.

48. Weimerskirch H, Guionnet T, Martin J, Shaffer SA, Costa DP: Fast and fuel efficient? Optimal use of wind by flying albatrosses. Proceedings Biological sciences, The Royal Society. 2000, 267: 1869-1874. 10.1098/rspb.2000.1223.

CHAPTER 15 REMOTE SENSING OF ECOLOGY, BIODIVERSITY AND CONSERVATION

Kai Wang, Steven E. Franklin, Xulin Guo and Marc Cattet

Keywords: Remote sensing, EBC (ecology, biodiversity and conservation), thermal infrared, small-satellite constellation, LIDAR, image classification, data fusion, integration of remote sensing (RS) and geographic information system (GIS)

INTRODUCTION

In general, ecological research refers to the investigation of organisms and their surrounding environment, including biotic and abiotic entities. Due to the multifaceted nature of biodiversity, it is difficult to simply express and measure biodiversity. Biodiversity should be related to not only the variation of life forms, but also the ecological complexes of which they are a part. Conservation has become an indispensable way of dealing with the accelerated native ecosystem conversion and degradation, which have a significantly negative effect on biodiversity. Remote sensing, the science of obtaining information via noncontact recording [1], has swept the fields of ecology, biodiversity and conservation (EBC). Remote sensing can provide consistent long-term Earth observation data at scales from the local to the global domain. In addition, remote sensing is not labor-intensive and time-consuming, compared with field-based observations. The review papers of Kerr and Ostrovsky and Turner *et al.*, published in the journal of "*Trends in Ecology and Evolution*", has been cited hundreds of times by scientists from around the world who are involved in remote sensing of EBC [2,3]. Turner *et al.* stated two categories of approaches, namely direct and indirect remote sensing approaches [3]. The direct approach refers to the direct observation of individual organisms, species assemblages, or ecological communities from airborne or satellite sensors, such as the application of high spatial resolution and hyperspectral sensors (e.g., [4]). Indirect approaches rely on environmental parameters derived from remotely sensed data as proxies. For example, habitat parameters, such as land cover, species composition, *etc.*, can be considered as a surrogate for precise estimates of potential species ranges and patterns of species richness [5]. The Foothills Research Institute Grizzly Bear Program (FRIGBP, formerly called Foothills Model Forest Grizzly Bear Research Program) has successfully applied this kind of approach in west-central Alberta (Canada) [6]. Kerr and Ostrovsky described ecological remote sensing in three main areas [2]. First, land cover classification, the physiographical characteristics of the surface environment, can be used to identify very specific habitats and predict the distribution of both individual species and species assemblages at a large spatial extent (e.g., [7]). Secondly, integrated ecosystem measurements offer the urgently needed measurements of functions at different spatial scales, including whole ecosystems, such as the derivation of leaf area index (LAI) and net primary productivity (NPP) mostly based on the normalized difference vegetation index (NDVI, e.g., [8]). Thirdly, change detection provides near-continuous, long-term measurements of key ecological parameters in order to monitor ecosystem through time and over significant areas, such as the application of climate change and habitat loss (e.g., [9]). Additionally, several quality review papers have contributed to this field, such as [10–14].

Most existing review papers too often discuss an issue from the viewpoint of ecologists or biodiversity specialists. For instance, Aplin reviewed the remote sensing of ecology as it relates to the significance of remote sensing in ecology, to spatial scale, and to terrestrial and marine ecological applications [11]. Gillespie *et al.* discussed the development of measuring and modeling biodiversity from space with a focus on species and land-cover classifications, modeling biodiversity, and conservation planning [14]. This review, on the other hand, focuses on the spaceborne remote sensing of

Citation: Kai Wang, Steven E. Franklin, Xulin Guo and Marc Cattet, Remote Sensing of Ecology, Biodiversity and Conservation, doi:10.3390/s101109647.

EBC from the perspective of remote sensing specialists, *i.e.*, it is organized in the context of state-of-the-art remote sensing technology, including instruments and techniques. Herein, the instruments to be discussed consist of high spatial resolution, hyperspectral, thermal infrared, small-satellite constellation, and LIDAR sensors; and the techniques refer to image classification, vegetation index (VI), inversion algorithm, data fusion, and the integration of remote sensing (RS) and geographic information system (GIS).

ADVANCED INSTRUMENTS IN REMOTE SENSING OF EBC

Based on the current status of remote sensing instruments, their existing applications in the literature, and future potential contributions to EBC, the aforementioned five types of instruments: high spatial resolution, hyperspectral, thermal infrared, small-satellite constellation, and LIDAR sensors, were selected. In order to avoid overlapping between high spatial resolution and hyperspectral sensors, the hyperspectral sensors discussed below will mainly refer to sensors with medium spatial resolution, such as Hyperion with 30 m spatial resolution. Radar sensors are not selected because their applications mostly concentrate on geology, ice and snow, marine surveillance, and agriculture. In addition, some uncertainties in radar remote sensing, such as the saturation issue under high vegetation biomass, hamper its applications on EBC.

High Spatial Resolution

Generally speaking, high spatial resolution, also called fine spatial resolution, is less than 10 m, and ranges from 0.5–10 m in the commercial domain for environmental research. IKONOS, QuickBird, OrbView-3 and SPOT-5 (Satellite Pour l'Observation de la Terre-5) are the commonly used systems (see [15] for the high-spatial resolution optical sensors). The benefit of high spatial resolution imagery is that it greatly increases the accuracy of identification and characterization of small objects at spatial scales which were previously only available from airborne platforms [3,14]. For example, Gillespie *et al.* provided several examples of accurately identifying plant species based on the high spatial resolution imagery [14]. Turner *et al.* have pointed out it is applicable and feasible to directly identify certain species and species assemblages at the scale of high spatial resolution [3]. In addition, high spatial resolution imagery can be employed to assess the accuracy of remote sensing precuts derived from moderate or coarse spatial resolution imagery. For instance, Wabnitz *et al.* assessed the accuracy of Landsat-based large-scale seagrass mapping against patterns detectable with very high-resolution IKONOS images [16]. However, the high spatial resolution imagery is still expensive to acquire from commercial satellites, at the price of approximately 3,000–5,000 US$ for 10 km^2 [14], although it has tended to decrease with the emergence of more sensors and the upcoming competition. Moreover, data coverage and security restrictions are still a significant hurdle before easily accessing high spatial resolution satellite data [17].

Due to the large amount of high spatial resolution sensors, the commonly-used IKONOS imagery was selected to display their typical applications in 2008 and 2009. First of all land cover, as the representative of basic landscape information, can be extracted quickly and reliably based on the high spatial resolution data. For example, the object-oriented classification of IKONOS-2 satellite images was utilized to explicitly recognize the transitional areas between tree crowns and tree shades (tree shadows), and then for the quantification of canopy cover [18]. Further, IKONOS imagery can be used to quantify and evaluate the spatial structure of critical habitats and how it affects endemic species, which is essential baseline information for biodiversity monitoring and management (e.g., [19]). In the context of marine applications, areas of coastline, with their fertile soil and unique flora and fauna which need to be highly protected, were planned for in a sustainable way through mapping the changes in land use of the area based on IKONOS imagery in the Cesme Peninsula (Turkey) [20]. Improving the science and conservation of coral reef ecosystems, such as the significant fish-habitat relationship, is often the objective of marine ecology, and also is an important facet in the application of IKONOS imagery [21]. Harborne *et al.* examined intra-habitat variability in coral-reef fish by mapping habitat heterogeneity, which is always considered a surrogate of biodiversity, in order to aid the design of networks of marine reserves [22]. Although high spatial resolution satellite remote sensing has been hailed as a very useful source of data, Nagendra and Rocchini pointed out that high spatial resolution remotely sensed data are one of the most potentially powerfully yet underutilized sources for tropical research on biodiversity, and stimulating discussion on the applications should be the first step in promoting a more extensive use of such data [17].

Hyperspectral

Hyperspectral data have the ability to collect ample spectral information across a continuous spectrum generally with 100 or more contiguous spectral bands. It is different from multispectral sensors which detect relatively few discrete bands [17]. Hundreds of spectral bands with 10–20 nm spectral bandwidths offer new possibilities to detect subtle differences between objects of interest. The best example is to discriminate fine-scale, species-specific land cover [3], such as vegetation categories or soil types [11], which make remarkable contribution to the study regarding biodiversity patterns. Moreover, Nagendra and Rocchini summarized that hyperspectral data have been successfully applied in recording information regarding critical plant properties (e.g., leaf pigment, water content and chemical composition), discriminating tree species in landscapes, and fairly accurate identification between different species [17]. What is more, spectral signatures acquired from atmosphere-corrected hyperspectral data can be directly compared to the existing spectral library (e.g., the Jet Propulsion Laboratory Spectral Library) in order to rapidly identify ground information useful in land-cover classification, characterization and change detection [3]. Similar to the situation with high spatial resolution imagery, the hyperspectral imagery encountered the same underutilization, and a high cost which may put it out of research for many ecologists [14], especially those in developing countries who eagerly need the data [17].

Shippert listed the existing hyperspectral sensors acquiring imagery from space, including the Hyperion sensor on NASA's EO-1 (National Aeronautics and Space Administration's Earth

Observing-1), the CHRIS (Compact High Resolution Imaging Spectrometer) sensor on the European Space Agency's PROBA (PRoject for On-Board Autonomy) satellite, and the FTHSI (Fourier Transform Hyperspectral Imager) sensor on the U.S. Air Force Research Lab's MightySat II satellite [23]. Of these sensors, the first-civilian and most commonly used data are derived from the Hyperion, which is operated by the EROS (Earth Resources Observation and Science) at a relatively low cost to the general public [23]. The EO-1, on which the Hyperion sensor is, was launched in November, 2000 as a one-year technology validation and demonstration in support of the LDCM (Landsat Data Continuity Mission; [24]). The Hyperion sensor, an upgrade from the LEWIS Hyperspectral Imaging Instrument (HSI), records visible light and other reflected electromagnetic energy in 220 spectral bands from 0.4 to 2.5 μm at a 30 m resolution [25]. Table 1 lists the Hyperion characteristics.

Table 1: Hyperion Imaging Spectrometer Characteristics (adapted from [26])

Characteristics	Values
Sensor Type	Push-broom imager
Wavelength Range	400–2,500 nm
Number of Spectral Bands	220
Spectral Resolution	10 nm
Spatial Resolution	30 m
Swath	7.5 km
Digitization	12 bits
Altitude	705 km
Repeat	16 day

The recent applications of Hyperion hyperspectral imagery mainly include ecology and biodiversity in forest, grassland [27], agriculture [28], and vegetation [29], fragmented ecosystem and ecosystem succession, coastal environment [30], *etc.*For example, vegetation types and densities were classified in support of the wildfire management, that is, fire propagation simulation models and fire risk assessment were based on a Hyperion classification map with 93% accuracy [31]. Foster *et al.*proposed hyperspectral imagery from EO-1 Hyperion is capable of mapping low-lying woody lianas, which are critical to tropical forest dynamics because of their strong influence on forest regeneration, disturbance ecology, and biodiversity [32]. Pignatti *et al.* analyzed the capability of Hyperion data for discriminating land cover in a complex natural ecosystem according to the structure of the currently used European standard classification system (CORINE Land Cover 2000), and the results showed the potential of the imagery up to the 4th level of the CORINE legend, even at the sub-pixel level, within a fragmented ecosystem [33]. Besides the application of land cover classification, the relationships between LAI and spectral reflectance were studied by [34] using narrowband (EO-1 Hyperion) and broadband (Landsat ETM+ [Enhanced Thematic Mapper Plus]) remotely sensed data in Sulawesi (Indonesia). Nagendra and Rocchini preliminarily discussed the strengths and drawbacks of hyperspatial (*i.e.*, high spatial resolution) and hyperspectral data [17]. Hyperspatial data was considered to be best suited for facilitating the accurate location of features such as tree canopies, but less suited to the identification of aspects such as species identity. However, conversely, hyperspectral data appear capable of identifying features with significantly increasing accuracy. Therefore, the integration of Hyperion and IKONOS imagery was proposed to differentiate the subtle spectral differences of land-use/land-cover types on household farms in the Northern Ecuadorian Amazon with an emphasis on secondary and successional forests, and the promising results supported the integrated use of hyperspectral and hyperspatial data [35].

Thermal Remote Sensing

Thermal remote sensing detects the energy emitted from the Earth's surface in the thermal infrared (TIR, 3 μm to 15 μm), which can be radiated by all bodies above absolute zero. Theoretically, TIR sensors measure the surface temperature and thermal properties of targets [36], which are essential for developing a better understanding, and more robust models, of land-surface energy balance interactions [37]. Moreover, TIR remote sensing is capable of uncovering the principles of ecological patterns of structure and function due to the development of ecological thermodynamics [37]. A thermal grey level image is generated based on relative radiant temperatures (a thermogram), and light tones correspond to warmer temperatures and dark tones to cooler temperatures [36]. TIR remote sensing plays an important role in observation of Earth surface characteristics, and is very useful for research regarding analysis of biophysical Earth processes, in particular landscape characterization and measurement of land surface processes [37]. The well-known sensors with TIR bands include the Advanced Very High Resolution Radiometer (AVHRR) onboard the Polar Orbiting Environmental Satellites (POES), the Landsat Thematic Mapper (TM) and ETM+, the Advanced Spaceborne Thermal Emission and Reflection Radiometer (ASTER) on the Terra Earth observing satellite platform, *etc.* [37].

TIR remote sensing has been developing since 1880, and has proven to be an integral part of understanding landscape characteristics [37], although it is relatively rarely used by ecologists [2]. However, interests have increasingly focused on the use of TIR remote sensing in EBC. For instance, biophysical variables were derived from thermal and multispectral remote sensing data and coupled with a Soil-Vegetation-Atmosphere-Transfer (SVAT) model [38]. Duro *et al.* pointed out the TIR region is an important source of information to study environmental disturbance because of the negative relationship between vegetation density and land surface temperatures [13]. Mildrexler *et al.* proposed a disturbance detection index using Moderate Resolution Imaging Spectroradiometer (MODIS) 16-day Enhanced Vegetation Index (EVI) and 8-day Land Surface Temperature (LST), and it was successfully applied to detect continental-scale disturbance events such as wildfire, irrigated vegetation, precipitation variability, and the incremental process of recovery of disturbed landscapes [39]. Another good use of TIR remote sensing data is to measure evapotranspiration, evaporation, and soil moisture. For example, Crow and Zhan analyzed the continental-scale performance of surface soil moisture retrieval algorithms depending on satellite passive microwave, scatterometer, and thermal remote sensing observations [40]. Petropoulos *et al.* reviewed Ts/VI (surface temperature/vegetation index) remote sensing based methods for the retrieval of land surface energy fluxes and soil surface moisture, and suggested one piece of future work should evaluate the accuracy of these methods under diverse environments [41].

Constellation of Small Satellites

A small satellite generally refers to its mass in the range of 1–500 kg and satellite constellation is defined as groups of satellites working in concert [42]. Since 1997, six symposia on small satellites have been organized by the International Academy of Astronautics (IAA) in Berlin, Germany. Kramer and Cracknell reviewed the development of small satellites in remote sensing [43]. With the launch of DMC (Disaster Monitoring Constellation, Table 2), the concept of the Earth-observation constellation of low-cost small satellites has been put into action. It is capable of obtaining multispectral images of any part of the world every day [24]. The DMC was initially proposed in 1996 and led by SSTL (Surrey Satellite Technology Limited), which is a world leader in high performance small satellites [42]. Wang *et al.* briefly introduced the characteristics of DMC imagery and its potential applications in environmental science [44]. Also, HJ-1 (Huan Jing-1, also called Environment-1, operated by China) is another outstanding constellation system. It is designed mainly for environmental protection and disaster monitoring, and the payload instruments onboard consist of a CCD (Charge-Coupled Device) camera, an infrared camera, a hyperspectral imager and an S-band SAR (Synthetic Aperture Radar, [45]).

Table 2: Disaster Monitoring Constellation (DMC) on Orbit (adapted from DMC International Imaging Ltd.)

Designation	Type	Imager	Launch	Waveband
Alsat-1	DMC	32m MS	2002	✓ MS
UK-DMC	DMC	32m MS	2003	NIR: 0.77–0.90 μm
Nigeriasat-1	DMC	32m MS	2003	Red: 0.63–0.69 μm
Beijing-1	DMC+4	32m MS/4m Pan	2005	Green:0.52–0.60 μm
Deimos-1	DMC	22m MS	2008	✓ Pan
UK-DMC2	DMC	22m MS	2008	0.50–0.80 μm

P.S. MS = Multispectral; Pan = Panchromatic

Besides the benefits in cost and operation, the constellation of small satellites has two obvious advantages in applications, *i.e.*, global surveying and increased revisit frequency [24]. It is relatively easy to obtain observation data across the world in a short time for constellation systems. The increased revisit frequency can not only satisfy the application of detecting rapid surface changes such as crop-growth monitoring and detecting intraseasonal ecosystem disturbance, but also promotes acquisition of good-quality imagery with limited cloud-contamination. Wang *et al.* discussed the issue of clouds and cloud shadows in the environmental remote sensing community, and advised looking for good solutions to the unavoidable problem in optical remote sensing [44]. The development of a constellation of low-cost small satellites is believed to make contributions to this issue at the sensor level. Only a few studies of EBC applied the imagery of small-satellite constellation, though Aplin has predicted the bright future of this kind of satellite imagery [11]. Qian *et al.* demonstrated that simulated HJ-1B satellite data performed better on smaller and cooler fires than MODIS or AVHRR data, and believe it will offer a great opportunity for fire detection [46]. The FRIGBP has started testing the applicability of DMC imagery for wildlife large-area habitat mapping in west-central Alberta (Canada) [44].

LIDAR

Light Detection and Ranging (LIDAR), also called Laser altimetry, is an active remote sensing technology that utilizes a laser to illuminate a target object and a photodiode to register the backscatter radiation [47,48]. The current LIDAR remote sensing can be categorized into two general groups: non-scanning LIDARs, and scanning LIDARs. The non-scanning LIDARs record pulsed ranging that measures the travel time between the transmitted and received signal backscattered from the object surface, and the scanning LIDARs register continuous wave ranging that is produced in a transmitted sinusoidal signal and carried out by modulating the laser light intensity [49]. According to the characteristics of LIDAR technology, it has been proven to provide horizontal and vertical information at high spatial resolutions and vertical accuracies [47]. For example, Miller stated that 5–30 cm range is the typical accuracies for LIDAR-derived vertical information [50]. Airborne LIDAR remote sensing systems such as LVIS (Laser Vegetation Imaging Sensor) have been used for bathymetry, forestry, and other applications [48,51,52]. For instance, Turner *et al.* briefly discussed the airborne LIDAR remote sensing for biodiversity science and conservation [3]; Lim *et al.* reviewed the application within forest structure (vertical information) [47], e.g., canopy and tree height, biomass, and volume; Goetz *et al.* claimed species distribution models have been improved through airborne LIDAR quantifying vegetation structure within a landscape [53]. LIDAR was underlined by [11] as one of the strong interests of the remote sensing community in ecology. Besides airborne LIDAR with the limitations of large data volumes, footprint size and high costs [54], spaceborne LIDAR has come through with the launch of the ICESat/GLAS (Ice, Cloud, and Land Elevation Satellite/Geoscience Laser Altimeter System), which is the first laser-ranging instrument for continuous global observations. The applications of the GLAS data in EBC, which are seldom reviewed, will be discussed below.

LIDAR focused on the forest vertical structure, especially forest canopy height and aboveground biomass estimation. Lefsky *et al.* estimated forest canopy height with an RMSE of 5 m (83% of variance explained) in varied forest types including evergreen needle leaf, deciduous broadleaf and mixed stands in temperate North America, and tropical evergreen broadleaf forests in Brazil [55]. Mangrove forests are considered as one of the most biodiverse and productive wetlands on Earth, and the mangrove height and aboveground biomass were measured and mapped based on SRTM (Shuttle Radar Topography Mission) elevation data, GLAS waveforms and field data [56]. Pflugmacher *et al.* compared GLAS height and biomass estimates with reference data from the Forest Inventory and Analysis (FIA) program of the U.S. Forest Service at a regional scale, and promising results were obtained [57]. Helmer *et al.* proposed the combination of Landsat time series and the GLAS to estimate the biomass accumulation of the Amazonian secondary forest, and the estimation agreed well with ground-based studies [58]. Duncanson *et al.* tested simulated GLAS data under tough conditions, e.g., areas with dense forests, high relief, or heterogeneous vegetation cover, and demonstrated the capability of GLAS waveforms as supplemental model input to improve estimates of canopy height [54].

ADVANCED TECHNIQUES IN REMOTE SENSING OF EBC

Similar criteria were applied to choose the remote sensing techniques discussed below, including promising algorithms or methods in image classification, vegetation index (VI), inversion algorithm, data fusion, and the integration of RS and GIS. Although these techniques are reviewed separately, they are frequently integrated in practice. For example, data fusion can be implemented to remotely sensed data before they are classified by advanced classifiers in order to improve classification accuracy.

Image Classification

Regardless of the variety of uses for remote sensing images, the first goal is to extract landscape information from the satellite images [59]. Image classification has been recognized as the most effective means to do so since mid-1800s, when humans first identified different types of land-use and land-cover in aerial photography [60]. Jensen discussed in detail the fundamental elements of image interpretation including grayscale tone, color, height and depth, size, shape, texture, pattern shadow, site, association and arrangement [1]. With the widespread of digital computers, special-purpose image classification algorithms have been used to extract land-use/land-cover and biophysical information directly from remotely sensed data [60]. In order to derive more accurate classifications, new approaches have increasingly emerged, and such approaches have made significant contributions to the science of EBC, examples of which would be support vector machines (SVMs), one-class classifier, object-oriented classification, and fuzzy classifications.

SVMs consist of many theoretically superior machine learning algorithms, and make use of optimization algorithms to find an optimal separating hyperplane (OSH) between classes based on training samples [61]. The hyperplane is called support vectors [62]. Foody and Mathur have demonstrated the robustness of SVMs through comparison with artificial neural networks (ANNs) and machine learning decision trees, especially for small training sets [63]. The SVM was selected by [64] to help understand the relationships among spectral resolution, classifier complexity, and classification accuracy obtained with hyperspectral sensors for the classification of forest areas. Ichii *et al.* applied SVM-based evapotranspiration estimation to refine rooting depths for ecosystem modeling in California [65].

Commonly, only one specific class is the foci of research interest [66]. Due to the fact that conventional multiclass classifier may be suboptimal in terms of the classification accuracy of the class of interest, a one-class-classification approach was suggested to focus tightly on the specific class. For example, Sanchez-Hernandez *et al.* applied the one-class classifier based on the support-vector data description (SVDD) to map fenland habitat in support of conservation activities [66]. An accuracy of 97.5% and 93.6% from the user's and producer's perspectives was obtained, and it performed much better than conventional maximum-likelihood classification. In the same year, the classifier was used to map and monitor coastal saltmarsh habitats of high conservation value under the European Union's Habitats Directive [67].

With the wide availability of high-spatial resolution satellite data, pixel-based classification algorithms seem not to be ideal to extract information desired from the data exhibiting high frequency components with high contrast and horizontal layover of objects [60]. Therefore, object-oriented classification algorithms have been developed to meet this need, and have established improved classification accuracy when compared with the traditional methods [5,60]. The basic processing units of object-oriented classification are segments, so-called image objects that represent a relatively homogenous unit on the ground [68]. Then classification was performed on image objects, and not on pixels. One of the most popular algorithms was developed to the software of Definiens' Developer (also called eCognition; [69]). Advantages of object-oriented classification are to make full use of meaningful statistic and texture calculation, uncorrelated shape information (e.g., length-to-width ratio, direction and area of an object, *etc.*) and topological features (neighbor, super-object, *etc.*), and the close relation between real-world objects and image objects [68]. Jensen *et al.* pointed out that the advantages include rapid process, and scale flexibility in which users can select different scale levels according to their images [60]. A variety of studies applied object-oriented classification into the science of EBC. For example, Collingwood *et al.* classified agricultural areas in Alberta grizzly bear habitats based on one of the object-oriented classification techniques – sequential supervised masking (SSM), in order to help ecologists understand the relationship between crop types and grizzly bear presence [5]. Wang *et al.* proposed that object-oriented classification may traverse the possible Landsat-gap on applications such as landscape pattern analysis or ecological models [44].

Traditionally, land cover information is assigned into a finite number of non-overlapping classes, and the classes are mutually exclusive [70], which is described as the one-entity-one-class method [71]. However, pixels may contain more than one class because of the heterogeneity and the limitation in spatial resolution of remotely sensed data, especially in medium and coarse spatial resolution imagery [70]. And the presence of mixed pixels could not be removed totally no matter how accurately map classes are defined [71]. Therefore, fuzzy classification, also called subpixel classification, arose in the context of the uncertainty associated to class mixtures. In fuzzy systems, every pixel is supposed to consist of multiple and partial memberships of all candidate classes [70]. Spectral mixture analysis (SMA) is one of the most popular and most effective approaches for dealing with mixed pixel problem [60,70]. For example, Lu and Weng used linear SMA to explore the relationship between urban thermal features and biophysical descriptors based on ASTER images [72]. Plourde *et al.*estimated species abundance in a northern temperate forest using SMA for better understanding changes in biodiversity, habitat quality, climate, and nutrient cycling [73].

Vegetation Index

Vegetation indices (VIs) are 'dimensionless, radiometric relative abundance and activity of green vegetation, including LAI, percentage green cover, chlorophyll content, green biomass, and absorbed photosynthetically active radiation (APAR)' [1]. Jensen summarized VIs benefit in maximizing sensitivity to biophysical parameters, normalizing or modeling external effects, normalizing internal effects, and assisting validation effort and quality control [1]. Additionally, VIs are simple to understand and implement, easy to quickly calculate, and useful to track temporal characteristics. To date, hundreds of VIs have been used in all kinds of applications of remote sensing. VIs can be roughly categorized into two groups, *i.e.*, biophysical indices and biochemical indices [74]. Biophysical indices represent those designed to link with vegetation biophysical characteristics including structure and condition. They can be grouped into simple ratio-based indices (e.g., Simple Ratio [SR]; [75]), soil-line-related indices (e.g., Soil Adjusted Vegetation Index [SAVI]; [76]), and chlorophyll-corrected indices (Ratio TCARI/OSAVI [Transformed Absorption in Reflectance Index/Optimized Soil Adjusted Vegetation Index]; [77]). Biochemical indices are those mainly employed to estimate vegetation biochemical properties such as Cellulose Absorption Index (CAI) and Lignin-Cellulose Absorption Index (LCAI) [78].

No doubt that NDVI is the most well-known vegetation index. Its use in EBC has been considerably reviewed by [2,13,14], *etc.* Nonetheless, other indices that are commonly used in the relevant applications are not taken seriously enough in the aforementioned review papers. For example, SR was validated to perform best in early and intermediate forest stages for the assessment of LAI based on ASTER data in East African rainforest ecosystems [79]. The modified soil adjusted vegetation index (MSAVI) was selected as the optimal vegetation index in a linear mixture model to map canopy fractional cover in tropical forests in the Amazonian state of Mato Grosso (Brazil) [80]. Haboudane *et al.* demonstrated that the existing VIs (e.g., NDVI, SAVI, Triangular Vegetation Index [TVI], and Modified Chlorophyll Absorption Ratio Index [MCARI], *etc.*) were either sensitive to chlorophyll concentration changes or affected by saturation at high LAI levels, whereas a modified triangular vegetation index (MTVI2) and a modified chlorophyll absorption ratio index (MCARI2) are proved to be the best predictors of green LAI [77]. Additionally, other recently proposed VIs such as WDRVI (Wide Dynamic Range Vegetation Index; [81]), L-ATSAVI (Litter-corrected Adjusted Transformed Soil Adjusted Vegetation Index; [74]), and VIUPD (Vegetation Index based on a Universal Pattern Decomposition; [82]). However, traditional measures such as the coefficient of determination and root mean square based on regression statistics, are not capable of evaluating the performance of VIs on the estimation of biophysical parameters because the sensitivity of a VI may change substantially with vegetation density [83]. Therefore, a statistical sensitivity function was developed to summarize the overall relationship between VIs and biophysical parameters instead of a constant [83].

Inversion Algorithms

Various process-oriented models are developed to characterize Earth environments because traditional methods based on simple statistical relationships are often sensor-dependent, and site-specific [84,85]. These models represent the in-depth understanding of physical processes deriving the Earth system, and are unquestionably useful in Earth observations in support of EBC [85]. Generally speaking, models can be run under two modes, namely inverse mode and forward mode. An inverse mode applies outputs to retrieve inputs that cause them, while a forward mode applies inputs to obtain resulting outputs. For example, Boyd and Danson suggested that a remote sensing model can be used to simulate the reflectance of forest canopies [84]. The forward mode treats data on the forest canopy variables as the inputs and the spectral signature as the output but the inverse mode is converse process, *i.e.*, the spectral signature is the input and estimates of the forest biophysical variables are the outputs. Obviously, the inverse model is more frequently used in remote sensing. The core of inverse model is inversion algorithms, which mostly follow the physical laws and establish cause-and-effect relationships [85]. In order to understand remote sensing signals and develop practical inversion algorithms to estimate land surface variables, physically-based models are advised to discuss the following three areas [86]: atmosphere (atmospheric radiative transfer modeling), land surface (surface radiation modeling), and sensor (sensor modeling). Liang grouped inversion algorithms into four categories: model simulation and statistical analysis, optimization algorithms, look-up table algorithms, and data assimilation [86]. Several recent examples are provided below to display the applications of inversion algorithms in EBC.

In order to monitor and model storm-water pollution, Park and Stenstrom proposed a Bayesian network approach, which falls into the category of model simulation and statistical analysis [87]. A leaf radiative transfer model called the LIBERTY

(Leaf Incorporating Biochemistry Exhibiting Reflectance and Transmittance Yields) was selected and incorporated with three pigments to better understand relationships between leaf biochemical, biophysical, and spectral properties [88]. A look-up table approach was developed to estimate LAI [89]. Migliavacca *et al.* assimilated remotely sensed vegetation index time series, such as MODIS NDVI, into a process-based model BIOME-BGC (Biome-BioGeochemical Cycles) to estimate the gross primary production (GPP) of agro-forestry ecosystems [90]. However, an intrinsic problem in inverse models is the process from inputs to outputs is often non-invertible, *i.e.*, more than one combination of inputs results in the same output of spectral signature. Liang stated that, because it is still a nonlinear, ill-posed problem to inverse land surface parameters, further research is required to focus on use of regularization [86].

Data Fusion

Each kind of imagery has its own benefits and drawbacks, which provide great potential to fully exploit increasingly sophisticated multisource data through data fusion. For example, MODIS imagery has significant advantage in temporal resolution (one day) but is very poor in spatial resolution (250, 500 or 1,000 m) for certain applications, whereas Landsat TM imagery performed very well in spatial resolution (30 m) but with 16-day revisit. Therefore, Hilker *et al.* developed Spatial Temporal Adaptive Algorithm for Mapping Reflectance Change (STAARCH) model to fuse high spatial- (Landsat) and temporal-resolution (MODIS) for mapping of forest disturbance [91]. A general definition of remotely sensed data (image) fusion is given as 'the combination of two or more different images to form a new image by using a certain algorithm' [92]. Since the late 1980s when data fusion emerged as a new topic [93], several comprehensive review papers have been published to review the data fusion techniques, such as [92–95]. In general, the fusion techniques can be categorized into two classes [92]: (1) colour-related techniques, such as colour composites (RGB), intensity-hue-saturation (IHS); (2) Statistical or numerical methods, such as principal component analysis (PCA), band combinations using arithmetic operators and others. Besides typical techniques, wavelet transform, SVM (support vector machine) and ANN (artificial neural network) represent the heart of new data fusion methods (e.g., [96–98]).

Data fusion has matured into a widely used application of EBC. Pan-sharpening technique, which is to integrate a panchromatic (Pan) image with high spatial resolution and a multispectral (MS) image with high spectral resolution [94] to produce a high spatial resolution MS image, is likely to be the first data fusion method to make installing to the commercial remote sensing software such as PANSHARP module in PCI Geomatica software. For example, Wunderle *et al.* pan-sharpened SPOT-5 imagery to classify stand age of western red cedar in British Columbia (Canada) [99]. Due to the complementary nature of optical and radar imagery, their both fusion is always at the leading edge of remotely sensed data fusion [44]. Huang *et al.* estimated the quantity and quality of coarse woody debris in Yellowstone post-fire forest ecosystem from fusion of SAR and optical data [100]. Optical (Landsat-5 TM) and SAR (RADARSAT-1 Wide 1) images were fused through the combination of PCA and IHS transforms to map geomorphological and environmental sensitivity index in the Amazonian Mangrove Coast (Brazil) [101].

Integration of RS and GIS

RS and GIS have a complementary nature and should develop interdependently. RS routinely provides extracted information from remotely sensed data at scales ranging from local to global and the purpose of GIS is to store, analyze and visualize spatial data [102]. Although Hinton has reviewed well the combined use of remotely-sensed data and vector GIS data [103], Merchant and Narumalani claimed the integration of RS and GIS has actually become increasingly apparent since Aronoff [104,105]. Merchant and Narumalani listed key factors to benefit the integration, including development of theory and analytical methods, advances in computing (hardware and software) and global positioning system (GPS) technology [104]. A state-of-the-art definition of the integration is given as 'the use of each technology to benefit the other, as well as the application of both technologies for modeling and decision support' [104]. Ehlers *et al.* proposed a three-level taxonomy of the integration [106]. First-level integration happens in the level of separate but equal data exchange between GIS and image analysis systems, e.g., displaying GIS (usually vector) data and remotely sensed (raster) data simultaneously. Second-level integration permits seamless tandem or combined raster-vector processing based on a common use interface. Certain RS or GIS software has capability of performing the second-level integration. For example, the aforementioned Definiens' Developer is capable of incorporating GIS data directly into image processing – image segmentation [69]. Third-level integration operates RS and GIS as a unified system, and finally generates an integrated model of the real world, e.g., accommodating raster and vector data in a hierarchical structure. Moreover, Gao pointed out GPS must be involved with the integration to build up seamless RS-GIS-GPS integration for geospatial information analysis [107]. Campbell, and Merchant and Narumalani summarized the contribution of RS to GIS, and GIS to RS [25,104]. The contribution of RS to GIS includes: (1) RS develops thematic layers for GIS, such as surface elevation (Digital Elevation Model [DEM]), land use and land cover mapping, biophysical parameters, feature extraction and landscape change; and (2) RS provides orthoimagery as base data, which plays key role in positioning, registration and geo-referencing. The contribution of GIS to RS consists of (1) mission planning; (2) ancillary data for geometric and radiometric correction, and image classification; and (3) collection, organization and visualization of reference data.

Foody demonstrated many commonly used examples of RS and GIS for biodiversity applications. The following review focuses on the promising applications of the integration in EBC in 2009 [102]. For example, an adaptable method integrating low-cost remote sensing imagery and GIS was developed to assess forest cover change and conversion in support of decision-makers in assessing regional and local land use and planning forest conservation measures [108]. Giriraj *et al.*applied data generated from RS and GIS to categorize habitats, and then determined the relationship between the habitat categorizations and species-distribution patterns in tropical rain forests of Southern Western Ghats (India) [109]. Dong *et al.*pointed out that the integration of high-resolution RS images and GIS technique is an effective way to analyze the landscape changes at river basin scale [110]. In the management of water resources, RS and GIS integration techniques were used to design sustainable development plan of area and locale watershed [111], river inundation impact reduction [112], rainwater harvesting for drinking [113].

CONCLUSIONS

Remote sensing plays an increasing role in EBC research, especially regarding large spatial and/or long-term temporal scales. Moreover, the use of remote sensing deepens with the support of state-of-the-art remote sensing products and technology. Certainly, it is impossible to make progress without the assistance of GIS and GPS. It is believed that remote sensing will develop in a path similar to that of computer science, which has penetrated all aspects of human life. EBC performs as a propeller to push up the naissance of advanced remote sensing instruments and techniques. For example, the object-based image analysis (OBIA) is maturing in hopes to answer the question "why are remote sensing and digital image processing still so focused on the statistical analysis of single pixels rather than on the spatial patterns they build up" raised by [114]. Blaschke summarized the status of OBIA for remote sensing through a comprehensive review several thousand abstracts [115]. However, with the popularity of remotely sensed data and commercial remote sensing packages, it is easy to obtain processed remote sensing products based on certain algorithms or modules. These products can be applied to answer questions in the field of EBC. But, it is noteworthy that these products may not be suitable or accurate enough to use. Therefore, it is still urgent to make EBC practitioners and remote sensing specialists communicate efficiently.

ACKNOWLEDGMENTS

This research was funded by the Natural Sciences and Engineering Research Council of Canada (CRDPJ 328937-05) and the Alberta Foothills Research Institute (FRI). Thanks go to the FRI team and the Foothills Research Institute Grizzly Bear Program funding partners. The first author also would like to acknowledge the scholarship supports of Department of Geography and Planning, and College of Graduate Studies and Research at the University of Saskatchewan.

REFERENCES

1. Jensen, JR. *Remote Sensing of the Environment: An Earth Resource Perspective*, 2nd ed; Pearson Prentice Hall: Upper Saddle River, NJ, USA, 2007.
2. Kerr, JT; Ostrovsky, M. From space to species: Ecological applications for remote sensing. *Trends Ecol. Evol* 2003, *18*, 299–305.
3. Turner, W; Spector, S; Gardiner, N; Fladeland, M; Sterling, E; Steininger, M. Remote sensing for biodiversity science and conservation. *Trends Ecol. Evol* 2003, *18*, 306–314.
4. Chambers, JQ; Robertson, AL; Carneiro, VMC; Lima, AJN; Smith, ML; Plourde, LC; Higuchi, N. Hyperspectral remote detection of niche partitioning among canopy trees driven by blowdown gap disturbances in the Central Amazon. *Oecologia* 2009, *160*, 107–117.
5. Collingwood, A; Franklin, SE; Guo, X; Stenhouse, G. A medium-resolution remote sensing classification of agricultural areas in Alberta grizzly bear habitat. *Can. J. Remote Sens* 2009, *35*, 23–36.
6. McDermid, G; McLane, A; Collingwood, A; Hird, J; Faraguna, A; Laskin, D; Linke, J; Cranston, J; Guo, X; Franklin, S. Remote sensing mapping and research update. In *Foothills Research Institute Grizzly Bear Program 2007 Annual Report*; Stenhouse, G, Graham, K, Eds.; Foothills Research Institute: Hinton, AB, Canada, 2008.
7. Imam, E; Kushwaha, SPS; Singh, A. Evaluation of suitable tiger habitat in Chandoli National Park, India, using multiple logistic regression. *Ecol. Model* 2009, *220*, 3621–3629.
8. Tagesson, T; Eklundh, L; Lindroth, A. Applicability of leaf area index products for boreal regions of Sweden. *Int. J. Remote Sens* 2009, *30*, 5619–5632.
9. Pellikka, PKE; Lotjonen, M; Sijander, M; Lens, L. Airborne remote sensing of spatiotemporal change (1955–2004) in indigenous and exotic forest cover in the Taita Hills, Kenya. *Int. J. Appl. Earth Obs. Geoinf* 2009, *11*, 221–232.
10. Nagendra, H. Using remote sensing to assess biodiversity. *Int. J. Remote Sens* 2001, *22*, 2377–2400.
11. Aplin, P. Remote sensing: Ecology. *Prog. Phys. Geogr* 2005, *29*, 104–113.
12. Fassnacht, KS; Cohen, WB; Spies, TA. Key issues in making and using satellite-based maps in ecology: A primer. *For. Ecol. Manage* 2006, *222*, 167–181.
13. Duro, DC; Coops, NC; Wulder, MA; Han, T. Development of a large area biodiversity monitoring system driven by remote sensing. *Prog. Phys. Geogr* 2007, *31*, 235–260.
14. Gillespie, TW; Foody, GM; Rocchini, D; Giorgi, AP; Saatchi, S. Measuring and modeling biodiversity from space. *Prog. Phys. Geogr* 2008, *32*, 203–221.
15. Toutin, T. Fine spatial resolution optical sensors. In *The SAGE Handbook of Remote Sensing*; Warner, TA, Nellis, MD, Foody, GM, Eds.; SAGE Publications Ltd: London, UK, 2009; pp. 109–122.
16. Wabnitz, CC; Andrefouet, S; Torres-Pulliza, D; Muller-Karger, FE; Kramer, PA. Regional-scale seagrass habitat mapping in the Wider Caribbean region using Landsat sensors: Applications to conservation and ecology. *Remote Sens. Environ* 2008, *112*, 3455–3467.
17. Nagendra, H; Rocchini, D. High resolution satellite imagery for tropical biodiversity studies: The devil is in the detail. *Biodivers. Conserv* 2008, *17*, 3431–3442.
18. Morales, RM; Miura, T; Idol, T. An assessment of Hawaiian dry forest condition with fine resolution remote sensing. *For. Ecol. Manage* 2008, *255*, 2524–2532.
19. Perotto-Baldivieso, HL; Melendez-Ackerman, E; Garcia, MA; Leimgruber, P; Cooper, SM; Martinez, A; Calle, P; Ramos Gonzales, OM; Quinones, M; Christen, CA; Pons, G. Spatial distribution, connectivity, and the influence of scale: habitat availability for the endangered Mona Island rock iguana. *Biodivers. Conserv* 2009, *18*, 905–917.

20. Gulgun, B; Turkyilmaz, B; Bolca, M; Ozen, F. An examination of the effects of land use changes on nature conservation rulings in Cesme peninsula, Turkey. *Environ. Monit. Assess* 2009, *151*, 457–476.

21. Mellin, C; Andrefouet, S; Kulbicki, M; Dalleau, M; Vigliola, L. Remote sensing and fish-habitat relationships in coral reef ecosystems: Review and pathways for systematic multi-scale hierarchical research. *Mar. Pollut. Bull* 2009, *58*, 11–19.

22. Harborne, AR; Mumby, PJ; Kappel, CV; Dahlgren, CP; Micheli, F; Holmes, KE; Brumbaugh, DR. Tropical coastal habitats as surrogates of fish community structure, grazing, and fisheries value. *Ecol. Appl* 2008, *18*, 1689–1701.

23. Shippert, P. Why use hyperspectral imagery? *Photogramm. Eng. Remote Sens* 2004, *70*, 377–380.

24. Goward, SN; Arvidson, T; Williams, DL; Irish, R; Irons, JR. Moderate spatial resolution optical sensors. In *The SAGE Handbook of Remote Sensing*; Warner, TA, Nellis, MD, Foody, GM, Eds.; SAGE Publications Ltd: London, UK, 2009; pp. 123–138.

25. Campbell, JB. *Introduction to Remote Sensing*, 4th ed; Guilford Press: New York, NY, USA, 2007.

26. Pearlman, J; Carman, S; Segal, C; Jarecke, P; Barry, P; Browne, W. Overview of the Hyperion imaging spectrometer for the NASA EO-1 mission. 2001. Available online: http://eo1.gsfc.nasa.gov/new/validationReport/Technology/TRW_EO1%20Papers_Presentations/19.pdf (accessed on 13 January 2010).

27. Guerschman, JP; Hill, MJ; Renzullo, LJ; Barrett, DJ; Marks, AS; Botha, EJ. Estimating fractional cover of photosynthetic vegetation, non-photosynthetic vegetation and bare soil in the Australian tropical savanna region upscaling the EO-1 Hyperion and MODIS sensors. *Remote Sens. Environ* 2009, *113*, 928–945.

28. Bannari, A; Khurshid, KS; Staenz, K; Schwarz, J. Potential of Hyperion EO-1 hyperspectral data for wheat crop chlorophyll content estimation. *Can. J. Remote Sens* 2008, *34*, S139–157.

29. Walsh, SJ; McCleary, AL; Mena, CF; Shao, Y; Tuttle, JP; Gonzalez, A; Atkinson, R. QuickBird and Hyperion data analysis of an invasive plant species in the. Galapagos Islands of Ecuador: Implications for control and land use management. *Remote Sens. Environ* 2008, *112*, 1927–1941.

30. Lee, Z; Casey, B; Arnone, R; Weidemann, A; Parsons, R; Montes, MJ; Gao, BC; Goode, W; Davis, CO; Dye, J. Water and bottom properties of a coastal environment derived from Hyperion data measured from the EO-1 spacecraft platform. *J. Appl. Remote Sens* 2007, *1*, 011502.

31. Keramitsoglou, I; Kontoes, C; Sykioti, O; Sifakis, N; Xofis, P. Reliable, accurate and timely forest mapping for wildfire management using ASTER and Hyperion satellite imagery. *For. Ecol. Manage* 2008, *255*, 3556–3562.

32. Foster, JR; Townsend, PA; Zganjar, CE. Spatial and temporal patterns of gap dominance by low-canopy lianas detected using EO-1 Hyperion and Landsat Thematic Mapper. *Remote Sens. Environ* 2008, *112*, 2104–2117.

33. Pignatti, S; Cavalli, RM; Cuomo, V; Fusilli, L; Pascucci, S; Poscolieri, M; Santini, F. Evaluating Hyperion capability for land cover mapping in a fragmented ecosystem: Pollino National Park, Italy. *Remote Sens. Environ* 2009, *113*, 622–634.

34. Twele, A; Erasmi, S; Kappas, M. Spatially explicit estimation of leaf area index using EO-1 Hyperion and Landsat ETM+ data: implications of spectral bandwidth and shortwave infrared data on prediction accuracy in a tropical montane environment. *GISci. Remote Sens* 2008, *45*, 229–248.

35. Walsh, SJ; Shao, Y; Mena, CF; McCleary, AL. Integration of Hyperion satellite data and a household social survey to characterize the causes and consequences of reforestation patterns in the northern Ecuadorian Amazon. *Photogramm. Eng. Remote Sens* 2008, *74*, 725–735.

36. Canada Centre for Remote Sensing. Fundamentals of Remote Sensing: A Canada Center for Remote Sensing Remote Sensing Tutorial, 2007. Available online: http://ccrs.nrcan.gc.ca/resource/tutor/fundam/pdf/fundamentals_e.pdf (accessed on 03 January 2010).

37. Quattrochi, DA; Luvall, JC. Thermal remote sensing in Earth science research. In *The SAGE Handbook of Remote Sensing*; Warner, TA, Nellis, MD, Foody, GM, Eds.; SAGE Publications Ltd: London, UK, 2009; pp. 64–78.

38. Gillies, R; Temesgen, B. Coupling thermal infrared and visible satellite measurements to infer biophysical variables at the land surface. In *Thermal Remote Sensing in Lands Surface Processes*; Quattrochi, DA, Luvall, JC, Eds.; CRC Press: Boca Raton, FL, USA, 2004; pp. 160–184.

39. Mildrexler, DJ; Zhao, MS; Heinsch, FA; Running, SW. A new satellite-based methodology for continental-scale disturbance detection. *Ecol. Appl* 2007, *17*, 235–250.

40. Crow, WT; Zhan, X. Continental-scale evaluation of remotely sensed soil moisture products. *IEEE Geosci. Remote Sens. Lett* 2007, *4*, 451–455.

41. Petropoulos, G; Carlson, TN; Wooster, MJ; Islam, S. A review of Ts/VI remote sensing based methods for the retrieval of land surface energy fluxes and soil surface moisture. *Prog. Phys. Geogr* 2009, *33*, 224–250.

42. Xue, Y; Li, Y; Guang, J; Zhang, X; Guo, J. Small satellite remote sensing and applications—History, current and future. *Int. J. Remote Sens* 2008, *29*, 4339–4372.

43. Kramer, HJ; Cracknell, AP. An overview of small satellites in remote sensing. *Int. J. Remote Sens* 2008, *29*, 4285–4337.

44. Wang, K; Franklin, SE; Guo, X; He, Y; McDermid, GJ. Problems in remote sensing of landscapes and habitats. *Prog. Phys. Geogr* 2009, *33*, 747–768.

45. Wang, X; Wang, G; Guan, Y; Chen, Q; Gao, L. Small satellite constellation for disaster monitoring in China. Proceedings of IEEE International Conference on Geoscience and Remote Sensing Symposium, Seoul, Korea, 25–29 July 2005; pp. 467–469.

46. Qian, YG; Yan, GJ; Duan, SB; Kong, XS. A contextual fire detection algorithm for simulated HJ-1B imagery. *Sensors*2009, *9*, 961–979.

47. Lim, K; Treitz, P; Wulder, M; St-Onge, B; Flood, M. LiDAR remote sensing of forest structure. *Prog. Phys. Geogr* 2003, *27*, 88–106.

48. Hyyppa, J; Wagner, W; Hollaus, M; Hyyppa, H. Airborne laser scanning. In *The SAGE Handbook of Remote Sensing*; Warner, TA, Nellis, MD, Foody, GM, Eds.; SAGE Publications Ltd: London, UK, 2009; pp. 199–211.

49. Wehr, A; Lohr, U. Airborne laser scanning—An introduction and overview. *ISPRS J. Photogramm. Remote Sens* 1999, *54*, 68–82.

50. Miller, SB. Photogrammetric Products. In *Manual of Photogrammetry*, 5th ed; McGlone, J, Ed.; ASPRS: Bethesda, MD, USA, 2004; pp. 983–1013.

51. Irish, JL; White, TE. Coastal engineering applications of high-resolution lidar bathymetry. *Coast. Eng* 1998, *35*, 47–71.

52. Drake, JB; Dubayah, RO; Clark, DB; Knox, RG; Blair, JB; Hofton, MA; Chazdon, RL; Weishampel, JF; Prince, SD. Estimation of tropical forest structural characteristics using large-footprint lidar. *Remote Sens. Environ* 2002, *79*, 305–319.

53. Goetz, S; Steinberg, D; Dubayah, R; Blair, B. Laser remote sensing of canopy habitat heterogeneity as a predictor of bird species richness in an eastern temperate forest, USA. *Remote Sens Environ* 2007, *108*, 254–263.

54. Duncanson, LI; Niemann, KO; Wulder, MA. Estimating forest canopy height and terrain relief from GLAS waveform metrics. *Remote Sens. Environ* 2010, *114*, 138–145.

55. Lefsky, MA; Keller, M; Pang, Y; de Camargo, PB; Hunter, MO. Revised method for forest canopy height estimation from Geoscience Laser Altimeter System waveforms. *J. Appl. Remote Sens* 2007, *1*, 013537.

56. Simard, M; Rivera-Monroy, VH; Mancera-Pineda, JE; Castaneda-Moya, E; Twilley, RR. A systematic method for 3D mapping of mangrove forests based on Shuttle Radar Topography Mission elevation data, ICEsat/GLAS waveforms and field data: Application to Cienaga Grande de Santa Marta, Colombia. *Remote Sens. Environ* 2008, *112*, 2131–2144.

57. Pflugmacher, D; Cohen, W; Kennedy, R; Lefsky, M. Regional applicability of forest height and aboveground biomass models for the Geoscience Laser Altimeter System. *For. Sci* 2008, *54*, 647–657.

58. Helmer, EH; Lefsky, MA; Roberts, DA. Biomass accumulation rates of Amazonian secondary forest and biomass of old-growth forests from Landsat time series and the Geoscience Laser Altimeter System. *J. Appl. Remote Sens* 2009, *3*, 033505.

59. Tseng, DC; Tseng, HT; Chien, CL. Automatic cloud removal from multi-temporal SPOT images. *Appl. Math. Comput*2008, *205*, 584–600.

60. Jensen, JR; Im, J; Hardin, P; Jensen, RR. Image classification. In *The SAGE Handbook of Remote Sensing*; Warner, TA, Nellis, MD, Foody, GM, Eds.; SAGE Publications Ltd: London, UK, 2009; pp. 269–281.

61. Huang, C; Davis, LS; Townshend, JR. An assessment of support vector machines for land cover classification. *Int. J. Remote Sens* 2002, *23*, 725–749.

62. Foody, GM; Mathur, A. The use of small training sets containing mixed pixels for accurate hard image classification: training on mixed spectral responses for classification by a SVM. *Remote Sens. Environ* 2006, *103*, 179–189.

63. Foody, GM; Mathur, A. A relative evaluation of multiclass image classification by support vector machines. *IEEE Trans. Geosci. Remote Sens* 2004, *42*, 1335–1343.

64. Dalponte, M; Bruzzone, L; Vescovo, L; Gianelle, D. The role of spectral resolution and classifier complexity in the analysis of hyperspectral images of forest areas. *Remote Sens. Environ* 2009, *113*, 2345–2355.

65. Ichii, K; Wang, WL; Hashimoto, H; Yang, FH; Votava, P; Michaelis, AR; Nemani, RR. Refinement of rooting depths using satellite-based evapotranspiration seasonality for ecosystem modeling in California. *Agric. For. Meteorol* 2009, *149*, 1907–1918.

66. Sanchez-Hernandez, C; Boyd, DS; Foody, GM. One-class classification for mapping a specific land-cover class: SVDD classification of fenland. *IEEE Trans. Geosci. Remote Sens* 2007, *45*, 1061–1073.

67. Sanchez-Hernandez, C; Boyd, DS; Foody, GM. Mapping specific habitats from remotely sensed imagery: support vector machine and support vector data description based classification of coastal saltmarsh habitats. *Ecol. Inform*2007, *2*, 83–88.

68. Benz, UC; Hofmann, P; Willhauck, G; Lingenfelder, I; Heynen, M. Multiresolution, object-oriented fuzzy analysis of remote sensing data for GIS-ready information. *ISPRS J. Photogramm. Remote Sens* 2004, *58*, 239–258.

69. *Definiens AG. Definiens Developer 7–user guide*, 2008. Available online: http://www.definiens.com/definiens-support-services_187_11_14.html (accessed on 14 July 2009).

70. Lu, D; Weng, Q. A survey of image classification methods and techniques for improving classification performance. *Int. J. Remote Sens* 2007, *28*, 823–870.

71. Rocchini, D; Ricotta, C. Are landscapes as crisp as we may think? *Ecol. Model* 2007, *204*, 535–539.

72. Lu, D; Weng, Q. Spectral mixture analysis of ASTER images for examining the relationship between urban thermal features and biophysical descriptors in Indianapolis, Indiana, USA. *Remote Sens. Environ* 2006, *104*, 157–167.

73. Plourde, LC; Ollinger, SV; Smith, ML; Martin, ME. Estimating species abundance in a northern temperate forest using spectral mixture analysis. *Photogramm. Eng. Remote Sens* 2007, *73*, 829–840.

74. He, Y; Guo, X; Wilmshurst, J. Studying mixed grassland ecosystems I: Suitable hyperspectral vegetation indices. *Can. J. Remote Sens* 2006, *32*, 98–107.

75. Schlerf, M; Atzberger, C; Hill, J. Remote sensing of forest biophysical variables using HyMap imaging spectrometer data. *Remote Sens. Environ* 2005, *95*, 177–194.

76. Huete, AR; Liu, H. An error and sensitivity analysis of the atmospheric and soil-correcting variants of the NDVI for the MODIS-EOS. *IEEE Trans. Geosci. Remote Sens* 1994, *34*, 897–905.

77. Haboudane, D; Miller, JR; Pattery, E; Zarco-Tejad, PJ; Strachan, IB. Hyperspectral vegetation indices and novel algorithms for predicting green LAI of crop canopies: modeling and validation in the context of precision agriculture. *Remote Sens. Environ* 2004, *90*, 337–352.

78. Serbin, G; Daughtry, CST; Hunt, ER; Brown, DJ; McCarty, GW. Effect of soil spectral properties on remote sensing of crop residue cover. *Soil Sci. Soc. Am. J* 2009, *73*, 1545–1558.

79. Kraus, T; Schmidt, M; Dech, SW; Samimi, C. The potential of optical high resolution data for the assessment of leaf area index in East African rainforest ecosystems. *Int. J. Remote Sens* 2009, *30*, 5039–5059.

80. Wang, C; Qi, J; Cochrane, M. Assessment of tropical forest degradation with canopy fractional cover from Landsat ETM plus and IKONOS imagery. *Earth Interac* 2005, *9*, 1–18.

81. Gitelson, AA. Wide dynamic range vegetation index for remote quantification of biophysical characteristics of vegetation. *J. Plant Physiol* 2004, *161*, 165–173.

82. Zhang, L; Furumi, S; Muramatsu, K; Fujiwara, N; Daigo, M; Zhang, L. A new vegetation index based on the universal pattern decomposition method. *Int. J. Remote Sens* 2007, *28*, 107–124.

83. Ji, L; Peters, AJ. Performance evaluation of spectral vegetation indices using a statistical sensitivity function. *Remote Sens. Environ* 2007, *106*, 59–65.

84. Boyd, DS; Danson, FM. Satellite remote sensing of forest resources: Three decades of research development. *Prog. Phys.Geogr* 2005, *29*, 1–26.

85. Liang, S. Recent developments in estimating land surface biogeographysical variables from optical remote sensing. *Prog. Phys.Geogr* 2007, *31*, 501–516.

86. Liang, S. Quantitative models and inversion in optical remote sensing. In *The SAGE Handbook of Remote Sensing*; Warner, TA, Nellis, MD, Foody, GM, Eds.; SAGE Publications Ltd: London, UK, 2009; pp. 64–78.

87. Park, MH; Stenstrom, MK. Using satellite imagery for stormwater pollution management with Bayesian networks. *Water Res* 2006, *40*, 3429–3438.

88. Di Vittorio, AV. Enhancing a leaf radiative transfer model to estimate concentrations and *in vivo* specific absorption coefficients of total carotenoids and chlorophylls a and b from single-needle reflectance and transmittance. *Remote Sens. Environ* 2009, *113*, 1948–1966.

89. Koetza, B; Baret, F; Poilve, H; Hill, J. Use of coupled canopy structure dynamic and radiative transfer models to estimate biophysical canopy characteristics. *Remote Sens. Environ* 2005, *95*, 115–124.

90. Migliavacca, M; Meroni, M; Busetto, L; Colombo, R; Zenone, T; Matteucci, G; Manca, G; Seufert, G. Modeling gross primary production of agro-forestry ecosystems by assimilation of satellite-derived information in a process-based model. *Sensors* 2009, *9*, 922–942.

91. Hilker, T; Wulder, MA; Coops, NC; Linke, J; McDermid, G; Masek, JG; Gao, F; White, JC. A new data fusion model for high spatial- and temporal-resolution mapping of forest disturbance based on Landsat and MODIS. *Remote Sens. Environ* 2009, *113*, 1613–1627.

92. Pohl, C; van Genderen, JL. Multisensor image fusion in remote sensing: concepts, methods and applications. *Int. J. Remote Sens* 1998, *19*, 823–854.

93. Gamba, P; Chanussot, J. Guest editorial: Foreword to the special issue on data fusion. *IEEE Trans. Geosci. Remote Sens*2008, *46*, 1283–1288.

94. Zhang, Y. Understanding image fusion. *Photogramm. Eng. Remote Sens* 2004, *70*, 657–661.

95. Dong, J; Zhuang, D; Huang, Y; Fu, J. Advance in multi-sensor data fusion: algorithms and applications. *Sensors* 2009, *9*, 7771–7784.

96. Dong, J; Yang, X; Clinton, N; Wang, N. An artificial network model for estimating crop yields using remotely sensed information. *Int. J. Remote Sens* 2004, *25*, 1723–1732.

97. Waske, B; van der Linden, S. Classifying multilevel imagery from SAR and optical sensors by decision fusion. *IEEE Trans. Geosci. Remote Sens* 2008, *46*, 1457–1466.

98. Hong, G; Zhang, Y; Mercer, B. A wavelet and IHS integration method to fuse high resolution SAR with moderate resolution multispectral images. *Photogramm. Eng. Remote Sens* 2009, *75*, 1213–1223.

99. Wunderle, AL; Franklin, SE; Guo, X. Age class estimation of western red cedar using SPOT-5 pan-sharpened imagery in British Columbia, Canada. *Geocarto Int* 2009, *24*, 47–63.

100. Huang, SL; Crabtree, RL; Potter, C; Gross, P. Estimating the quantity and quality of coarse woody debris in Yellowstone post-fire forest ecosystem from fusion of SAR and optical data. *Remote Sens. Environ* 2009, *113*, 1926–1938.

101. Souza, PWM; Goncalves, FD; Rodrigues, SWP; Costa, FR; Miranda, FP. Multi-sensor data fusion for geomorphological and environmental sensitivity index mapping in the Amazonian mangrove coast, Brazil. *J. Coast. Res* 2009, *2*, 1529–1596.

102. Foody, GM. GIS: Biodiversity applications. *Prog. Phys. Geogr* 2008, *32*, 223–235.

103. Hinton, JC. GIS and remote sensing integration for environmental applications. *Int. J. Geogr. Inf. Sci* 1996, *10*, 877–890.

104. Merchant, JW; Narumalani, S. Integrating remote sensing and geographic information systems. In *The SAGE Handbook of Remote Sensing*; Warner, TA, Nellis, MD, Foody, GM, Eds.; SAGE Publications Ltd: London, UK, 2009; pp. 257–268.

105. Aronoff, S. *Remote sensing for GIS managers*; ESRI Press: Redlands, CA, USA, 2005.

106. Ehlers, M; Edwards, G; Bedard, Y. Integration of remote sensing with geographic information systems: a necessary evolution. *Photogramm. Eng. Remote Sens* 1989, *55*, 1619–1627.

107. Gao, J. Integration of GPS with remote sensing and GIS: reality and prospect. *Photogramm. Eng. Remote Sens* 2002, *68*, 447–453.

108. Briggs, NA; Sader, SA. Tracking forest change and development using low-cost remote sensing imagery and GIS integration. *North. J. Appl. For* 2009, *26*, 148–155.

109. Giriraj, A; Murthy, MSR; Ramesh, BR; Dutt, CBS. A method for assessing evergreen habitats using phytodiversity and geospatial techniques in tropical rain forests of Southern Western Ghats (India). *Ecol. Res* 2009, *24*, 749–760.

110. Dong, X; Liu, L; Wang, J; Shi, J; Pan, JH. Analysis of the landscape change at river basin scale based on SPOT and TM fusion remote sensing images: A case study of the Weigou River Basin on the Chinese Loess Plateau. *Int. J. Earth Sci*2009, *98*, 651–664.

111. Chowdary, VM; Ramakrishnan, D; Srivastava, YK; Chandran, V; Jeyaram, A. Integrated water resource development plan for sustainable management of Mayurakshi watershed, India using remote sensing and GIS. *Water Resour. Manag*2009, *23*, 1581–1502.

112. Kurnaz, S; Salahova, S; Rustamov, RB; Zeynalova, M. River inundation impact reduction based on space technology application. Proceedings of the 4th International Conference on Recent Advances in Space Technologies, Istanbul, Turkey, June 2009; pp. 265–272.

113. Jasrotia, AS; Majhi, A; Singh, S. Water balance approach for rainwater harvesting using remote sensing and GIS techniques, Jammu Himalaya, India. *Water Resour. Manag* 2009, *23*, 3035–3055.

114. Blaschke, T; Strobl, J. What's wrong with pixels? Some recent developments interfacing remote sensing and GIS. *GIS-Zeitschrift für Geoinformationssysteme* 2001, *14*, 12–17.

115. Blaschke, T. Object based image analysis for remote sensing. *ISPRS J. Photogramm. Remote Sens* 2010, *62*, 2–16.

INDEX

M

N

P

Q

R

S

T

U

V

W